Advances in Spatial Science

Springer
Berlin
Heidelberg
New York
Barcelona
Budapest
Hong Kong
London
Milan
Paris
Singapore
Tokyo

Titles in the Series

C. S. Bertuglia, M. M. Fischer and *G. Preto* (Eds.)
Technological Change, Economic Development and Space
XVI, 354 pages. 1995. ISBN 3-540-59288-1

H. Coccossis and *P. Nijkamp* (Eds.)
Overcoming Isolation
VIII, 272 pages. 1995. ISBN 3-540-59423-X

L. Anselin and *R. J.G.M. Florax* (Eds.)
New Directions in Spatial Econometrics
XIX, 420 pages. 1995. ISBN 3-540-60020-5

H. Eskelinen and *F. Snickars* (Eds.)
Competitive European Peripheries
VIII, 271 pages. 1995. ISBN 3-540-60211-9

J. C.J.M. van den Bergh, P. Nijkamp and *P. Rietveld* (Eds.)
Recent Advances in Spatial Equilibrium Modelling
VIII, 392 pages. 1996. ISBN 3-540-60708-0

P. Nijkamp, G. Pepping and *D. Banister* (Eds.)
Telematics and Transport Behaviour
XII, 227 pages. 1996. ISBN 3-540-60919-9

D. F. Batten and *C. Karlsson* (Eds.)
Infrastructure and the Complexity of Economic Development
VIII, 298 pages. 1996. ISBN 3-540-61333-1

T. Puu
Mathematical Location and Land Use Theory
IX, 294 pages. 1997. ISBN 3-540-61819-8

Y. Leung
Intelligent Spatial Decision Support Systems
XV, 470 pages. 1997. ISBN 3-540-62518-6

C. S. Bertuglia, S. Lombardo and *P. Nijkamp* (Eds.)
Innovative Behaviour in Space and Time
X, 437 pages. 1997. ISBN 3-540-62542-9

A. Nagurney and *S. Siokos*
Financial Networks
XVI, 492 pages. 1997. ISBN 3-540-63116-X

M. M. Fischer and *A. Getis* (Eds.)
Recent Developments in Spatial Analysis
X, 434 pages. 1997. ISBN 3-540-63180-1

R.H.M. Emmerink
Information and Pricing in Road Transportation
XVI, 294 pages. 1998. ISBN 3-540-64088-6

Piet Rietveld · Frank Bruinsma

Is Transport Infrastructure Effective?

Transport Infrastructure and Accessibility: Impacts on the Space Economy

With 50 Figures
and 99 Tables

Springer

Professor Dr. Piet Rietveld
Dr. Frank Bruinsma

Vrije Universiteit
Faculty of Economics
De Boelelaan 1105
NL-1081 HV Amsterdam, The Netherlands

ISBN-13: 978-3-642-72234-9 e-ISBN-13: 978-3-642-72232-5
DOI: 10.1007/978-3-642-72232-5

Cataloging-in-Data applied for
Die Deutsche Bibliothek - CIP-Einheitsaufnahme
Rietveld, Piet:
Is transport infrastructure effective? : transport infrastructure and accessibility impacts on the space economy; with 99 tables / Piet Rietveld; Frank Bruinsma. - Berlin; Heidelberg; New York; Barcelona; Budapest; Hong Kong; London; Milan; Paris; Singapore; Tokyo : Springer, 1998
(Advances in spatial science)

Softcover reprint of the hardcover 1st edition 1998

Hardcoverdesign: Erich Kirchner, Heidelberg
SPIN 10655441 42/2202-5 4 3 2 1 0 - Printed on acid-free paper

Preface

When in 1989 the authors started research on infrastructure, they did not foresee that this would lead to a long-term involvement in this area. Our beginning happened to coincide with the publication of David Aschauer's article on public capital and productivity, which induced a large flow of publications in this field. Infrastructure has indeed been a hot topic in policy and research during the past decade. It is surprising, however, that the number of monographs on spatial and economic impacts of infrastructure has remained very limited. The aim of this book is to contribute to the literature in a consolidated way.

A distinguishing feature of our book is that we analyze infrastructure impacts using various methods (both modelling and non-modelling) at a variety of spatial levels (from local to international). Other special features are that we make ample use of 'accessibility' as a bridge concept between the areas of infrastructure and the economy. Finally, we not only treat transport infrastructure projects as given, as is the usual approach in infrastructure impact research, but we also analyze the factors influencing infrastructure supply.

We have adopted a mainly non-technical approach throughout most of the book. This means that it can also be used by readers without a strong background in statistics, modelling or micro-economics.

A considerable part of the book is clearly related to papers already published in various journals and edited books. The essential sections of the introductory Chapters (1-4) however, have been written specifically for this book.

The empirical part of the book has both a national (Dutch) and an international (European) orientation, however, the emphasis is not on the specific region or country itself, but rather on the existence and spatial distribution of impacts of infrastructure projects (urban belt ways, international express ways, railways, etc). Therefore, we believe that the book is also valuable for those not particularly interested in the specific regions examined.

It was about seven years ago after completing a book that one of the authors insisted he would never write a book again because of the large amount of work involved and the frustrations with the word-processors of that time. The present book shows that this claim was premature. Improved word processors have indeed helped soften some of the hardships of writing, and infrastructure appeared to be a challenging theme to address.

We want to express our gratitude to several colleagues who contributed to specific sections of the book. In alphabetical order we thank Jaap Boonstra, Maarten Hilferink, Joost van Nierop, Peter Nijkamp, Gerard Pepping, and Sytze Rienstra for their valuable contributions. The maps in the book have mostly been produced by our colleagues Wim Douven, Michel Grothe, and particularly Niels Vlaanderen, who have great expertise with Geographical Information Systems (GIS).

Not only are we thankful to colleagues working in our own department; the following colleagues throughout Europe made their data available for us: Nadine Cattan, DATAR, Ulf Erlandsson, Javier Gutiérres, Wilhelm Meester, Piet Pellenbarg, Klaus Spiekermann, and Michael Wegener. Michel Wegener and Chris Jensen-Butler and Bjarne Madsen were so kind to allow us to use parts of their empirical studies as examples of international-oriented infrastructure impact research.

The results of some research projects are already presented in journals or as chapters in edited books. We thank the publishers Carfax (Regional Studies, Urban Studies), KNAG (TESG), Nomos, Pion (Environment & Planning A), Springer (Annals of Regional Science), and Wiley for their permission to rewrite and use the material once again for this book.

We also thank the editorial board of Advances in Spatial Science (especially Folke Snickars), as well as Marianne Bopp from Springer for their encouragement to complete this project. Finally, we thank Sue Davis who corrected our English.

Amsterdam, Easter 1998
Piet Rietveld
Frank R. Bruinsma

Contents

Part B: European Accessibility Studies

Part C: Regional and Urban Impact Studies

1 Introduction

1.1 Introduction

Economic development is not only the result of a proper combination of private production factors such as labour and capital, but also of infrastructure in general and transport infrastructure in particular. Improving infrastructure leads to a higher productivity of private production factors. Conversely, a neglect of infrastructure leads to a lower productivity of the other production factors. A related effect is that transport infrastructure investments lead to changes in transport costs. They may therefore affect trade relationships and hence the location of production factors. This implies differentiated growth of regions or countries. For these reasons, transport infrastructure projects play an important role in the political agenda of many countries.

The discussion of the economic impacts of transport infrastructure has taken place at various spatial levels: urban, regional, national and international. In each spatial range specific issues play a role. At the urban level important issues are the consequences of congestion and infrastructure for public transport on spatial development. At the regional level infrastructure is often considered as a policy instrument to stimulate lagging regions. At the national level the balance between private and public investments to arrive at an optimal growth path is an important issue. At the international level improvement of transport infrastructure is considered as a way to reduce trade barriers and enhance the competitiveness of national economies.

Many countries witnessed a considerable decline of infrastructure investments during the 1980s. The budgetary problems of most national governments are an important explanation for this decline. Another reason is that environmental problems gained a more prominent role in political decision making. The environmental problems associated with transport led to a greater critical attitude towards large transport infrastructure projects among voters.

The political interest in the economic effects of transport infrastructure investments is high. Pressure groups such as organizations of firms or Chambers of Commerce and regional governments generally take for granted that the economic benefits of transport infrastructure investment are large. National governments often use the argument of high economic benefits of infrastructure to counter the critique of environmentalists who are concerned about environmental impacts. The political interest is related to the equity aspects of infrastructure. In a *spatial* sense, equity aspects are important because transport infrastructure has a spatial structure such that certain grids in space are more strongly influenced than others. This is the immediate consequence of the node and link structure of networks. Some places will benefit more than others, or - implying a stronger degree of conflict - some places will benefit, but other places will only experience disadvantages. From a *stakeholder* point of view, equity is important because the distribution of costs and benefits among actors is unbalanced. Usually those who benefit do not coincide with those who pay the costs (either as a tax-payer or as victim of environmental effects).

One of the reasons why (transport) infrastructure has attracted much attention as an instrument to strengthen competitive positions of countries and regions in Europe is that the creation of the Common Market has led to the disappearance of a number of policy instruments available to national governments. For example, taxation and subsidies will be subject to the same constraints in the various European Union member states. Infrastructure investments have become one of the important instruments with which national economies can compete with each other. A related reason which explains the importance attached to transport infrastructure is that the process of European economic integration is expected to lead to an increase in the growth of the economies and even more so in the international trade flows. This will induce a relatively rapid growth in the demand for infrastructure, especially international infrastructure links. Examples can be found in large-scale border crossing infrastructure projects such as the Channel tunnel, bridges and tunnels connecting Denmark with Germany and Sweden, tunnels through the Alps, and international high-speed rail connections.

1.2 The Scope of the Book

Is transport infrastructure effective? The title chosen for this book somewhat resembles David Aschauer's (1989) well-known article (is public expenditure productive?). There are two differences, however. First, our book is more specific: 'transport infrastructure' is a much narrower concept than 'public expenditures'. In another respect our title indicates a broader interest because 'productive', as used by Aschauer, implies a focus on productivity, whereas 'effective' may relate to various other economic objectives.

The aim of our work is to shed light on the spatial economic effects of transport infrastructure investments; both theoretical and empirical approaches will be used for this purpose. The aim has both scientific and social relevance. From a scientific viewpoint, the issue of transport infrastructure impacts is challenging: it concerns a field of research with various theoretical perspectives sometimes leading to conflicting empirical results. The social relevance is evident given the significant ongoing debates in many countries about infrastructure projects.

From a policy point of view, transport infrastructure in general and new transport infrastructure investments in particular have received considerable attention. There are at least three topics to be mentioned of particular interest to politicians:

- accessibility impacts;
- economic and locational impacts;
- environmental impacts.

The first two topics are dealt with in this book. The fact that we do not include the environmental aspects of transport does not mean that we consider them as unimportant. To keep the size of the book manageable we decided to exclude them. Some attention will be given to environmental issues in this introductory Chapter, however (section 1.7). In the next sections a short introduction on the above mentioned topics will be given. We start with a brief discussion of impact studies in general, followed by a more detailed discussion of accessibility and locational impacts of infrastructure.

1.3 Policy Analysis of Infrastructure Projects: Impact Studies

Compared with other types of public expenditures, investments in transport infrastructure have a high degree of visibility. In the case of infrastructure projects it is easy to show taxpayers (voters) what the public sector did with the public funds in the general interest. This may elicit a tendency among policy makers to give priority to such projects above other types of measures. Therefore, studies on infrastructure impacts often have a strong relevance for policy debates. Certain public sector actors and immediate beneficiaries may use impact studies to justify particular infrastructure projects. Opponents of the projects try to reduce the claims that the project in question has large beneficial effects.

A policy analysis to structure the debate and improve the quality of decision processes on transport infrastructure projects contains several elements including:

- definition of relevant effects;
- generation of alternatives;
- estimation of effects of infrastructure projects;
- weighing the effects.

The definition of relevant effects entails the changes in welfare positions of all actors. Part of the effects will be in monetary terms, part will be in other terms such as travel time, environmental effects, safety, etc. Since infrastructure projects may have effects that spread quite broadly in a thin way, one has to delimitate the sets of effects and actors that are included and those that are not included.

In multicriteria decision analysis methods have been developed to assist in a systematic generation of alternatives (Voogd et al., 1990). Such methods are useful, but one should not overlook the fact that the generation of alternatives is often a creative process where certain constraints that seem to be binding can be removed. In addition, the number of alternatives in infrastructure projects is often very large, so that methods are needed to select the most promising ones (or a representative subset of all alternatives).

Once alternatives have been defined, the question is how to determine the various economic and non-economic effects of infrastructure. With the economic impacts, discussions especially concern the broader programme effects: how do economies adjust in the long-run to the development of infrastructure networks? Looking back at past infrastructure projects it would be fair, however, also to pay attention to the uncertainties about costs. Costs of infrastructure projects seem to be systematically underestimated (Button, 1993). Other relevant effects of infrastructure concern environmental effects. It is interesting to observe that in many countries environmental impact assessments have been institutionalized, whereas the status of an analysis of economic effects is lower. And also the quality check of the impact studies tends to be stronger for environmental effects compared with economic effects. This has obviously stimulated the development of environmental impact assessment methods compared with economic impact assessments of infrastructure projects.

As already indicated above, it is one thing to determine the effects of infrastructure projects, the valuation of these effects remains another issue. The classical approach to valuation is social cost-benefit analysis. It implies a valuation of the various effects in monetary terms. Especially with environmental and safety effects such a valuation is often difficult.

In this book we decided to concentrate on the third aspect of policy analysis: the uncertainty about the economic impacts per se. This does of course not imply that we consider the other issues to be trivial. But we find that this subject deserves a focused analysis because of its substantial complexity.

An important reason for the uncertainty about transport infrastructure project effects is the very long life time of such projects combined with the fact that some impacts of infrastructure improvement only materialize after a rather long period. It may take a long time before the relevant actors have adjusted to the possibilities and challenges implied by the new infrastructure. It is therefore often difficult to assess how developments would have been without a particular project. Another difficulty is that with major infrastructure programmes, the size of the impacts may be so large that ordinary analytical tools that rely heavily on marginal changes cannot easily be applied.

An important distinction in the research of infrastructure impacts is the one between ex-ante and ex-post studies. The first deals with the problem of giving predictions of impacts of projects that have not yet been completed. The second one addresses the impacts of projects that have already been completed. Policy makers are usually interested in ex-ante studies to discover whether proposed projects can be justified. On the other hand, academic research will tend to carry out ex-post studies to learn whether the claims of proposed projects have become true. Indeed the two approaches are linked because ex-post studies can be used for ex-ante purposes. A basic issue is of course to what extent one can learn from past experience; every project can be claimed to be unique so that it is not easy to predict its effects based on past experience.

The question then becomes what makes a project unique. A relevant issue is the life cycle of an infrastructure type. During the earlier phase of the introduction of an infrastructure type one may expect larger impacts than in the saturation phase when the infrastructure density is already high. The economic and institutional environment are also important in this respect. During a phase of rapid economic growth, an infrastructure project may have larger impacts than during a period of economic stagnation. Institutional aspects are also important. The institutional changes that occurred in eastern Europe after 1989 imply that it is dangerous to rely on regularities of the period before that year. This also has important implications for the methods to be used. The more stable the environment, the easier it will be to apply models based on earlier years for ex-ante purposes. In the extreme reverse case of a unique project or unique new circumstances, it is much more dangerous to rely on ex-post models. In that case various types of expert judgement and stated preference approaches can be used. Combinations of the two approaches are possible as well (see also Chapter 4).

1.4 Accessibility Impacts

Accessibility is sometimes used as a 'miracle' word to explain the necessity for transport infrastructure (investments) in political debates on all governmental levels.

At the European level the European Commission is striving for the development of Trans European Networks (TENs) of highways and high-speed rail lines. The call for TENs is inspired by equity and efficiency principles (cf Vickerman, 1995). In the new European Common Market all members should have access to this huge market to benefit from the advantages offered by this common market. So every country should be well-connected to this common market by the TENs. The national borders are not only barriers for international trade because of different languages, laws, cultures etc., but also because in the past, infrastructure networks tended to have an intra-national bias. The low infrastructure densities on border lines must be overcome by the construction of those TENs.

In the last decades transport infrastructure construction often took place to support declining regions or regions already lagging behind. Providing access for regions to the national market by infrastructure investments was expected to lead to a rise in regional employment in the short-term, and offer a secure base to support regional economic growth in the longer-term. In the following Chapters we will investigate to what extent these expectations conform with theory and empirical data.

The third level of spatial aggregation on which accessibility plays a role is the local or urban level. As important as the good connections between regions and cities is the accessibility within the urban area itself. For the daily activities like for instance, commuting, business travel, shopping, daily deliveries etcetera, a good road and public transport network is necessary. Especially cities - as the nodes in transport networks - have difficult coping with increasing transport flows; congestion problems are endangering accessibility of central zones. Orbital motorways are constructed to reduce the pressure on the inner city road network and also high-quality public transport systems - urban light rail or hybrid rail systems - are developed.

As we will show in the next Chapters, accessibility is a concept which is defined and used in many different ways (cf. Song, 1996; Handy & Niemeier, 1997). A careful treatment of the concept is needed. We show that accessibility, when defined properly, can play a key role in the analysis of infrastructure impacts.

1.5 Infrastructure and Location Decisions of Firms

The role of transport infrastructure in (re)location processes of firms is often regarded to be important. This holds true for the (re)location of various activities such as production plants, offices and service providers. Direct transport infrastructure factors, like access to the highway network or the availability of parking lots and also indirect factors such as just-in-time deliveries and distance to suppliers or customers play a role here.

However, a number of questions can be raised about the real importance of transport infrastructure for firms:

- in which phases of the relocation decision process is transport infrastructure important?
- can economic development be steered in desired directions by infrastructure construction?
- does transport infrastructure have an impact on the 'image' of a firm or region?

To start with the first question, should it be impossible that only after the decision to relocate is taken transport infrastructure is regarded to decide where to relocate? It is arguable that in the initial phase the decision to relocate is taken because of shortcomings at a company's present location. We postulate the hypothesis that push factors such as the inability to expand the company at the present location, or the inability to fulfil environmental preconditions made by the government to protect residential areas against pollution, are more decisive in the relocation decision than pull factors that show the attractiveness of another location.

In the second question we mention the possibility for spatial planners to steer economic developments towards desired locations by the development of transport infrastructure. However, how far is their reach? Does the development of transport infrastructure and industrial sites only attract either firms that had already decided to relocate and starting firms; or do firms relocate because of the possibilities that an improved transport infrastructure network has to offer them? When infrastructure is effective only for firms that already had decided to relocate along with a few newcomers, the steering impact is rather marginal in the short-run: the percentage of relocating firms is rather low compared to the total population of firms.

A somewhat different question concerns the image effect of transport infrastructure. In the literature there are a number of examples of buildings such as the new Arc de Triomphe in the La Defense quarter of Paris, which act as an image determinant (Poppelaars, 1993, Funck, 1996 and Rienstra & Rietveld, 1997). Could the same be said of certain transport infrastructure projects? There are indications that transport infrastructure can act as an image determinant on different levels of spatial aggregation.

Firms sometimes state that they located near an entry of a public transport system to show their commitment to infrastructure policy by supporting public transport. However, closer investigation reveals that neither their employees nor their costumers make regular use of the public transport system. They predominantly commute/travel by car. Another example is the common desire to have a sight-location near a highway or a railway station. Those examples of transport infrastructure and their impact on the image of the company are found everywhere. There are also more unique transport infrastructure projects which have a strong impact on the image of companies. An example is the Schiphol airport near Amsterdam. The Schiphol area is an absolute top location with the highest price level in the Netherlands. Nevertheless, the majority of the 85.000 jobs in the area has no relation what so ever with aviation (Kramer, 1990). Why are companies located on those expensive locations when much cheaper locations are available nearby? Just because 'it' is happening there? To show that the firm is a dynamic company?

More researched in-depth is the question whether transport infrastructure can be used to improve the image of a region. The introduction of a new transport mode will have a positive impact on the image of the region. The impact of transport infrastructure which is already mature or in a saturation phase may be expected to be marginal.

The impact and perception of transport infrastructure and the location behaviour and image of firms is rather complex. In the empirical part of the book questions of the type raised here will come to the fore.

1.6 Transport Infrastructure: Supply or Demand Driven?

Most of the research on transport infrastructure is addressing the impacts of infrastructure on the economy: how do economic actors adjust to the opportunities offered by new infrastructure? The supply of infrastructure is usually not studied, or is studied in a normative way: 'if social cost benefit analysis indicates that a project has a sufficiently high rate of return, it should be carried out'. However, as indicated by for example, Mueller (1989), such a normative approach presents an incomplete picture, because the public sector considers many other factors in its policies.

A more explicit treatment of the behaviour of the supplier of transport infrastructure is therefore called for. What factors determine the behaviour of suppliers of infrastructure? How do these suppliers define their roles? These questions have important implications for policy debates. Some would argue that it is the role of the public sector to take the *lead* in the economy and supply infrastructure to stimulate economic development. Others would argue that the public sector is not a leading sector in the economy, but instead *responds* to bottlenecks and political claims.

In Chapter 3 we will present a perspective where both roles of the public sector are included. One may argue that at the local level the supply of large transport infrastructure projects can be assumed to be exogenous: it is mainly determined by forces above the local level. However, at a higher spatial level infrastructure supply is no longer exogenous because the range of options of the public sector is clearly determined by the functioning of the economy. It is especially here that the full range of interrelationships between the public sector and the other sectors has to be considered, thus implying the need to take into account crowding out effects on factor and product markets. Special attention to the role of public sector supply will be given in Chapters 12 and 13.

1.7 Infrastructure and the Environment

Among the reasons why transport infrastructure projects attract so much attention is that environmental issues are important. In some respects, the transport sector has been successful in achieving environmental improvements such as a substantial reduction of lead and NO_x emissions by passenger cars (Verhoef, 1996). However, a shift towards more environmentally friendly modes of transport for both passengers and freight (rail, inland navigation) has not been realized. Even the reverse seems to hold. For example, aviation, a transport mode with large environmental effects, is growing in importance as a long-distance transport mode.

Environmental aspects play a role in infrastructure projects in several ways (Banister & Button, 1993). A first distinction is between the existence and the use of infrastructure. The existence of infrastructure may lead to barrier effects, visual intrusion and landscape deterioration. The use of infrastructure also leads to emissions of various pollutants and noise.

Another viewpoint is that certain infrastructure projects are built to avoid environmental effects at other places. In that case not only are the environmental effects of the project itself in need of a resolution, but so too, do the environmental effects intended to be avoided at other places demand attention. For example, high-speed rail development has been advocated among others because it may function as a substitute for aviation and car, which are less environmentally friendly. A critical issue is then to what extent the substitution effect (less pollution by passengers travelling by car or plane) compensates for the generation effects of more pollution of passengers per train who are travelling longer distances because of the improvement of the rail infrastructure. These examples make clear that environmental impacts of transport infrastructures may be difficult to assess.

Environmental quality in a city may in some cases be of immediate consequence for its economic performance. For example, firms do pay attention to environmental aspects in locational decisions as will be shown in Chapter 4. Therefore, the distinction between economic and locational impacts on the one hand and environmental impacts on the other is not absolute.

Other issues in the field of environmental aspects of transport relate to problems of valuing the effects (see also section 1.3) and the development of financial instruments (taxes) to induce infrastructure users to reduce the environmental effects to optimal levels. For various reasons it is difficult to adjust taxes so that they exactly reflect the (marginal) external costs of transport. Therefore, often, second best solutions, such as a fuel tax are used.

We conclude that environmental issues of transport infrastructure are an important and interesting field of research. In the present study we have decided to concentrate on economic impacts to achieve a more focused analysis. The reason that we nevertheless mention the environmental aspects here is that they play a large role in transport infrastructure debates.

1.8 Aim and Structure of the Book

The aim of the book is the analysis of the economic effects of transport infrastructure investments at various spatial levels: urban, regional, national and international. The analysis will have both a theoretical and an empirical character. Accessibility will function as a key concept in both the theoretical and empirical parts of the book.

The means by which we try to attain our goal can perhaps best be illustrated by an outline of the structure of the book (see Figure 1.1.).

We discuss *definitions and measurement* issues in Chapter 2. Special attention is paid to the concept of accessibility because it plays a central role in our theoretical framework and our empirical applications.

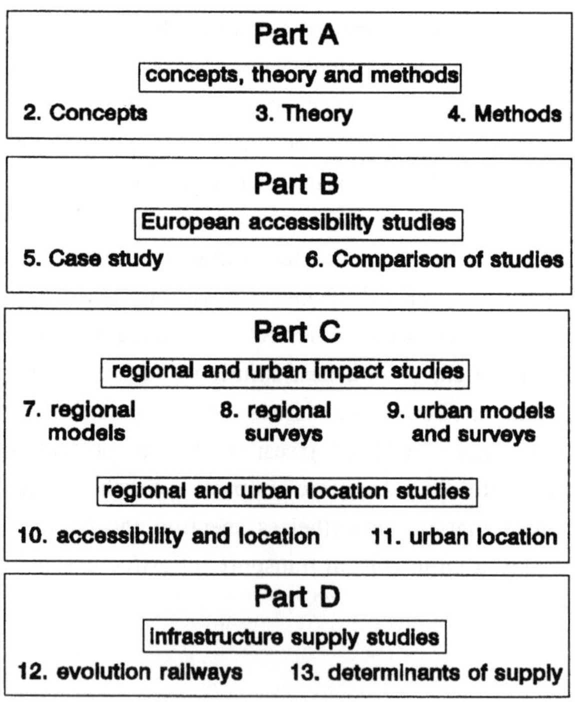

Figure 1.1 Outline of the structure of the book

A number of *theoretical perspectives* are discussed in Chapter 3. Here we discuss theories that not only address the contribution of infrastructure to productivity, but also the spatial location of economic activity.

Methods of infrastructure impact research are surveyed in Chapter 4. We pay attention to both modelling and non-modelling approaches at different spatial levels of aggregation.

After these foundations we proceed with a list of empirical studies.

Chapters 5 and 6 address the key notion of *accessibility*. Chapter 5 gives an empirical study of the accessibility of major European cities in multi modal infrastructure networks. The impacts of changes in these networks on the relative accessibilities are also studied. In Chapter 6 a comparison is conducted of a number of related accessibility studies recently carried out at the European level.

In Chapters 7, 8 and 9, the results of *infrastructure impact studies* for the Netherlands are presented. In Chapters 7 and 8 the focus is on the regional level of spatial aggregation. Regional models (Chapter 7) and entrepreneurial surveys (Chapter 8) are used to investigate the economic impacts of the A1 highway in the eastern part of the Netherlands. In Chapter 9 an office price model and an entrepreneurial survey are used to determine the impact of the Amsterdam orbital motorway at the urban level of spatial aggregation.

Chapters 10 and 11 are devoted to *location studies*. In Chapter 10 the inter-urban location preferences of entrepreneurs are explained by a number of variables of which the accessibility via the road network appears to be the most important one. In Chapter 11 the intra-urban location, the relocation pattern and the locational desires of entrepreneurs are analyzed for seventeen different location profiles. Moreover, trade-offs among various locational factors are analyzed.

In Chapters 12 and 13 the interrelationship between *infrastructure supply* and economic development is discussed. In Chapter 12 a network design model is used to analyze the historical development of the Dutch railway network. In Chapter 13 we explain the differences in infrastructure supply among European regions.

Finally, Chapter 14 rounds up by giving a summary of the most important conclusions found throughout the book. By using accessibility as a central concept in all three fields of research - impact studies, location studies and supply studies - some new insights can be attained to answer the question: 'Is transport infrastructure effective?'

PART A

Concepts, Theory and Review of Empirical Research

2 Concepts and Data

2.1 Introduction

Infrastructure is a well-known term used by many experts and laymen. But a closer examination of the term reveals that it is not easy to provide a satisfactory definition. These semantic difficulties will be addressed in section 2.2. Based on this we give further reflections on the public character of infrastructure and the role of the public sector in providing infrastructure services (section 2.3). Various ways of measuring infrastructure are highlighted in section 2.4. This is followed by some data on infrastructure investments in various countries in section 2.5. A basic limitation of these investment figures is that they do not really address the *services* provided by infrastructure. These services are discussed more fully in section 2.6, where we focus on the concept of accessibility and provide a list of ways to operationalize this concept. Section 2.7 concludes.

2.2 Defining Infrastructure

A large number of definitions for infrastructure (sometimes called *social overhead capital*) have been given in the literature. For example, Nurske (1952) mentions properties such as: 'provide services basic to any production capacity' and 'large and costly installations'. Hirschman (1958) and Biehl (1993) emphasize that infrastructure is capital that provides *public* services.

Diamond and Spence (1989, p. 14) define infrastructure as 'the collective and integrative basis for economic activity'. This definition contains four elements: (1) capitalness, (2) collective nature, (3) integrative role, and (4) support of economic activities (either directly or indirectly).

In some of these definitions the properties of capitalness and publicness are mentioned. The first element is important since it helps to distinguish infrastructure (defined in terms of a *stock*) from public goods (defined in terms of a *flow* per time unit). The second element of the definition concerns the notion of publicness. This again has a multidimensional meaning. Musgrave and Musgrave (1984), for example, mention two elements:

- *non-rivalness*: one person's use of the services provided by public capital does not deprive others from using them;
- *non-excludability*: if one person consumes services provided by public capital, it is impossible to restrict others from consuming them.

Table 2.1 gives a review of the various possible combinations of rivalness and excludability (see also World Bank, 1994). The Table makes clear that a strict use of the notion of public goods implies that only a small range of goods are included into this public goods category. Many types of goods often considered as 'infrastructure' only have one of the properties of non-rivalness and non-excludability, or even lack both properties.

Table 2.1 Combinations of rivalness and excludability of goods.

	excludable	non-excludable
rival	Private goods (e.g. telecommunications, airport services rail services	Common property goods (e.g. urban roads, ground water)
non-rival	Club goods (e.g. inter-urban highways)	Public goods (e.g. rural roads, radio broadcasting)

In addition to the aspects of rivalness and excludability, other properties are sometimes mentioned in the literature (cf. Youngson, 1967; Diewert, 1986; Lakshmanan, 1989; Snickars, 1989). Biehl (1993) for example, mentions the following properties:

- *indivisibility*: providing infrastructure to the first user leads to very high cost, but the additional costs to serve more users are very low;
- *non-substitutability*: it is often very costly, or even impossible to replace infrastructure by a private production factor;
- *immobility*: once an investment in infrastructure has been undertaken, it is very difficult to move it to another location;
- *polyvalence*: infrastructure can often be used as an input to a large number of production processes, and it may also be used by both consumers and producers (the opposite case is called monovalence).

Again, it appears that not all types of capital usually considered as infrastructure share these properties. For example, firms may find it profitable to generate their own power rather than buy it from the public electricity net (contra: non-substitutability). Therefore, it should be admitted that it is not possible to give an unambiguous list of properties according to which it can be judged whether a particular type of capital is part of infrastructure as this term is used in common parlance.

A common element of what is usually called infrastructure seems to be that there is strong public involvement in investment and/or exploitation. For example, Biehl (1993) defines infrastructure as 'that part of the overall capital stock of national or regional economies that, due to its publicness, is normally not provided by free markets at all or only inefficiently'. However, this way of defining infrastructure also has its handicap, because every country has its own way of determining government's role in investment decisions of various types. This would make intercountry comparisons of infrastructure impossible, because for example in one country, exploitation of expressways would be a public sector activity, whereas in other countries it is taken care of by the private sector.

We conclude that many analysts in the field have tried to formulate a number of properties in order to delimitate the concept of infrastructure. However, it is very difficult to formulate this list of properties in such a way that it coincides with what is commonly understood by 'infrastructure'. Therefore, we will refrain from giving a definition of infrastructure in terms of a list of properties. Here we follow Button (1996), who proposes avoiding a formal definition and treating it as 'what most people consider it to be'.

As an alternative to the above approach, where infrastructure is defined in terms of a number of *properties*, one can also follow a *nominal approach* by defining infrastructure as a set of *things*. For example, the Penguin Dictionary of Economics describes infrastructure as 'roads, airports, sewage and water systems, railways, the telephone and other public utilities' (Bannock et al., 1992, p.215). A more elaborate list is given by Biehl et al. (1986):

- transport infrastructure;
- communications infrastructure;
- energy supply infrastructure;
- water infrastructure;
- environmental infrastructure;
- education infrastructure;
- health infrastructure;
- special urban (local) infrastructure;
- sports and tourist facilities;
- social infrastructure;
- cultural facilities;
- natural endowment.

It is clear that such lists will differ in length and detail. Some researchers advocate a narrow definition of infrastructure by deleting the social, cultural and/or environmental elements of the above list. For example, Hirschman (1958) mentions that the hardcore of the concept can probably be restricted to transportation and power.

As far as transport infrastructure is concerned, Biehl (1993) distinguishes:

- roads;
- railways (lines, stations);
- waterways;
- airports;
- seaports;
- pipelines;
- telecommunications.

This list will lead to few objections in most respects, but some analysts may prefer to place telecommunications as a separate infrastructure category (see also Table 2.2 in section 2.3).

2.3 Infrastructure and Publicly Provided Capital

From the above discussion it appears that infrastructure and public capital are only partially overlapping concepts. As illustrated in Figure 2.1, there are parts of infrastructure that are in many countries not considered as part of the public capital stock (for example telecommunications, oil pipelines), and there are public capital goods such as defense materials and the equipment of civil servants that are usually not considered as infrastructure.

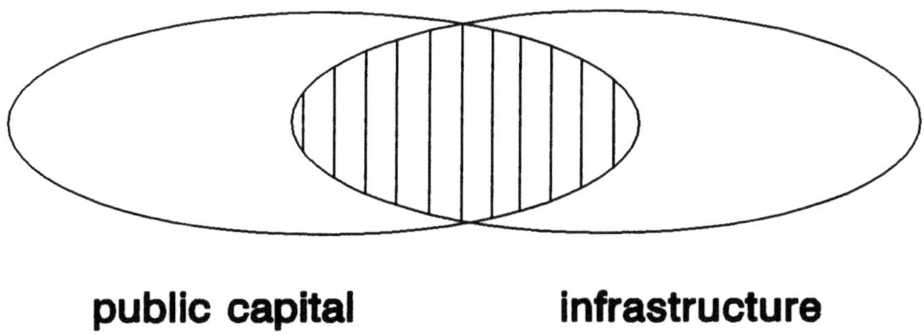

Figure 2.1 Partial overlap between public capital and infrastructure

The above distinction between the public capital stock and infrastructure given in Figure 2.1 is also quite relevant in the context of transport. We observe in many countries a shift away from the public sector to the private sector. The background of this shift is provided by Table 2.2. This Table indicates that according to criteria such as potential for competition, potential for cost recovery and relevance of equity considerations, there are for several types of infrastructure and related services no intrinsic reasons why they should be provided uniquely by the public sector. High suitability for private sector supply can be observed for example for port and airport services, rail freight and passenger services and long-distance telecommunications services. At the other end of the range we have rural and urban roads, where the public sector is clearly the most suitable supplier. A good number of infrastructure categories is in the middle of the range so that one may expect various countries to adopt different strategies. A change in view on the role of the public sector in economic development as can be observed during the last ten years will then easily lead to a reduction of public sector involvement in infrastructure supply.

Table 2.2 Feasibility of private sector delivery for various infrastructure components

	1[a]	2	3	4	5	6
Telecommunications						
- Local services	medium	private	high	medium	low	2.6
- Long distance and added value	high	private	high	few	low	3.0
Transport						
- Railbed and stations	low	club	high	medium	medium	2.0
- Rail freight and passenger services	high	private	high	medium	medium	2.6
- Urban bus	high	private	high	many	medium	2.4
- Urban rail	high	private	medium	medium	medium	2.4
- Rural roads	low	public	low	many	high	1.0
- Primary and secondary roads	medium	club	medium	few	low	2.4
- Urban roads	low	common property	medium	few	high	1.8
- Port and airport facilities	low	club	high	few	high	2.0
- Port and airport services[b]	high	private	high	few	high	2.6

1 Potential for competition [high = 3, medium = 2, low = 1]
2 Characteristics of good services [private = 3, club/common property = 2, public = 1]
3 Potential for cost recovery from user charges [high = 3, medium = 2, low =3]
4 Public service obligations (equity concerns) [few = 3, medium = 3, many = 1]
5 Environmental externalities [low = 3, medium = 2, high = 1]
6 Marketability index as an average of ratings across each row
a Due to either absence of scale economies or sunk costs, or existence of service substitutes
b Including cargo handling, shipping, and airlines

Source: World Bank (1994)

The above distinction between public sector capital and infrastructure capital has consequences for the way in which infrastructure can be measured. For analyses at the macro-economic level there are often no aggregate financial data available, because infrastructure is not a concept used in national accounts. Therefore, in many studies one uses investments of the public sector as a substitute. However, as pointed out above, this concept only partially coincides with the data one needs. It includes, for example, defence related investments, but investments in airports or telecommunications are not included. It is important to keep this in mind when interpreting empirical results on infrastructure investments, which will be given in the section 2.5.

2.4 Ways of Measuring Infrastructure

Data on infrastructure can be presented in various ways (see Table 2.3). A first distinction is between physical and monetary units. Physical units (for example, number of kilometres of roads, canals or railways, number of runways, etc.) are often used. However, more specific information is often needed for an appropriate interpretation (single track versus double track, electrified versus non-electrified; capacity of ships allowed in canals, number of locks, maximum possible speeds, etc.). Such distinctions often make simple aggregate sums such as total length of railway system rather uninformative. A general trend in Europe, for example, is that this length is decreasing as a consequence of closing down strongly underutilized tracks. In terms of length this often dominates the new investments in new tracks, double tracks or electrification. In such a case the decrease in aggregate track length does not reflect the increase in capacity which actually takes place. This problem of aggregation is even more difficult in the case of different infrastructure types. This may be one of the reasons why one often uses data on infrastructure stocks in terms of monetary values, because there is no aggregation problem when all variables are measured in the same unit of measurement. Monetary data may however be plagued by other problems. For example, when construction costs in one region (or country) are much higher than in another region, data in monetary terms will give a misleading impression of relative infrastructure endowment. In such a case one would need a 'quantity index' to correct for price differences.

Another dimension of infrastructure measurement raised in the Table concerns the difference between stocks and flows. When physical units are used, the stock and the flow represent the capacity and the change in the capacity (cases I and II), respectively. Note that change in capacity may be measured in gross terms (by only considering extensions) and in net terms (by taking into account that existing capacity may be discontinued).

Table 2.3 Types of data on infrastructure

	stock	flow
physical	I	II
monetary	IV	III

Monetary data on capital stocks are usually only available in terms of flows: the annual investments (III). These are typically *gross* investment data, which means that they include expenditures on maintenance. This part is quite substantial. For example, in countries with a well-developed highway system, some 50 % of road investments will relate to maintenance activities.

Measuring infrastructure stocks in monetary terms (IV) is not as easy as one might think. Necessary ingredients are data on capital investments of type III. These have to be corrected for price changes over the course of time. Depreciation has to be taken into account. Since investment data usually are only available during a limited number of years, one needs an estimate of the value of the capital stock in the first year. Because of these difficulties it is no surprise to see that when data in monetary units are used these are usually flow data.

For spatial analyses one will often be in the position to compare infrastructure levels among regions. When regions differ in size, a standardization of infrastructure figures would be helpful. Two basic ways of standardization are often used: standardization by population and by area. When regions differ strongly according to population density, the choice of standardization measure has a clear influence on the result.

Table 2.4 Motorway density in various European regions (1993)

region	population density (persons per square km)	motorway density (km per 1.000 square km)	km. motorway per 1.000 capita
Urbanized regions			
- Ile de France (FR)	910	43.4	0.05
- South East (UK)	653	33.8	0.05
- West Nederland (NL)	605	61.5	0.10
- Nordrhein Westfalen (GE)	520	63.1	0.12
- Lombardia (IT)	373	22.7	0.06
Rural regions			
- Sud (IT)	152	13.5	0.09
- Noroest (SP)	96	4.6	0.05
- Mecklenburg Vorpommern (GE)	80	10.2	0.13
- Voreia Ellada (GR)	59	1.3	0.02
- Ireland	52	0.5	0.01

Source: Eurostat (1997)

As shown in Table 2.4, highly developed regions have a high density of motorways per 1.000 square km compared with rural regions. However, if we were to use per capita figures, the picture may be less evident: in rural regions such densities are often comparable to the densities in highly urbanized regions, or at least show a less poignant contrast. It is not so straightforward to say which of the two standardization methods is most appropriate. When population is uniformly distributed within a region, motorway *density* reflects the average distance of any place to the nearest point on the network as well as the average detour users have to make when they want to travel from one place to another in the network. Motorway length *per capita* would in general be a less relevant variable: a long length per capita is of little use when the average distance to the network is quite long. An exception occurs when congestion is a relevant issue. In this case road length per capita may reflect something about absence/presence of congestion.

A similar comparison can be carried out for point infrastructure as, for example, railway stations. Railway station *density* provides information about the average distance from all places in a region to the nearest railway station. Railway stations *per capita* reflect something about potential problems in capacity within the facility, which may be useful when there is the risk of congestion.

We conclude that when infrastructure levels have to be compared, we tend to prefer the use of densities (in the absence of congestion), but we immediately note, that these densities are still very poor proxies for the quality of infrastructure services. For example, much depends on the specific spatial structure of the network and the spatial distribution of demand. These issues will be further discussed in section 2.6, but first we will present data on infrastructure investments in various European countries.

2.5 Trends in Infrastructure Investments

2.5.1 Europe

As is shown in Table 2.5, most European countries have witnessed a decline in infrastructure investments. However, in recent years in nearly all countries there is a slight increase in gross investment in transport infrastructure measured as the share in Gross National Product. In 1992 in Italy and Sweden only the level is above the 1975 level. The same pattern of decrease and a recent increase is found when gross investment in transport infrastructure is measured as the share in Gross Fixed Capital Formation. The only difference is that according to this measure, in five countries the 1992 share is above the 1975 level. Apparently, the Gross National Product has grown at a faster rate than Gross Fixed Capital Formation in most countries.

In Table 2.6 the share of rail, road and waterway in the total infrastructure investments is given for the period 1975-1992. A first remark is that only the Netherlands, Belgium and - to a lesser extent - Germany and France invest in their waterways. However, the share is low compared with the share received by roads and railways. In all countries the majority of the investments are in the road network. In most countries the share of the road network in total infrastructure investments is above the 75 percent level. This also holds true for France, where during the past decades large investments took place in the high-speed rail network. There are only a few European countries which show a comparable low share of the rail investments. Apparently, nearly all the rail investments in France were directed towards the high-speed rail network.

Table 2.5 Gross investments in transport infrastructure (roads, rail, canals) as a percentage of Gross National Product and of Gross Fixed Capital Formation

	Gross National Product					Gross Fixed Capital Form.				
	'75	'80	'85	'90	'92	'75	'80	'85	'90	'92
Austria	2.3	2.1	1.4[1]	---	---	8.7	8.2	6.4[1]	---	---
Belgium	1.9	1.8	0.9	0.6	0.7	8.7	8.5	5.9	3.1	3.8
Denmark	1.3	1.2	1.4	0.9	1.0	7.0	6.3	7.4	5.3	6.7
Finland	1.9	1.9	1.0	1.1	1.4	6.4	7.5	4.2	4.2	7.8
France	1.3	1.1	0.8	0.9	---	5.9	5.3	4.0	4.4	---
Germany	1.8	1.3	1.1	---	---	8.6	5.8	5.4	---	---
Greece	0.8	0.7	0.8	---	---	3.6	2.9	4.0	---	---
Ireland	0.6	0.8	0.9	---	---	3.0	2.6	4.6	---	---
Italy	1.1	0.7	0.9	1.1	1.6	5.6	3.5	4.1	5.4	5.9
Luxembourg	2.2	1.8	1.2	---	---	8.2	6.8	8.3	---	---
Netherlands	1.5	1.1	0.9	1.1	0.7	7.4	5.2	4.6	5.3	3.3
Norway	1.8	1.3	1.0	1.0	---	5.1	5.3	4.4	5.4	---
Portugal	0.9	0.5	0.5	0.7	0.8	4.6	1.8	2.5	2.5	3.1
Spain	1.5	0.7	0.7	1.5	1.4	6.1	3.4	3.7	6.2	6.7
Sweden	1.1	0.8	0.9	0.9	1.2	5.6	3.9	4.5	4.4	7.0
Switzerland	2.2	1.8	1.4	1.5	1.6	9.2	7.4	5.9	5.7	6.6
Turkey	---	1.0	1.2	---	---	---	5.4	6.3	---	---
United Kingdom	1.0	0.5	0.7	0.8	0.9	4.9	3.1	4.1	4.1	5.5

1 = 1984

Source: ECMT (1988, 1997)

Table 2.6 Share of rail, road and waterway in total infrastructure investments in the period 1975 - 1992

	Rail					Road					Waterway				
	’75	’80	’85	’90	’92	’75	’80	’85	’90	’92	’75	’80	’85	’90	’92
Austria	10.0	19.2	22.8	--	--	88.9	80.3	76.2	--	--	1.1	0.6	1.0	--	--
Belgium	22.8	23.4	24.0	23.6	25.8	69.2	72.2	52.9	65.9	68.3	8.0	4.4	23.0	10.5	6.0
Denmark	14.2	15.7	11.2	15.7	16.3	85.8	84.3	88.8	84.3	83.7	0	0	0	0	0
Finland	26.8	22.3	14.3	15.3	17.3	72.0	76.4	84.1	83.8	81.1	1.2	1.2	1.6	0.9	1.6
France	11.4	16.7	13.5	18.3	--	85.9	82.0	85.7	81.2	--	2.7	1.3	0.8	0.5	--
Germany	26.5	16.4	23.4	--	--	69.2	79.9	71.9	--	--	4.2	3.7	4.6	--	--
Greece	11.7	20.9	19.4	--	--	88.3	79.1	80.4	--	--	0	0	0	--	--
Ireland	17.1	11.9	3.2	--	--	82.9	88.1	96.8	--	--	0	0	0	--	--
Italy	10.5	21.5	28.5	24.1	22.8	89.2	77.9	71.3	75.6	77.1	0.3	0.5	0.2	0.2	0.1
Luxembourg	11.9	20.0	19.3	--	--	88.1	80.0	80.6	--	--	0	0	0.1	--	--
Netherlands	7.4	15.9	22.7	14.0	18.8	85.9	78.9	74.3	79.4	73.8	6.8	5.2	3.0	6.6	7.4
Norway	12.4	13.3	14.1	8.7	--	87.6	86.7	85.9	91.3	--	0	0	0	0	--
Portugal	8.7	13.3	21.2	20.4	20.3	91.3	86.7	78.8	79.6	79.7	0	0	0	0	0
Spain	31.8	39.1	19.8	20.7	14.7	68.2	60.9	80.2	79.3	85.3	0	0	0	0	0
Sweden	22.5	13.5	17.2	19.8	27.9	77.5	86.5	82.8	80.2	72.1	0	0	0	0	0
Switzerland	18.7	13.0	16.6	23.4	22.1	80.4	86.4	82.9	76.5	77.8	0.9	0.6	0.6	0.1	0.2
Turkey	--	1.0	35.2	--	--	--	99.0	64.8	--	--	--	0	0	--	--
United Kingdom	15.8	19.2	18.3	15.1	19.2	84.0	80.6	81.5	84.7	80.6	0.2	0.2	0.2	0.1	0.2

Source: E.C.M.T. (1988, 1997)

2.5.2 The Netherlands

In this section we will discuss trends on infrastructure investments in a particular country where available data allows a long-term perspective (the Netherlands). As already indicated in section 2.3, data on *public investments* are the main source of information available. Table 2.7 shows that in the Netherlands before the second world war, the share of public investments in total public expenditures has been considerable: 35-40 percent. After the war, the share of public investments declined to 20-30 %. In the 1980s the share even dropped to 13-14 %. Before world war II, the public sector undertook 25-30 % of total investments. After the war, the share fluctuated for a long time around 18 %, but fell to about 11-15 % in the 1980s.

Table 2.7 Public investments in relation to other macro-economic indicators, The Netherlands, 1921-1994

	Public invest. as a % public expen-ditures	Public invest. as a % total invest.	Public invest. as a % GNP	Total invest. as a % GNP	Growth rate GNP (%)	Growth rate employ-ment (%)
1921-24	35	23	4.6	20	5.2	0.5
1925-29	38	24	4.4	18	4.4	2.0
1930-34	41	32	5.9	19	-1.4	-1.4
1935-39	37	31	4.6	15	3.6	2.5
1950-54	21	18	3.6	20	4.7	0.9
1955-59	28	17	4.0	24	3.6	0.4
1960-64	23	18	4.5	25	5.6	1.3
1965-69	24	19	4.8	25	5.4	0.6
1970-74	21	18	4.2	23	4.4	0.0
1975-79	17	16	3.5	21	2.4	1.4
1980-84	14	15	2.9	19	0.6	-0.1
1985-89	13	11	2.4	21	2.2	2.7
1990-94	15	14	2.7	20	2.3	0.6

Source: CBS (1995)

The Table can also be used to give a first impression on the primary theme of this monograph: the link between public investments (infrastructure) and the economy. The main message is that based on these macro-economic data, no simple immediate relationship can be observed. For example, the peak period of public investments (1930-34) is characterized by negative GNP growth. During the 1980s public investments decreased, but GNP growth increased. The connection with employment is also unclear. Relatively high levels of public investments in the 1950s and 60s are accompanied by very modest levels of employment growth. The period 1985-89 combines the lowest level of public capital investments with the highest level of employment growth. Of course macro-economic relationships between public investments and the economy are much more complex than the simple comparisons made here, but the first impression is still important: there are no simple immediate links between public investments on the one hand and GNP or employment growth on the other.

2.5.3 Trends in Private Versus Public Capital Formation

Table 2.7 shows large shifts in the ratio of public and private investments. The long-run relationship between public and private investments can be illustrated by means of Figure 2.2. The balanced growth path in this Figure indicates how public and private capital can be combined to reach a maximum level of national production.

The course of this growth path is primarily determined by technological developments and by relative prices. Hirschman (1958) has suggested that in reality deviations from the balanced growth path will occur, especially because public capital is characterized by indivisibilities. The actual development is sketched by the curved line. According to Figure 2.2, we are now facing a situation of obvious bottlenecks in public capital.

Another explanation for the cyclical pattern in the relationship between private and public capital formation is provided by political economy theories of decision making in the public sector. For example, Frey (1983) makes clear that the ways politicians, bureaucrats and voters interact will easily lead to policy cycles with alternating periods of high and low infrastructure capital formation.

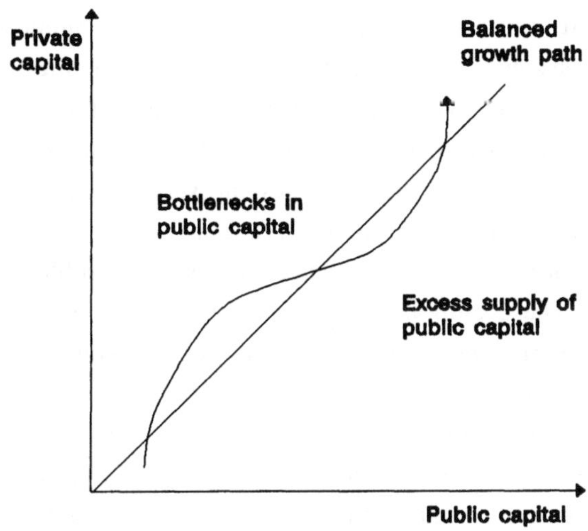

Figure 2.2 Excess capacity and bottlenecks in public capital

Several factors can be mentioned that explain the long run decline of the share of public sector investments, in total investments observed in many countries over the last 20 years:

- The financial position of the public sector has been weak in many countries. Public investments are less difficult to postpone than other categories of public expenditures, and have therefore suffered most from the policy to reduce the budget deficit.
- There is excess capacity for certain types of infrastructure, owing to large investments in the 1960s and 1970s. Complementarity with the private sector means that investments will only take place after excess capacity has disappeared.
- The 1980s were characterized by changing opinions about the role of the state in the economy and the society as a whole. The idea of a retreating government offers little inspiration for ambitious investment programmes initiated by governments.
- The potentially negative external effects of many public investments (especially those on the environment) have received more attention on the political agenda in the 1980s compared with the decades before.

- Technological developments lead to a shift in the need for infrastructure towards the private sector (telecommunications services, airline system, etc.). This means that falling public investments may be compatible with an adequate level of infrastructure. It implies that the overlap between public sector investments and infrastructure is not fixed, but changes in such a way that the overlap becomes smaller (see Figure 2.3). This notion is supported by a life-cycle approach of infrastructure types (cf. Grübler & Nakicenovic, 1991). These authors demonstrate that in the long-run, clear cycles of infrastructure investments take place with a canal cycle that began in the first part of the 19th century. It was followed by a railway cycle in the second half of the 19th century. The 20th century has witnessed the road cycle. In the present decades aviation and telecommunications have become quite dynamic.

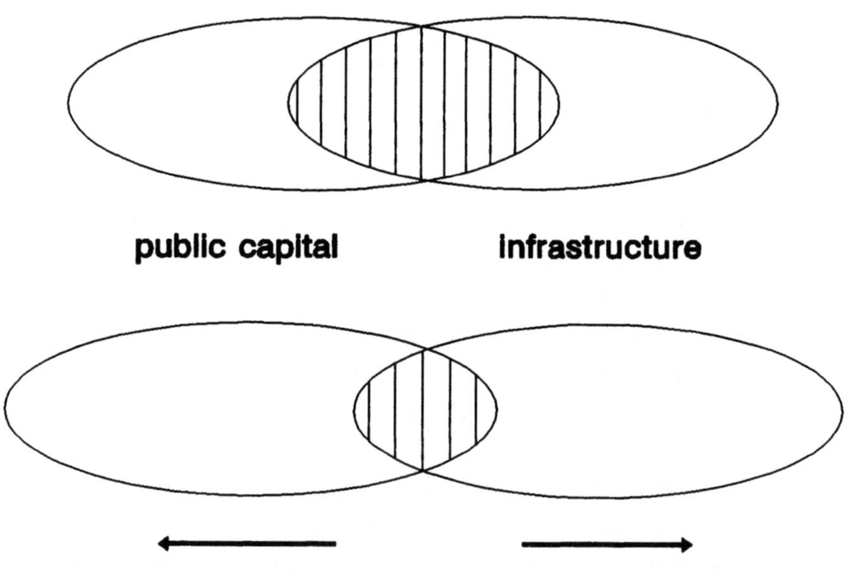

Figure 2.3 Shift in overlap between public investments and infrastructure

The last line in Table 2.7 indicates, however, that the long-run decline in the share of public sector investments has come to a halt in the 1990s. Some factors contributing to this seem to be that public investments in the road system have grown as a response to increasing congestion and that the financial problems of the public sector have become less pressing. In addition, the environmental and congestion problems related to road transport have led to a new emphasis on rail investments. The level of public sector investments will most probably remain higher than that of the 1980s, because in the Netherlands as well as in other European countries, large construction projects have begun for long-distance links in the context of trans-European networks.

2.6 The Accessibility Concept

2.6.1 Definitions of Accessibility

In section 2.4 we found that to study infrastructure services, data on infrastructure stocks and flows are not sufficient. In this section we will discuss the concept of accessibility in order to obtain a more precise view of the potential use of infrastructure (see Bruinsma & Rietveld, 1998a). Indeed, the accessibility concept will play a crucial role in this monograph. Therefore, we pay special attention to ways of measuring it. A general definition of accessibility is that it is the potential of opportunities for interaction (see Hansen, 1959 and Martellato et al., 1998). Related definitions use terms like 'ease of spatial interaction' 'potentiality of contacts with activities or supplies' or more precisely: 'attractiveness of a node in a network taking into account the mass of other nodes and the costs to reach those nodes via the network'.

Accessibility for whom and to what? These basic questions indicate that accessibility is specific per type of user and per trip purpose. Therefore, when measuring accessibility, these aspects should be specified. An implication is that accessibility becomes trip purpose specific (see also 2.6.2).

In practice a large number of possible operationalizations is used. In Table 2.8 we list a number of possibilities. Related surveys have been given by Pirie (1979), Song (1996), Handy and Niemeier (1997), and Miller (1998). An axiomatic approach to accessibility is given by Weibull (1980).

When we compare these definitions of accessibility, it appears that for most

Table 2.8 Alternative operationalizations of accessibility

operation-alization	definition	assumptions/ remarks	example
acc1	A node has access to a network if a link exists between the node and the network	Accessibility actually means access or connectivity; accessibility is a binary variable: 1 or 0	The city of Bonn is connected to the German autobahn network
acc2	The accessibility of a node with respect to a network is the distance one has to travel to the nearest node on the network (neg.)	If accessibility defined according to acc1 equals 1, acc2 attains its most favourable outcome (acc1=1 implies acc2=0)	The distance of village A to the nearest point of entry of the national express-way system is 16 km. The distance from Gent to Brussels airport is 60 km
acc3	The accessibility of a node in a network is the total number of direct connections with other nodes		From Rotterdam airport one can fly to 12 destinations without changing planes
acc4	The accessibility of a node in a network is the total number of links connected to this node		From Hannover the railway lines extend in four directions
acc5	The accessibility of a node to an-other node is measured as the travel cost between these nodes (neg.)	This definition considers accessibility in a strictly bilateral way without summa-tion across destinations	It takes 2.5 hours to fly from London to Lisbon; the costs of a round-trip are USD 460
acc6	The accessibility of a node in a network is the weighted average travel cost between the particular node and all nodes in the network (neg.)	Weights may relate to the masses of the nodes, or to the total number of trips made to the nodes	The average distance from Vienna to all major cities in Europe weighed by pop-ulation size is 880 km; when weighed by the shares in the total number of trips it is 350 km

Table 2.8 Alternative operationalizations of accessibility (continued)

acc7	The accessibility of a node in a network is the expected value of the maximum utility of a visit to any node	Utility of a visit to a certain node is assumed to depend on: -the mass of the node -the travel costs of a trip to the node -a stochastic term	The accessibility of Milan for road transport in Europe is 56 compared with Frankfurt 100 (index)
acc8	The accessibility of a node in a network is (proportional to) the spatial interaction between the node and all other nodes	The spatial interaction between nodes may be directly measured or computed by means of a spatial interaction model	See acc7
acc9	The accessibility of a node in a network is the total number of people one can reach from the node within a certain transport cost limit	The transport costs limit can be formulated in any dimension: distance, travel time, etc.	From Copenhagen one can reach 80 million people within a time interval of 4 hours
acc10	The accessibility of a node is the inverse of the balancing factor in a singly or doubly constrained spatial interaction model.	This interpretation has been given by several authors (e.g., Wilson, 1982, Hamerslag, 1980)	See acc7
acc11	Accessibility is measured by means of expert judgement	No formal definition is given	The five European cities with the best accessibility are: A, B, C, D and E

Remark: accessibility is defined as a positive concept; the higher the value attained the better. Only in 3 cases (acc2, acc5, acc6) is the definition in negative terms so that an inverse transformation is needed.

of them the information we need concerns spatial data such as the location of nodes, the length of links, and data on transport costs (travel time, fares, etc.). However, for some of them we require additional information. These cases are acc7, acc8 and acc10, where we would need data on spatial interaction patterns, or parameters of models describing these spatial interactions.

A rather general formulation of a spatial interaction model that can serve as a basis for accessibility studies is:

$$T_{ijk} = a_k \, X_i \, X_j \, f_k(c_{ij}) \tag{1}$$

T_{ijk} denotes travel demand from i to j for trip purpose k; a_k is a constant, reflecting intensity of travel for trip purpose k. X_i and X_j denote the size of origin and destination j. The deterrence function $f_k(c_{ij})$ gives the dependence of the number of trips for purpose k on (generalized) transport costs c_{ij}. Using this formulation of a spatial interaction model as a starting point for measuring accessibility of i means that a summation takes place across all destinations j. The accessibility is subsequently based on T_{ijk}/X_i to make it independent on the mass of the origin. When in addition accessibility is measured as an index (for example, the highest value equals 100) the constant c_k may also be dropped.

In the case of acc8 a frequently used definition is: the accessibility of a city is measured as the weighted sum of the population in all cities where weights are equal to the 'travel time decay'(cf. Keeble et al., 1982):

$$Acc_i = \sum\nolimits_j pop_j \;/\; (\text{travel time}_{i,j})^a \tag{2}$$

According to this measure, the location of a city at one hour's travelling distance contributes more to the accessibility than if that city were to be located four hours away. If the travel time parameter a is set equal to 1 (this is the parameter value used most often in empirical studies), the ratio of the weights is 4:1.

In the case of acc7, accessibility is given a basis in utility theory by using a utility function as a starting point (see, for example Williams & Senior, 1978, Bröcker, 1989, Miller, 1998):

$$u_{ij} = v_j - bc_{ij} + e \tag{3}$$

where u_{ij} denotes the net utility of a visit from i to j, v_j denotes the utility without transportation costs c_{ij}. The parameter b represents utility per unit transport costs, and e is a stochastic component. Suppose that residents of location i choose the destination of their trip by maximizing the expected net utility of the trip with i as the origin (denoted as A_i). Then when e is Weibull distributed, stochastic utility theory leads to the following result for A_i

$$A_i = \ln \sum_j \exp(v_j\text{-}bc_{ij}) = \ln \sum_j \exp\ v_j \exp(-bc_{ij}) \tag{4}$$

Clearly, the definition of A_i is very close to the one given for acc8 when v is the log of population size. In the case of acc9 one would need a standard for a transport costs limit (maximum 1 hour travel time for commuting trips; maximum 30 minutes travel time for shopping trips, etc.).

A good number of these accessibility measures has also been surveyed by Song (1996). He uses a set of alternative definitions of accessibility and tests which one gives the best statistical explanation of population densities in the Reno-Sparks metropolitan area in Nevada (US). He concludes that the most commonly used measure (equation (1) with parameter a=1) is not statistically surpassed by any other measure. This is also the specification that will be used several times in the present monograph (Chapters 5, 6, 7 and 10).

2.6.2 Further Variations in the Accessibility Concept

In fact, many more definitions can be given than those presented here. Variations may relate to a number of factors listed below.

- *The way the mass of a node is measured.* In most applications the mass of a node is represented by its population size. However, various other mass indicators can be used such as employment or GDP. When one studies accessibility for specific production activities, the masses may also be related to more specific variables, such as the volumes of sales in cities, or the amounts of inputs received from certain input nodes.
- *The use of a uni- versus a multimodal approach.* Many studies are unimodal. They address accessibility according to a particular transport mode. However, in certain cases one cannot do without a second mode, i.e., when a mode has incomplete coverage of space: not every city has an airport, inland navigation is not possible in certain regions, and not every municipality has

an access point to the national highway system. In this case one usually adds a supporting mode with a spider structure connecting all points in space to the nearest node in the network considered. A more drastic step is the introduction of a truly multimodal network with transshipment points. In the case of multimodality one can proceed in several ways. One approach would be to treat each modality separately so that one ends up with several accessibility indicators. Another approach is to construct an accessibility indicator based on the composite costs of transport modes. This can be done by using a log-sum approach (see Williams, 1977, and Vickerman et al., 1998). A step further would be to allow for intermodal transport chains.

- *The way the transport costs are measured.* Usual proxies of transport costs are distance and travel time. When one would use a generalized transport cost approach one would add up distance dependent costs (fares, or the variable costs of vehicle use) and time dependent costs (travel time times the value of time). Waiting times would also have to be taken into account, as well as the costs of inconvenience when a certain transport mode has a low frequency causing activities to be rescheduled. An important implication of this line of thought is that since values of time differ among different groups of people, accessibility would also have to be specified for different groups. For low income people with ample time, the relevant costs when measuring accessibility would be the direct monetary costs of transport. On the other hand, for busy high income people, the relevant transport costs relate to travel time. A final cost component which is usually neglected in accessibility studies concerns the costs of uncertainty in unreliable networks. As a side remark it is interesting to note that in particular discussions about transport policies, 'accessibility' is interpreted as 'lack of congestion'. For example, in the Dutch transport policy, accessibility is closely connected to the achievement of minimum congestion probabilities at certain links of the expressway network (Dutch Ministry of Transport, 1989).
- *The choice of a particular trip purpose.* Accessibility is often studied in view of a certain type of activity. This not only has an implication for the way masses are defined (see above), but also affects the spatial interaction function used in acc8. Thus, the spatial interaction formula should be trip purpose specific: for example, business trips tend to be more distance dependent than shopping trips.
- *The choice of the point in time.* When networks are congested, the choice of the point in time is important. This issue is especially consequential for

national and regional accessibility studies.

- *The choice of the spatial interaction function.* As already mentioned above, different specifications can be used for the spatial interaction function. Two commonly used forms are the power function (acc8) and the exponential function (acc7). The shapes are similar, but a difference is that the power function may achieve very high values for short distances. Fotheringham and O'Kelly (1989) note that the exponential function tends to be more suitable for analysis at short distances and the power function tends to be more appropriate for longer distance spatial interactions.

2.6.3 Problems in the Measurement of Accessibility

Some problems that are important in the measurement of accessibility are listed below. They relate to:

1. Dimension of measurement of accessibility.
2. Choice/demarcation of the nodes.
3. Demarcation of total area.
4. Treatment of internal accessibility.

ad 1. Dimension of measurement of accessibility. Most of the accessibility concepts presented have a clear dimension and are easy to interpret. However, there are some exceptions: acc7, acc8 and acc10 require an index value, for example by comparing different nodes (Paris = 100), or different years (1990 = 100) in order to provide the user with an appropriate frame of reference.

ad 2. Choice/demarcation of the nodes. Three different approaches can be identified in the demarcation of the nodes.

First, the analysis can be confined to a set of nodes which do not completely cover the entire area. For example, as is often done, the nodes represent the major cities in a country. Since administrative boundaries are not useful here, one faces the problem of how to demarcate the cities. Especially when accessibility is studied in an international context this is often a bottleneck, because of a lack of commonly used measurement procedures among countries.

A second approach would be to use nodes as point representations of regions (surfaces) so that the total area is covered. Here again we have the problem of demarcation. In the case of the European Union, one has an advantage in that a commonly accepted regionalization in terms of NUTS regions can be used.

A third approach, which has become possible from the emergence of GIS (geographical information systems), is the use of grids. These grids have a large advantage in that demarcation problems can be avoided because the spatial units are exogenously given. However, when one finally wants to formulate conclusions about the accessibility of cities such as Paris or London, one has to decide on which grids this should be based and how the grid scores should be aggregated.

ad 3. Demarcation of total area. The total area taken into account in an accessibility study is obviously an important choice. Of course, the demarcation used depends on the purpose of the accessibility study. The smaller the sensitivity to transport costs, the larger the study area should be. In many studies the spatial demarcation coincides with the borders of a country. This tends to lead to low accessibilities of locations near borders. In view of the ongoing process of European integration, such a demarcation along a national boundary may be questioned. Cities that are peripheral in a national context suddenly appear to have high accessibility in an international setting.

When accessibility is studied at the European level, the question becomes how to demarcate 'Europe'. Some studies only take into account the European Union members. Indeed, the choice for the European Union has advantages because of the availability of a standardized data set and because the economic and policy importance of the European Union justifies a special treatment of the European Union. Nevertheless, even if one wants to focus on accessibility of European Union members, it would be strange to neglect their accessibility to non-European Union countries. For the accessibility of Greece, for example, it is much more important to include its neighbour countries than member-countries like Ireland and Denmark.

ad 4. Treatment of internal accessibility. The internal accessibility of a spatial unit may have a substantial impact on the final outcome in the case of accessibility concepts acc7, acc8 and acc10 (see also Frost & Spence, 1995). The reason is that the functional forms used lead to high weights for the internal accessibility. The problem is that usually the data about the local transport network are weak. For example, one would like to know whether the average travel time between all origins and destinations in a metropolitan area is 30 minutes or 75 minutes. This uncertainty has far-reaching consequences for the accessibility scores of cities, especially the larger ones.

It would be tempting to avoid this problem by ignoring the internal accessibility of a city. This would lead to counter intuitive outcomes, however, with high scores for smaller cities near large cities and low scores for the large cities themselves. For example, Orleans would receive a high accessibility score because of its proximity to the big mass city of Paris, but in the accessibility of Paris itself, its own mass would be neglected. The use of small areas in the case of a grid system clearly leads to a smaller dependence on the internal accessibility, and hence provides a good way of avoiding this problem.

2.6.4 Broader Comparison of Some Accessibility Measures

An obvious advantage of the average distance based indicator (acc6) is that it is easy to interpret. However, a disadvantage is that it depends rather strongly on the demarcation of the total area of study. For example, when Russia is not taken into account, the average distance from Rotterdam to all European cities is 1100 km. However, when Russia is included, the figure becomes 1600 km, and when Asia is included it becomes 6000 km. Note that in this case the relative differences between European cities also become small. When the choice of the total area of study is clear because of a priori reasons (for example, in the case of an island such as Great Britain; or when the study area is given by administrative boundaries) this is not a serious problem. But in many cases there are no compelling a priori reasons for such a demarcation. Thus, accessibility measured as the (inverse function of) average distance has the unattractive property that it becomes rather sensitive to the demarcation of the area of research. 'Too narrow' a demarcation means that relevant nodes are missing, while when the demarcation is 'too broad', the outcomes of the accessibility indicator are strongly influenced by irrelevant nodes.

One possibility of solving this problem is to apply a weighting by means of the shares in the total number of trips. This has the attractive property that when areas are added which are irrelevant in the sense that no trips are made to those areas, the accessibility indicator is not affected. However, this weighting scheme also has an unattractive feature: it may lead to high outcomes for isolated cities where the large majority of the trips is made in the immediate vicinity. This would for example, imply a high accessibility for an isolated city like Tromso in Northern Norway.

The gravity and utility based definitions of accessibility (acc7 and acc8) do not suffer from such a strong dependence on the demarcation of the area of research. The reason is that adding an irrelevant destination does not affect the accessibility values of the other nodes, and therefore is harmless. At the same time the gravity or utility based approach does not give rise to the problems mentioned above in which isolated cities may receive good scores. To stay with the example given above, Tromso will receive a low accessibility value (according to definitions acc7 and acc8), because its large distance to major European cities prevents it from benefiting from the large masses present there. However, there is another problem with acc7 and acc8 which deserves our attention: they may depend strongly on the way their own mass is treated; this subject has already been discussed above.

Thus, we conclude that an interesting trade-off exists between average distance based indicators (acc6) and interaction based indicators (acc7 and acc8). The first ones are sensitive to the demarcation at the outer side of the area of study; the latter ones depend rather strongly on the way the internal distance problem is handled.

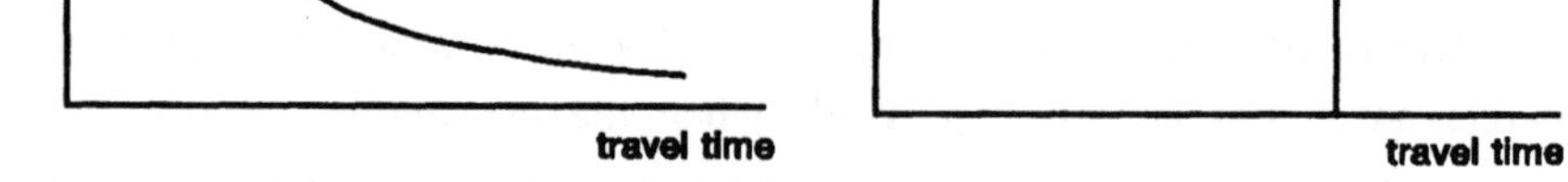

Figure 2.4 Travel time decay in acc8 and acc9

A difference between the approaches of acc8 - which indicates the *potential* accessibility - and acc9 - which indicates the *daily* accessibility - concerns the treatment of travel time decay (see Figure 2.4). In acc8 there is a gradual decay: a halving of travel time leads to a doubling of the weight for the pertaining city pair. Very remote cities do receive a positive weight, although it may be very small. The contribution of a city to its own accessibility may be considerable for large cities. It explains part of the relatively high rankings of cities such as Istanbul and Athens, but also of Paris and London.

In acc9 on the other hand, travel time decay does not occur until the travel time exceeds a critical level (about 6 hours) such that it is no longer possible to spend 4 hours at the location of a visit. No further differentiation is used within these ranges of travel times. For example, a major improvement of the link between two cities so that travel time is reduced from 5 hours to 2 hours does not lead to an improvement of the accessibility of these two cities according to acc9.

2.6.5 Link between Infrastructure Improvement and Accessibility

Improvement of infrastructure will in general lead to an improvement of accessibility of locations. There are some exceptions, however.

First, in congested road systems an improvement of a link or the addition of a new link may lead to higher average transport costs because of the possible discrepancy between user equilibrium and system optimum (the so-called Braes paradox, see for example Sheffi, 1985). This paradox may be of some relevance in congested urban road networks. A second exception is that improvements in an infrastructure network usually imply disadvantages for particular points in space. For example, the upgrading of a highway will lead to a reduction in the number of highway access points so that for some locations the highway system may be less accessible.

An inspection of the various definitions of accessibility in Table 2.8 reveals that most of them will be able to capture changes in the network, although the level of refinement varies. For example, distance based measures will not be able to capture speed improvements. In addition, frequency improvements are only captured when waiting times and scheduling inconveniences are taken into account.

When we have a closer look at the locations where a network improvement takes place, we note that accessibility indicators acc1 up to acc4 have a rather local orientation. An improvement on a link between A and B is only important for A and B, not for other nodes. With the other accessibility concepts this is rather different. For example, the construction of the Channel tunnel will not only affect regions immediately connected such as South East England and Northern France, but also other regions. The particular ways in which this happens may be rather different, however. For example, with an average travel time index (acc6), the absolute improvement in travel time between Paris and London is equally large compared with Madrid and London. In the case of accessibility concepts acc7 and acc8, Paris would benefit much more than Madrid, because the weight of London will be higher for Paris compared with Madrid in the accessibility formula.

Another property of accessibility measures concerns their *symmetry*. Suppose a link is improved connecting a large city and a small city. To what extent does this lead to symmetric improvement of accessibilities of both cities?

Symmetry holds for the measures acc3 up to acc5. But for the measures acc6 up to acc10, it no longer holds true. For these measures we will find that an improvement of travel costs of the link between a large node and a small node will lead to a larger increase with the small node than with the large node. The transport cost improvement is the same for both nodes (assuming symmetry in transport costs), but the weights applied are different, leading to a larger increase in accessibility for the small city compared with the large city. This lack of symmetry seems to be larger for the average travel costs function compared with the other two functions. The lack of symmetry occurs both in absolute and relative terms. In relative terms the discrepancy is even larger.

2.7 Conclusion

It appears impossible to give a definition of infrastructure that entirely matches what is commonly understood by it in daily parlance. Therefore, we decide to use a nominal approach (see section 2.2).

In the discussion of definitions we paid special attention to the property of publicness. An important point made in this Chapter is that transport infrastructure is often publicly provided, but that also the private sector will play its role. There are important sectors of infrastructure (telecommunications, airports) where the private sector is a main provider. From an historical perspective the overlap between the concepts of infrastructure and public capital tends to become wider.

In our empirical work in Chapters 5-13 we will mainly deal with infrastructure types such as roads and railways. These are strongly dominated by the public sector. This means that we will focus on the public sector as the provider of infrastructure. Thus, for most of the empirical part of the book the leading question concerns the private sector's response to *publicly* provided infrastructure.

Data on infrastructure and public capital indicate that long-run cycles occur, not only for individual types of infrastructure, but also for infrastructure (or public capital) as a whole. During the last decades we witnessed a decrease, but this decrease seems to have come to a halt.

Special attention has been given to the concept of accessibility because it sheds more light on the potential use of infrastructure. A substantial number of operationalizations appears to be possible, some of which are related to spatial interaction models. We find that particular measures of accessibility are based on the choice of, among others: the way the mass of a node is measured, the use of a uni- versus multimodal approach, the way transport costs are measured, choice of a particular trip purpose, point in time, and specification of deterrence function. Therefore, it should not come as a surprise when the results of different accessibility studies are conflicting (comparisons will be carried out in Chapters 5 and 6). Some frequently observed problems in measuring accessibilities concern: the choice and demarcation of nodes (for example cities), the demarcation of the total area of study, and the treatment of internal accessibility. We discussed a number of ways to alleviate these problems. Apparently some of the measures are more sensitive to problems of measurement than others. This provides an important criterion for the selection of accessibility measures.

After this Chapter on definitions and measurement we will now first give a review of relevant theories on the relationship between infrastructure and the economy.

3 Theory: Infrastructure and the Space Economy

3.1 Introduction

This Chapter is devoted to a discussion of theoretical perspectives on infrastructure and the space economy. Our starting point is Table 3.1, according to which transport infrastructure investments have both temporary and non-temporary effects on the economy. A major temporary effect concerns the stimulation of employment and income during the construction phase via the demand side.

Table 3.1 Temporary and non-temporary effects of transport infrastructure investments

	demand side	supply side
temporary effect	construction effects; crowding out	-
non-temporary	operations and maintenance	effect on productivity and location of activities

This effect can be studied by means of input-output analysis and there is little controversy about it, except for the fact that one should pay attention to the question of how the infrastructure is financed. A tax increase or an increase in interest rates due to government borrowing on the capital market would have a negative impact on consumption or investments which would counter the initial demand stimulating effect of government spending. Such crowding out effects are often ignored in regional or urban studies where the infrastructure project is considered 'small' compared with the size of the national economy. However, if local projects are financed by means of local financial resources (local taxes), one should of course take into account the impacts of these taxes on investment behaviour of firms in the area concerned. Another demand related effect of infrastructure occurs in the field of operations and maintenance; this effect is non-temporary. Although maintenance is not an activity which strikes the imagination of the general public as much as new infrastructure projects might, it is nevertheless an important activity (OECD, 1986) with high rates of return on investments.

In this book we will focus on the programme effects, i.e. the permanent effects at the supply side. Figure 3.1 shows a number of possible effects which are relevant for an economic analysis of transport infrastructure investments: generalized transport costs, GDP, employment, environment, welfare, and equity.

Transport infrastructure investments lead to changes in *generalized transport costs* via shorter distances or higher speeds, which give rise to reductions in fuel, capital and labour costs. Such changes will have impacts in the transport system in the form of mode choice, choice of time of day (in the case of congested networks) and the generation and attraction of trips per zone.

The reduction in generalized transport costs lead to an increase in *productivity* in firms. This increase in productivity will manifest itself in the form of a change in value added, which will in turn lead to a growth of gross domestic product (GDP) in the region or country concerned.

Another effect of the change in transport costs is the change in *accessibility* of locations (see also Chapter 2). An important feature of transport infrastructure is that its services are distributed unequally in space. Accessibility is an appropriate concept for representing these differences in service levels.

Effects on *employment* of infrastructure investment take place among others via substitution and complementarity relationships between labour, private capital and infrastructure. They also occur by differences in growth rates of economies in regions or zones due to the differences in advantages they receive from changes in infrastructure networks.

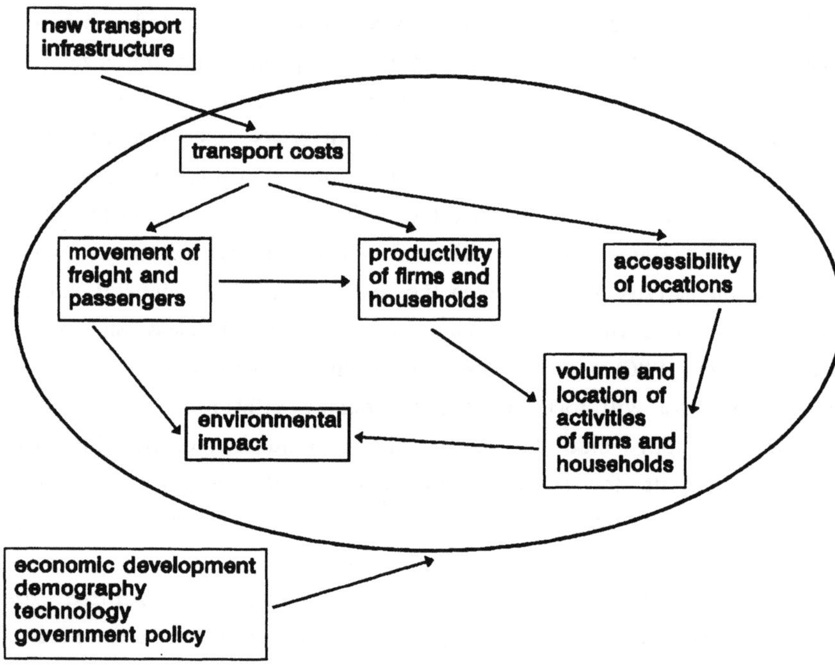

Figure 3.1 Direct relationships between transport infrastructure and spatial development

There will be several types of *environmental* impacts of road infrastructure. Development of a major highway will make road transport more competitive compared with rail and inland navigation. The resulting modal shift is disadvantageous from the perspective of transport related emissions. In addition, it induces more transport in the form of longer trips. Highway construction may also affect natural areas and lead to visual intrusion. Highway construction does have some positive environmental effects, however. It leads to changes in route choice so that usually local environmental effects are reduced in urban areas. In addition, highways may imply more continuous traffic flows having favourable effects on energy use. The balance of these effects depends strongly on local circumstances.

The economic importance attached to environmental aspects appears among others from the fact that firms consider it as one of the relevant aspects of locational profiles of regions. Investments of the public sector in environmentally friendly infrastructure may thus lead to an improvement of the competitive position of cities, regions or countries.

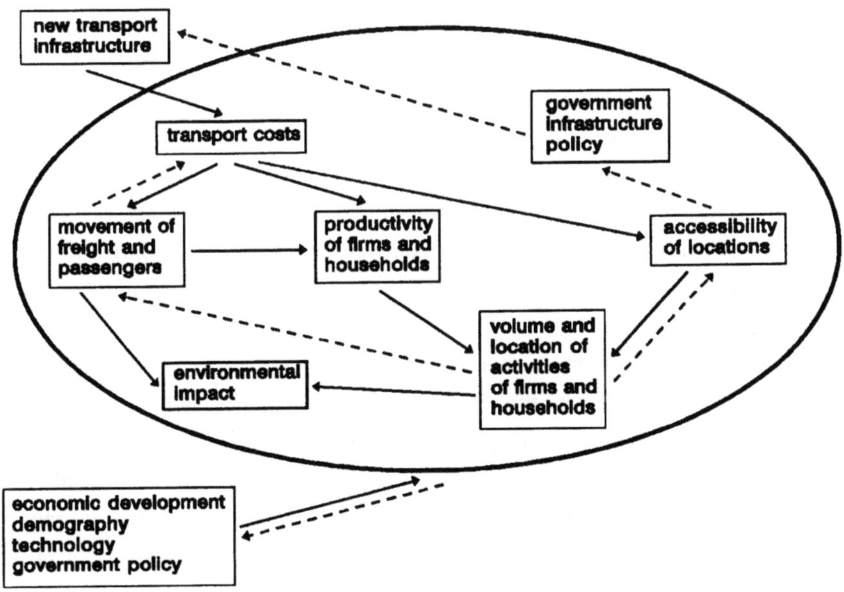

Figure 3.2 Direct and indirect relationships between transport infrastructure and spatial development

Figure 3.1 does not give a complete picture of the relationships involved. A more complete view is presented in Figure 3.2, where various indirect relationships are added. Among these feed-backs are the link between transport flows and transport costs (via congestion), and the link between the functioning of the transport system and government infrastructure policy. This latter link makes explicit that infrastructure is not only a cause of changes in the economy, but also a consequence. This subject will be discussed further in Chapters 12 and 13.

The impact of infrastructure on the welfare of people is not simply the sum of the items covered in Figures 3.1 or 3.2. In many cost-benefit analyses of road infrastructure projects in urban areas, it appears that the major benefits occur in the form of reductions of travel time of household members, an item which is not included in any of the above points. These reductions give people better opportunities for enjoying more leisure time, for increasing their range of consumption activities by travelling to destinations farther away, or for working longer. Only in the latter case is there a direct link with the development of the GDP.

A theoretical foundation for the analysis of the impacts of infrastructure on welfare is provided by social cost-benefit analysis (cf. Mishan, 1976). In several countries this type of analysis is used as a tool to support *go-no go* decisions about infrastructure. It is important to realize, however, that policy objectives such as the stimulation of temporary or permanent employment only play a subordinate role in the way social cost-benefit analysis is usually carried out. There is a remarkable gap between the interest of governments to stimulate employment by infrastructure projects on the one hand, and on the other hand the way cost-benefit analysis is usually carried out.

A final point of interest is that *equity* aspects are very important in infrastructure investments. First, the benefits of projects are usually distributed in a rather uneven way among people and firms, depending on their location. In addition, it often happens that infrastructure investments which lead to an improvement of accessibility of most people in an area, have a negative impact on accessibility for a minority of people. Another equity aspect is related to the financing of the project. Those who benefit most from a project are not necessarily the ones who pay most for the provision of the infrastructure services. Environmental effects are also often distributed in a spatially very uneven way. For example, the regions being most disturbed by high-speed rail lines are usually not the ones where the largest benefits materialize.

Equity aspects are very important in the political decision process. These aspects are ignored, however, in standard social cost-benefit analysis. This is one of the reasons why in some countries cost-benefit analysis is used in combination with multicriteria analysis to allow an explicit treatment of equity issues.

Generalized transport costs play a central role in Figure 3.1. It is important to note that generalized transport costs not only depend on the quality of the infrastructure provided, but also on the prices charged for its use. The analysis of infrastructure impacts always relies on the assumptions made about the tariffs for its use. One extreme case is the provision of infrastructure services free of charge (all costs are covered by receipts of the public sector which are not directly related to the use of the infrastructure). This implies a redistribution of the burden of the public sector, which is usually not neutral with respect to the total level and composition of economic activity in a region or country. The other extreme takes place where prices are charged for the use of infrastructure which coincide with those which would emerge if the infrastructure were to be entirely funded and maintained by the private sector.

In the next sections we will discuss some theoretical approaches to the analysis of infrastructure impacts on the economy. In addition to the positive effects of infrastructure on productivity and welfare, attention will be paid among others to possible adverse impacts of infrastructure on employment, which are sometimes overlooked in policy debates. We will discuss subsequently the role of transport infrastructure in:

* production processes (section 3.2)
* interregional/international trade (section 3.3.1)
* industrial location (section 3.3.2)
* distribution systems (section 3.4.1)
* housing/labour markets (sections 3.4.2)
* transport infrastructure and land prices (section 3.4.3)
* monopolistic/monopsonistic price setting (section 3.4.4)

3.2 Transport Infrastructure as a Production Factor

Transport infrastructure can be considered as a stock of a certain type of capital available to a region or a country. A general formulation of a production function for sector i in region r, with various types of infrastructure is[1]:

[1] An alternative approach of studying the role of infrastructure in production processes is to use cost functions (Seitz, 1993, 1995a); see also Chapter 4.

$$Q_{ir} = f_{ir}\,(L_{ir}, K_{ir}; IA_r, \ldots, IN_r) \qquad (1)$$

where:

Q_{ir}	value added in sector *i*, region *r*
L_{ir}	employment in sector *i*, region *r*
K_{ir}	private capital in sector *i*, region *r*
$IA_r, \ldots, IN_r$	infrastructure of various types in region *r*

Among the types of infrastructure distinguished are: transportation, communication, energy supply, water supply, education, health services, etc. As far as transport infrastructure is concerned, it is not easy to take into account its *network* properties in the production function approach. One thing one can do is to identify various types of transport infrastructure according to their spatial range: intraregional, interregional and possibly international.

A related problem with infrastructure is that its impact may transcend the boundaries of regions. A certain region may benefit from a university or airport, even though these facilities are not located in the region itself. This problem of spatial spill-overs may be solved by using the concept of *accessibility* of certain types of infrastructure in the production function (see e.g. Johansson, 1992).

What are the services offered by transport infrastructure which increase productivity? First, the improvement of transport infrastructure enables a reduction of the costs for the collection of inputs and the distribution of outputs. This is not only a matter of direct transport costs, but also of possible reorganizations of logistical systems (see section 3.4). Second, a reduction of transport costs implies increased opportunities for the exploitation of economies of scale implying a productivity effect. Third, improvement of transport infrastructure leads to better functioning markets. Some examples are given in section 3.4: monopolistic and monopsonistic tendencies will be curbed, and on labour markets better matches are feasible, thus implying higher labour productivities.

These are convincing arguments that transport infrastructure plays a role in production functions. Yet, as noted among others in Gramlich (1994), analysts finding a significant contribution of infrastructure to production must be aware of a quite different interpretation: there may be an inverse causality problem. Rich regions have more resources to invest in public infrastructure implying a positive correlation between infrastructure and production. This causality issue, which is also addressed in the last part of this book, deserves more attention than it usually receives in the literature.

Production functions have been applied at various spatial levels: national, regional, metropolitan. Their orientation can thus be characterized as macro to meso. Production functions give an aggregate view of the contribution of infrastructure to productivity. Note that this approach does not cover all welfare aspects of infrastructure supply: the impact on the consumers is not taken into account. Due to the aggregate nature of the production functions it can hardly be used to give an ex-ante prediction of specific projects. Only when the network properties of infrastructure are adequately represented in the production function may this approach become useful for this purpose.

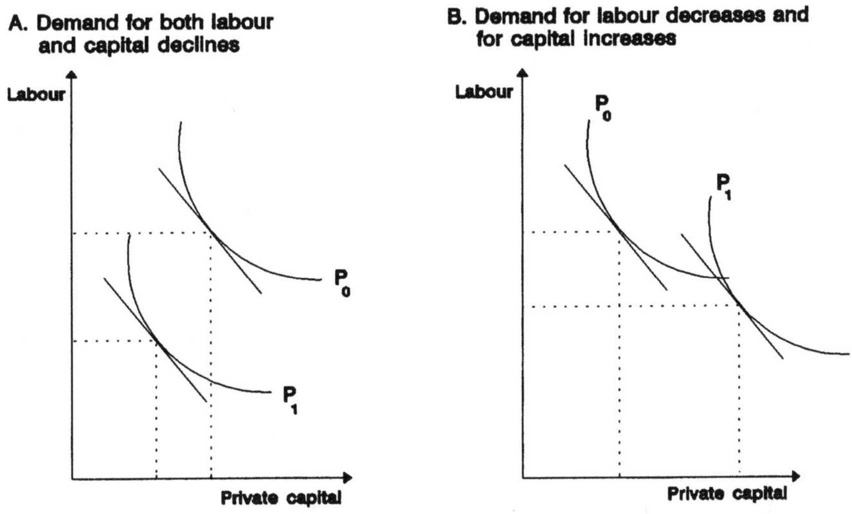

Figure 3.3 Allocation of labour and capital before and after improvement of infrastructure

According to the production function approach, an increase in the infrastructure stock leads to a shift in productivity: *it does not lead to a permanently higher growth rate.* Such a permanent increase in the growth rate might be present, however, when transport infrastructure investments lead to a higher level of knowledge production, which would affect the growth level according to endogenous growth theory.

Production functions as shown above can be used to derive demand functions for labour and private capital. With respect to labour demand, two effects can be distinguished. First, an increase in the availability of infrastructure leads to a shift in the optimum allocation of labour L and private capital K at a given level of production. This shift may be such that the total level of both L and K needed to produce the same volume of output is decreased (Figure 3.3A). But it may also occur that demand for L increases and demand for K decreases (Figure 3.3B) or vice versa. In all cases, the total costs of private production factors will decrease.

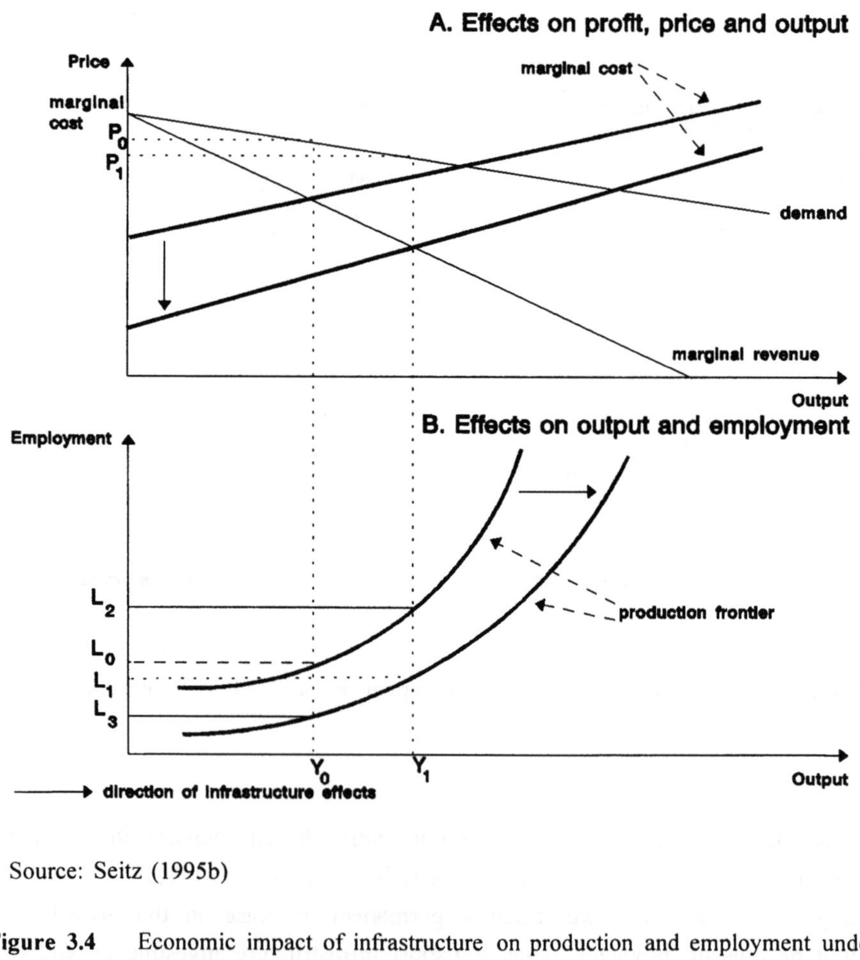

Source: Seitz (1995b)

Figure 3.4 Economic impact of infrastructure on production and employment under monopoly

In other words, the direction of the first effect of infrastructure improvement on labour demand is not certain. A second effect is that due to the decrease in the costs of production, output can be expanded. This will under the usual assumptions lead to an increase in the demand for private production factors.

Another way to illustrate these results is suggested by Seitz (1995b). Figure 3.4A indicates that given an elastic demand function, when marginal costs decrease, a monopolist will respond to an infrastructure improvement by expanding production and reducing prices. Figure 3.4B shows what will happen with labour demand. It is assumed that the infrastructure is labour saving: the production frontier has moved outward. If output would have remained unchanged, we would have observed a decrease of employment from L_0 to L_3. On the other hand, if the infrastructure would have been labour neutral (i.e. no shift in the production frontier), employment would increase from L_0 to L_2. The combination of the effects results in a modest decrease in labour demand from L_0 to L_1. Note that the figure could also have been constructed in such a way that it would illustrate an increase in employment.

3.3 Transport Infrastructure, Interregional Trade and Industrial Location

3.3.1 Transport Infrastructure and Interregional Trade

The standard model of interregional trade is illustrated in Figure 3.5. Demand and supply functions are given for two regions. Equilibrium levels of prices in both regions are rather different. Export takes place from region 1 to region 2 when transportation cost is less than the difference in equilibrium price for a certain good in both regions. Compared with the situation without trade, an additional surplus is created consisting of areas A (in region 1 producer surplus increases more than consumer surplus falls) and B (in region 2 consumer surplus increases more than producer surplus falls). Thus, both regions receive net benefits from trade according to the model, although one has to note that consumers in region 1 have to pay a higher price, whereas producers in region 2 lose part of the market.

Improvement of infrastructure leads to a decrease in transportation costs and hence to an increase in transportation volumes. The equilibrium price in region 1 will increase and the price in region 2 will decrease. Thus, in region 2, consumers benefit from the improvement in infrastructure, whereas producers are negatively affected. In region 1 it is the other way around. In employment terms, region 1 benefits, but region 2 is hurt by the improvement of transportation infrastructure. The Figure illustrates the classical case that a sector sheltered by isolation (high transport costs) is threatened by a decrease of transport costs implying an increase of competitive forces.

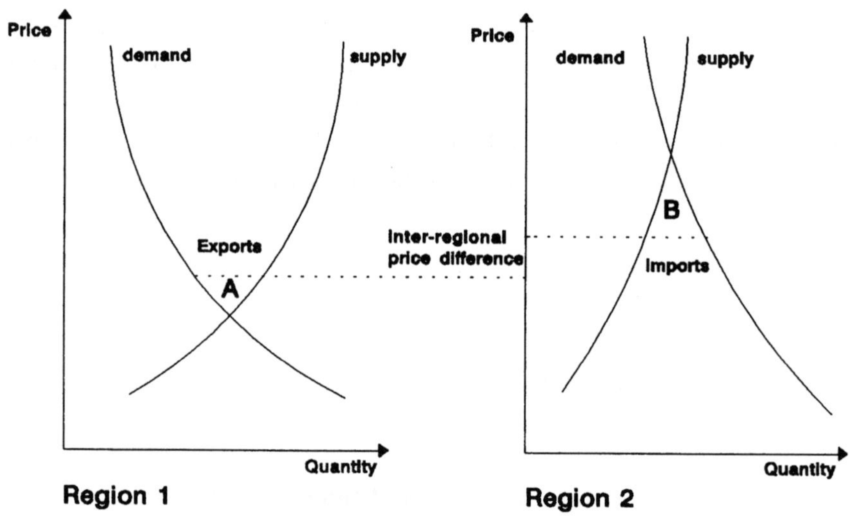

Figure 3.5 Supply and demand in two regions

Economies of scale in production play an important role in interregional/international trade, since they reinforce specialization tendencies. In the presence of economies of scale, regions with an initial advantage may benefit much more from a reduction of transport costs than other regions (Krugman, 1991), leading to a process of 'cumulative causation' (Myrdal, 1957).

A nice example of the trade-offs between transport costs and economies of scale is given by Krugman (see Table 3.2). Production costs in the core (10) are assumed higher than in the periphery (8) because of differences in factor costs. Because of economies of scale, production in the two locations would be even more expensive (12). Since the core market is larger than the periphery market, the total transport costs are higher when production takes place in the periphery compared with the case that production takes place in the core. Total transport costs are zero when production takes place in both core and periphery. With high transport costs, international trade is not profitable and production will take place in both core and periphery (note that 12 is smaller than both 10+3 and 8+8). However, with medium levels of transport costs, total costs of production and transport are lowest when production takes place in the core: 10+1.5 is lower than both 8+4 and 12+0. Here we observe a process of spatial polarization when infrastructure improves. However, a further improvement of infrastructure would lead to a shift of the production to the periphery, because with zero transport costs it would only be production costs that count.

Table 3.2 Hypothetical effects of transport infrastructure improvements on manufacturing

Production in:	production costs	shipping costs: high	medium	low
Core	10	3	1.5	0
Periphery	8	8	4	0
Both	12	0	0	0

Source: Krugman (1991)

This example is obviously simple and partial, but it nevertheless explains important tendencies in the location of manufacturing activity due to infrastructure improvements (see Table 3.3). With high transport costs production will be dispersed; medium transport costs lead to spatial polarization and low transport costs lead to a shift towards the periphery.

Table 3.3 Tendencies in the location of the manufacturing sector as a function of transport costs

transport costs:	high	medium	low
optimal location:	both	core	periphery

Source: based on Krugman (1991)

The model sketched above is a partial equilibrium model. It deals with the market for only one good. General equilibrium models are more suitable for analyzing the effects of changes in infrastructure, but they are of course more complex. Figure 3.6 (which has been taken in adapted form from Pluym & Roosma, 1984) presents some of the main effects when more than one sector is considered. In this case the net effects are difficult to predict.

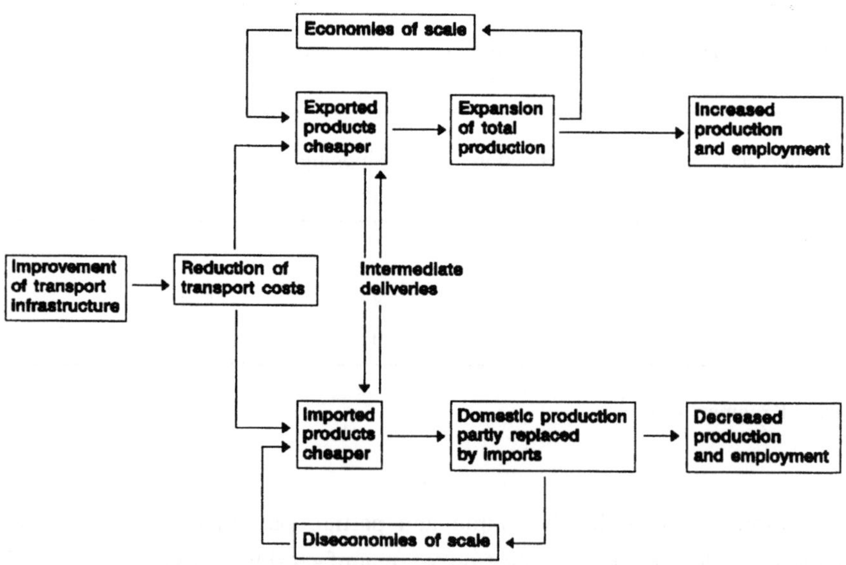

Figure 3.6 Effects of improvement of transport infrastructure on the location of economic activity

Labour market flexibility is important in overcoming the potential frictions on the labour market which may be the consequence of changes in transport costs. With flexibility, workers in a certain sector who lose their jobs due to changes in transport costs (lower part of Figure 3.6) may find work in an expanding sector as shown in the upper part of the Figure. Without such flexibility, one might have a combination of high total unemployment in one sector and a high level of vacancies in the other. Thus, a flexible labour market is important to realize the potential employment gains in regions which are affected by transport infrastructure investment.

In terms of the types of infrastructure considered, it can be noted that in the productivity approach, the *intra*regional aspects of the services of infrastructure are emphasized, whereas in the trade approach the *inter*regional aspects come to the fore.

The basic notion in this section is that since a transport link connects two regions, both regions may experience positive and negative effects. As will be shown in section 3.5, the generalization towards three or more regions leads to additional insights. Clearly, the materials provided in this section are most relevant for footlose economic activities. When activities are firmly connected to certain locations, relocation effects are obviously not relevant. The effect on the transport sector deserves further attention. An increase in trade between two regions implies a larger level of transport activities on the link between both regions, the region with the best (cheapest) transport firms getting the bigger share of the increase. Another part of the transport activities concerns nodes (terminals, ports) rather than links. Because of the complementarity of link and node related transport services, a decrease in link costs leading to an increase in transport flows will also lead to higher demand for node services in both regions. Since node services in both regions are complements, we arrive at a different conclusion here compared with the case illustrated in Figure 3.5 where substitution is assumed.

3.3.2 Transport Infrastructure and Industrial Location

After the meso level analysis of (shifts in) interregional trade, we will have a closer look at the underlying locational decisions of firms. A list of location factors (based on Schmenner, 1982, and Chapman & Walker, 1992) is given in Table 3.4. Transport costs, accessibility and infrastructure play a definite role here. Two points deserve attention: for each sector the relevant markets and suppliers are different. This would call for a differentiated treatment of accessibility in location studies. Second, the quality and availability of information is limited. Locational decisions take place in the realm of search,

Table 3.4 Major location factors; information quality and availability

Factor	Quality of information
Access to markets	
Distribution cost to existing markets	Estimate
Future trends in sales (including additional sales generated by new plant)	'Guesstimate'
Access to supplies	
Assembly costs	Estimate
Future trends in supplies (including new suppliers)	'Guesstimate'
Competitive considerations	
Location of competitors	Quantifiable
Reaction of competitors	Non-quantifiable
Transport facilities	
Availability, quality and cost of services	Estimate
Utilities and services	
Availability, quality and cost	Estimate
Labour factors	
Availability, quality and cost	Estimate
Taxes and financing	
Prevailing tax rates and obligations	Quantifiable
Tax incentives	Quantifiable
Site factors	
Purchase price and development costs	Quantifiable
Environmental considerations	
Prevailing regulations	Quantifiable
Ease and speed of compliance	Non-quantifiable
Community factors	
Cost of living	Quantifiable
Community facilities	Estimate
Community attitudes	Non-quantifiable

Source: Chapman and Walker (1992), Schmenner (1982)

risk and uncertainty where perceptions of qualities of alternative locations (instead of their objective features) will play a role (see Chapman & Walker, 1992). The issue of perceptions will be discussed among others in Chapters 5 and 10.

Firm location patterns are strongly influenced by their life cycle. In the first phase when a firm starts to exist, it is mainly dependent on the residential location of the starting entrepreneur. Infrastructure does not play an important role here. It does play a role however, in the success of firms. Successful firms need space for expansion, which is often not available at the original location. It is in relocation decisions that infrastructure may be expected to play a more important role.

A high percentage of firms are not independent, but rather are part of a larger corporate enterprise. In this case it is not fruitful to study locational decisions from an individual plant perspective. Instead, a corporate enterprise perspective should be used which yields a differentiated view for production plants, distribution facilities, R&D activities, and head offices. Each of these has its own infrastructure requirements. Infrastructure for freight traffic is usually important for production plants and distribution centres. Changes in infrastructure networks may not only lead to changes in the location of a distribution centre, but also within the distribution structure, including the number of layers in distribution networks. Head offices and research and development rely much more on urbanization economies and on facilities for passenger transport.

3.4 Other Approaches to Measure Effects of Infrastructure Improvements

3.4.1 Spatial Organization of Distribution

Above we noted that infrastructure improvement leads to a reduction in transport costs and hence affects trade flows. A closer look reveals that infrastructure investments may have various effects in the way production, transport and distribution are organized. Infrastructure provision may in principle affect choices concerning matters such as: route choice (including port choice), mode choice, location of distribution centres, number of levels in distribution structure, choice of logistical strategies, etc. McKinnon (1996) mentions three types of 'reorganizational benefits' of transport infrastructure supply that are of particular importance: market expansion, spatial concentration and tighter scheduling.

Market expansion may be quite important in less developed regions where the improvement of infrastructure makes the exploitation of natural resources feasible. It may also have large effects when island economies get fixed links. Note further that due to various factors (such as limits on driving hours of truck drivers, and the wish of business travellers to return home on the same day) in some cases even a modest improvement of infrastructure may induce a substantial reduction in transport costs, thus implying improved opportunities for reaching markets farther away.

Spatial concentration is the result of an adjustment of the optimal plant size as a result of a trade-off between size dependent production costs and transport costs of inputs and outputs. These scale effects also play an important role in warehousing.

Examples of tighter scheduling are the just-in-time principle in manufacturing and quick response principles in the retail sector. The application of these principles presupposes the existence of a well developed and reliable infrastructure. These principles induce a decrease in the stocks and an increase in freight traffic, although as McKinnon (1996) notes, due to various consolidation and collection schemes, this increase is smaller than often thought.

The general conclusion is that improvement of transport infrastructure not only leads to larger trade flows as discussed in Section 3.3, but also to a more transport intensive way of organization of production and distribution having distinct effects on the spatial distribution of production and distribution activities.

3.4.2 Housing and Labour Market

It is conventional wisdom that improvement of infrastructure leads to better functioning of labour markets. More workers can be recruited within reasonable commuting distances leading to a reduction of unemployment and vacancies due to spatial frictions and a better match between the demand and supply side of the labour market. In the long-run, however, the effects are more diffuse. Urban economic theory predicts that a decrease in transport costs will lead to a shift in settlement patterns towards a more diffuse pattern of land use (Fujita, 1989). This will result in an increase of commuting distances. This means that next to a reduction of spatial frictions which indeed implies an improvement of productivity of firms (see section 3.2), there is also a welfare effect for households, because they can now live in a more spacious dwelling in a nicer environment. There is ample evidence that commuting distances in many countries have increased considerably as a response to the improvement of transport infrastructure. These longer distances are not so much the consequence of new recruitments at longer distances, but are much more the result of voluntary moves of households who relocate to other dwellings (see, for example Rouwendal & Rietveld, 1994). It is therefore probably that the welfare improving effects for households are larger than the productivity improving effects for firms.

3.4.3 Transport Infrastructure and Land Prices

To what extent will the benefits of transport infrastructure investments be reflected by changes in land values? This issue has been addressed among others in urban economic theory (see, for example Mohring, 1961, Arnott and Stiglitz, 1981, and Fujita, 1989). The effect of a uniform decrease of transport costs per unit distance on the total land rents depends strongly on the substitution elasticity between land and other consumption goods. If demand for lot size would not depend on land rents, a decrease of transport costs per unit distance would imply a decrease in land rents. But when lot size would be very responsive to land rents, a decrease of transport cost per unit distance would imply an increase in aggregate rents. Therefore, Arnott and Stiglitz (1981) conclude: 'Aggregate land rents may, in general, rise, remain the same, or fall in response to a transport improvement'. An important implication is that it would be misleading when one infers the benefits of transport improvements from the changes in aggregate land rents.

The above results have been derived for a *uniform* fall of transport costs per unit distance. But in real world settings, transport infrastructure improvements are not uniform: some zones clearly benefit, others do not benefit, but instead experience a decline in relative terms. This case is studied by Mohring (1993), who considers the effects of an introduction of a radial expressway on land prices in a monocentric city. The expressway will lead to an expansion of the city in the area it serves; the rest of the city will shrink. Mohring concludes that when the reduction in transport costs made possible by the expressway is modest, the major effect will be an increase of land rents in the area affected. If the reduction of transport costs would be substantial, its impact area would of course expand, but the fraction of the total travel cost savings that appears as a rent increase becomes smaller. Even negative impacts on total land rents cannot be ruled out. The effects of transport infrastructure improvements on land rents depend considerably on behavioral reactions in terms of trip frequencies and demand for space per dwelling. Therefore, there is no clear-cut relationship between the change in land rents and the change in transport costs. Mohring (1993) concludes that increases in land rents in zones directly affected by the infrastructure project substantially understate benefits unless these benefits are small either absolutely (because the improvement is modest), or relatively (because the impact zone is part of a homogeneous area of much larger size). We will pay attention to land rents in Chapter 9.

Table 3.5 Distribution of benefits and costs of transport infrastructure improvement among users and landowners (figures between brackets can be positive or negative)

	household	firm	landowner	government	total
infrastructure benefit	(1)	(2)			(1+2)
land benefit	(3)	(4)	(5)		0
tax:					
-infrastructure use	(6)	(7)		(8)	0
-property			(9)	(10)	0
project cost				(11)	(11)
total	(1+3+6)	(2+4+7)	(5+9)	(8+10+11)	(1+2+11)

The above results indicate that changes in land rents in general do not elucidate about the benefits of transport infrastructure improvement. Nevertheless, they are relevant when one wants to consider distributional issues between users and owners of land. A useful scheme for the study of the redistribution of the benefits of transport infrastructure improvement is given by Morisugi et al. (1993) (see Table 3.5).

The Table shows that benefits of the transport infrastructure accrue to households (1) and firms (2) using it. In a second step a zero sum redistribution takes place between users and landowners: (3)+(4)+(5)=0. In a similar way taxes related to the use of the infrastructure and to property lead to a redistribution of benefits between users and landowners on the one hand, and government on the other. In aggregate terms the total net benefits equal (1)+(2)+(11), where (11) denotes the project costs. The bottom line of the Table gives the distribution of net benefits among the four parties distinguished: households, private firms, landowners and government.

3.4.4 Monopolistic (Monopsonistic) Price Setting

A final point of interest is that improved infrastructure leads to an increase in the number of suppliers (or buyers) in spatial markets. When transport costs are very high, one not only has the disadvantage (as already indicated in section 3.3) that economies of scale in production cannot be exploited, but also that suppliers begin to function as monopolists. Improvement of transport infrastructure means that spatial market areas start to overlap. This has a favourable effect on consumer welfare, because it reduces the probability of collusive behaviour between suppliers.

Large improvements may occur in the more extreme case that areas are dependent on only one trader, as may be the case in isolated regions in developing countries. This phenomenon has already been mentioned by Von Thünen (1821), who describes that buyers of agricultural products in rural areas may act as monopsonists because due to the high transport costs it becomes too expensive for farmers to seek alternative buyers. This has also been vividly described by Johnson (1970), who indicates that high transport costs and lack of market places in medium sized cities will lead to monopsonistic price setting. For further discussions on the impacts of transport costs on monopolistic price setting, we refer to Jara-Diaz (1986) and Blum (1997).

3.5 Distributive and Generative Effects

An important feature of transport infrastructure is that improvement of infrastructure will generally lead to both 'distributive' and 'generative' effects. Distributive effects occur when positive impacts of infrastructure improvements are compensated for by negative impacts elsewhere in the economy (both in a sectoral and in a spatial sense). Generative effects, on the other hand, refer to the net (welfare) improvement that accrues to the total spatial system affected by the investment. This distinction is important, particularly because the generative effect of an investment may be overestimated when the study area is too narrowly defined.

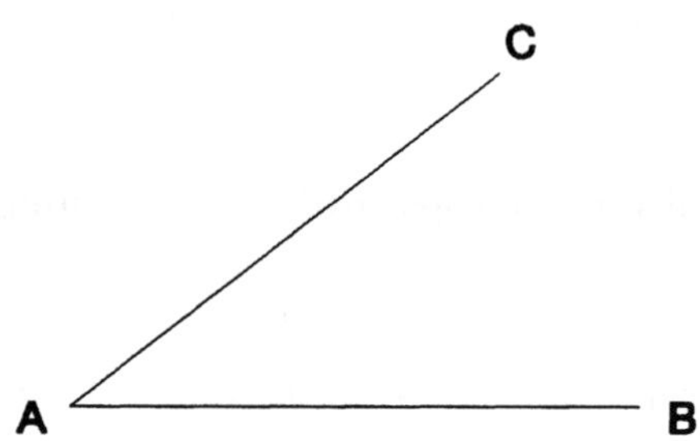

Figure 3.7 Simple network with three nodes

A major reason for this feature of infrastructure is its network character. Improvements (and construction) of links will usually not only affect that link itself, but will also affect other links in the network, and therewith employment and welfare in the nodes not directly served by the improved link. To see how this works, consider an extremely simple network with one origin (A) and two possible destinations (B and C). This is illustrated by Figure 3.7. For the determination of the benefits of an improvement in link 1 between A and B, it is necessary to distinguish three groups of users: those who do not change their behaviour; those who will switch from choosing destination C before the improvement, to B afterwards, due to its increased accessibility; and newly generated traffic. These groups together determine the change in total traffic for both routes. Especially when the network was originally congested, none of the groups will generally be of zero size.

The group of users that switches from the AC link to the AB link represent a distribution effect. They reflect the improvement of accessibility of B relative to C. If one were to only measure the effects of the link improvement in A and B itself without observing what is happening in C, one would certainly overestimate the effects of link improvement[2].

This overestimation may be mitigated in situations of congestion since a switch of users from the BC link to the AB link will have a positive effect on the performance of the BC link. This leads to newly generated traffic, not only on the AB link, but also on the BC link when the latter link would have been congested in the zero situation.

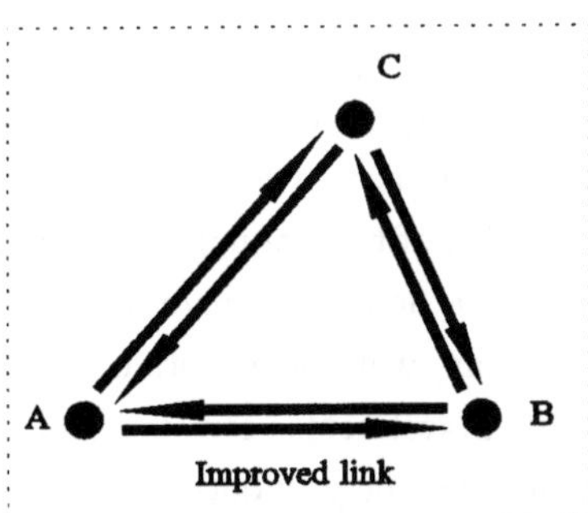

Origin \ Destination	A	B	C
A	-	+	0
B	+	-	0
C	-	-	0

Figure 3.8 Response of trade shares to transport cost reductions in a three-region network owing to an improvement on link A-B

---

[2] This discussion clearly ties in with the literature on induced traffic (see, for example SACTRA, 1994, and Kroes et al., 1996). Improvement of a certain link in a network often leads to substantial increases of its use. Careful analysis usually reveals that major parts of extra traffic on the links are due to route and mode changes (cf van Vuren, 1997). In the case of congestion, expansion of the capacity of a link will induce a 'return to the peak' effect, which is another reason why road improvement may have disappointing effects on speeds.

After 'closing' the above network by adding link BC, we find the network given in Figure 3.8, which can be used to explore the network implications of infrastructure improvements a bit further. Based on a model of interregional trade presented by Amano and Fujita (1970), in this spatial configuration an improvement on link A-B can be shown to have an unambiguously negative effect on the trade share of region C. For the regions A and B, the effect on trade shares is not clear, as this depends on the extent to which the loss on the home market can be compensated for by an increased penetration on the other region's market. Note that this model is concerned with trade *shares* only; when total production changes, and especially economies (or dis-economies) of scale occur, the picture becomes even more complicated.

It may be clear from these examples that even in very simple networks, improvements in infrastructure will generally have an impact not only on the links and in the nodes or regions directly involved, but will also often have effects on other links in the network and on other regions and nodes. The distinction between 'distributive' and 'generative' effects is therefore indeed important. Moreover, these interdependencies imply that the impacts of certain infrastructure improvements not only vary over space and sectors within the area directly concerned, but may often carry over to (much) larger areas. Generally, areas directly involved will benefit (although this certainly need not hold for all interest groups within these areas; compare Figure 3.3), whereas the indirect effects for other areas may well be negative (compare Figure 3.8). This implies that infrastructure investments can be seen as a strategic game, where improvements in one area may induce the need for other areas to follow.

When deciding on its infrastructure policies, a central government faces the dilemma that accessibility is both an absolute and a relative concept. Therefore, favouring certain regions may have negative impacts on other regions (relatively and absolutely speaking), and trade-offs between economic efficiency and (spatial and sectoral) equity in infrastructure policies are often unavoidable.

3.6 External Benefits to Transport?

From the above list of (spatial) effects it is clear that changes in transport may bring about a large number of changes in the economy, many of them having welfare improving effects. An important question is to what extent these are taken into account in an appropriate way in standard cost-benefit analyses. Are they fully reflected by the consumer and producer surpluses as shown in Figure 3.5? A classical result is indeed that with a competitive equilibrium, the producer and consumer surpluses in the underlying goods markets are equal to the transport consumer surplus (see, for example Blum, 1997). Blum shows however, that in the context of price discrimination of a monopolistic supplier, and with monopolistic competition, this equality no longer holds true. Thus, in certain monopolistic contexts, the transport consumer surplus will underestimate the total benefits of infrastructure improvement.

Another issue is whether there are additional benefits of transport that are not captured by the transport consumer surplus. For example, Willeke (1992) claims that there are substantial additional benefits owing to infrastructure (road) improvement, whereas Rothengatter (1994) claims that such benefits - if they exist at all - would be small. The political importance of this debate lies in the fact that road transport is paying substantial Pigouvian taxes in many countries to correct for its external costs. Therefore, if there would exist external benefits to (road)transport, this would imply that there is a case for subsidies to infrastructure use, or at least a reduction in the taxation levels.

It is indeed true that transport infrastructure provision has effects that are much wider than only take place in the transport sector itself. Without infrastructure our economies would collapse. This, however, is not the issue at stake here. Two important aspects should be considered.

First, we should not focus on the *average* contribution of transport to our welfare, but on the *marginal* one. When discussing a subsidy for infrastructure use because of positive externalities, the relevant question is not how large the welfare gain is when we compare an economy without and with (road)transport; instead the relevant question is what is the marginal change in social welfare given a marginal change in the number of kilometres driven? Compare the difference between the average and marginal utility of drinking water: the first is close to infinity, because life is impossible without it, the last is close to zero because nobody cares about another glass of water.

Second, it should be checked in the procedure of estimating benefits of infrastructure improvements via consumer surpluses as shown above, whether there is a neglect of certain benefits which occurs outside the transport sector. For most of the examples given above, it is clear that there is no need to worry. For example, in the trade case dealt with in section 3.3.1, it can be shown that the total benefits due to a reorientation of production and consumption are fully reflected by the area under the demand curve for transport.

External benefits would imply that the total benefits of the use of infrastructure are larger than the immediate user's willingness to pay. Thus, there must be another party who gains without there being a payment by this beneficiary to the actor using the transport infrastructure. A closer look indicates that in the relevant cases discussed in section 3.3.1, such a payment always takes place, hence there are no externalities involved.

However, some additional cases can be listed where positive marginal externalities might be present. We will discuss them in some detail below.

- *Plane (or car-) spotters.* These consumers indeed receive a welfare increase from observing planes or cars without a financial transfer. The economic importance of this phenomenon is negligible.
- *Trips with a social purpose.* Many trips give positive utility to those who are visited. The visitor pays the costs of the visit, and the person visited does not compensate him for it. As indicated by Verhoef (1996), these trips are characterized by reciprocity (the compensation of the visit is a counter visit), or by altruism (the visitor includes the utility of the person visited into his utility). In the first case there is no external element involved; in the second there may be an external benefit, i.e., when the person visited feels that his utility does not receive sufficient weight in the visitor's utility. In principle, the person visited could reveal his willingness to pay to the visitor by offering a compensation (for example, in the form of a good meal), so why would the government subsidize the visitor's transport costs? Thus, this does not provide a convincing case of a positive externality. Note further that some people may dislike visits from certain persons so that one would arrive at a negative externality. It is clear that this example is too weak and unimportant to serve as a case for a subsidy to social visits in (road) transport, let alone to (road) transport trips at large.
- *Emergency services (police, ambulance, fire-brigade).* These services may save lives and large riches; they are mentioned by Rothengatter (1994) as one

of the few possible exceptions to the rule that positive externalities do not exist in the use of transport infrastructure. Without a proper transport system these services cannot function. Does this imply that these services deserve a subsidy? Note that demand for these service trips is (almost) price inelastic: the willingness to pay for it is very high, so that for the relevant price ranges of transport costs the demand can be considered as fixed. In most institutional settings the services are offered by the public sector and the consumer does not pay. Expenses are usually covered by taxes. The risk that the fire-brigade will not arrive because its budget does not allow it to pay for the use of the road is negligible. We conclude that the willingness to pay for these emergency services is high; this will be reflected by a relatively large contribution to aggregate consumer surplus implying a strong case for the construction of the infrastructure. However, there is no externality involved so that this is not a solid case for a subsidy on infrastructure use. A paradoxical result of making road use cheaper would be that it may increase congestion, thus implying a deterioration of the quality of emergency services.

- *Image effects related to infrastructure.* For example, port A becomes a more attractive transfer seaport when it has a railway connection to the hinterland; even when the railway will not be used (because existing water and road connections with the hinterland are appropriate and cheaper), it improves the status of A so that its market share as a transfer port increases. If the status of hinterland connections really matters in the decision of container lines to include port A into their operations, an investment into a railway line may indeed be profitable. From a welfare economic point of view, it is most probably not a good investment, because it is difficult to see how it will yield additional consumer surplus. Since the status effect is assumed to be independent of the volume of traffic on the line, there is no reason to subsidize the use of the line. There are external elements here, but they do not relate to the *use of the infrastructure*, and consequently there is no need to bother about subsidizing its use. This is an extreme example of a status effect. A more realistic example would be that the railway line will be heavily underutilized, but again there is no reason in that case to subsidize its use.

We conclude that no clear and significant case of a positive externality of infrastructure use has been identified. An important source of confusion seems to be that transport has many (positive) effects outside the transport sector itself; however, these effects are properly represented in the consumer surplus of the transport demand curve. There is no need to worry that there would be an underinvestment or underutilization of transport infrastructure accordingly. This is not a solid basis for a subsidy to infrastructure use.

It appears that the major issue should not be whether cost-benefit analysis overlooks certain (external) benefits of infrastructure. The real problem is that cost-benefit analysis should be based on properly estimated demand functions for transport. The demand functions should reveal the various indirect effects discussed in the preceding sections. Priority is needed for research to improve integrated ex-ante impact studies of transport infrastructure. This does not imply that cost-benefit analysis itself is without problems, but the problems seem to be located elsewhere: the way negative externalities are valued and the possibility for taking (spatial) equity problems into account.

3.7 Costs and Benefits Versus Other Effects of Infrastructure Investment

Cost-benefit analysis provides a useful framework for analyzing infrastructure impacts in a policy oriented context concerning investment decisions and pricing. However, cost-benefit analysis also often meets criticism. Some of these are indeed difficult to overcome (lack of knowledge of how to value external effects, how to deal with equity issues). Another criticism of cost-benefit analysis is that it is too soft in the sense that its outcomes depend strongly on welfare effects such as time gains of persons travelling; instead one would like to concentrate on the effects of infrastructure on the business sector. The basis of this criticism seems to be that only effects measurable via the GDP are important. In our opinion this is a dangerous view that does not sufficiently recognize the importance of infrastructure for the consumer.

Another issue concerns the employment effects of infrastructure investments. It is striking that especially local and regional governments base their arguments in favour of employment effects. Obviously it is easier to communicate to the general public that a project will generate 500 jobs than that its internal rate of return is 9 %. A question is to what extent projects may be expected to lead to identical rankings when cost-benefit analysis is used, compared with employment based rankings. The answer is that there is little reason to expect that the two criteria run parallel. The first reason is that the employment approach ignores the positive effects on consumers, who often play a dominant role on the benefit side. The second reason is that productivity gains as studied by means of the production function approach (section 3.2) may have a negative effect on employment. The background is that substitution effects would lead to a decrease in demand for employment. One cannot be sure a priori whether such a substitution effect will be compensated by a positive output effect. This depends on the extent to which the productivity increase leads to a decrease in prices of the various goods produced and on their price elasticities.

The concern about employment can be understood from the situation of chronic unemployment experienced by many economies during the past 25 years. Indeed, in this case the use of the current wage does not reflect the opportunity costs of labour. Many more workers would be prepared to work for the current wage. Therefore, a lower shadow price for labour should be employed in cost-benefit analysis. This holds true for both the temporary employment generated during the project and the permanent employment effect (if there is any) after the project. Our conclusion is that in situations of chronic unemployment, cost-benefit analysis should be based on a shadow price of labour, which is lower than the market price. If this is not taken care of, it underestimates the net benefits of infrastructure supply. However, in situations of more balanced labour markets there is less reason to focus on employment. From a welfare economic viewpoint, employment is not a goal per se and it makes no sense to increase employment without taking into account its welfare implications in terms of leisure and consumption opportunities. In economies plagued by overheated labour markets, one would even prefer to see infrastructure projects of a labour saving nature.

Clearly, in situations of chronic unemployment as experienced in many economies, the use of the wages observed on the market leads to an underestimate of the net benefits of projects. This provides an argument for paying special attention to them in impact studies. Since equity aspects and (intangible) environmental effects also have to be dealt with, we arrive at the conclusion that multicriteria analysis as a tool for a systematic analysis of conflicting criteria is a welcome complement to the standard cost-benefit approaches (see van Pelt, 1994).

We conclude that employment effects are too narrow as the sole basis for a policy analysis of infrastructure projects. Nevertheless, employment effects remain important, especially in periods/regions with chronic unemployment.

In addition, the study of changes in the volume and location of employment provides deeper insights into the adjustments of spatial economies to changes in infrastructure networks. For an appropriate evaluation of infrastructure projects, the study of employment effects as an intermediary variable remains a necessary ingredient.

3.8 Conclusion

Because of the derived demand for transport, the effects of transport infrastructure investments are felt in many sectors of the economy and in several fields of household behaviour. One of the reasons why it is so difficult to determine these effects is that they are often diffuse in a spatial and sectoral way; another reason is that it may take a long time before they materialize.

Several theoretical perspectives have been discussed to deal with infrastructure impacts, the major ones being production function theories and interregional trade theories, with the former focussing on intraregional effects and the latter on interregional effects. An important result is that changes in infrastructure may give rise to reallocations and spatial redistributions so that equity aspects are relevant. The trade-off between transport costs and economies of scale in production leads to interesting patterns of polarization and de-polarization in space depending on the present quality of the transport system.

Three main indicators have been used in research and policy making to study the effects of infrastructure: productivity, employment and cost-benefit ratios. These indicators are only partially overlapping, and each has its relevance in policy making, so that we need a multidimensional framework for a proper impact assessment. In the present book however, we will mainly focus on the employment effects of infrastructure.

4 A Review of Empirical Infrastructure Impact Research

4.1 Introduction

After having discussed the theoretical perspectives used in infrastructure impact research, we now turn to a survey of results of empirical studies. For this purpose we begin with a short review of research methods used. A wide variety of methods for studying infrastructure impacts exists. It is important to pay special attention to these methods, because the results of infrastructure impact research depend to some extent on the types of methods used. We give a short review of methods in section 4.2, where both modelling and non-modelling approaches are distinguished.

Based on the theoretical approach presented in Chapter 3, various fields of infrastructure research can be identified. The first one concerns infrastructure impacts on productivity and costs. The second one is the role of infrastructure in industrial location. Here, two spatial ranges are usually identified: the interregional (international) level, and the intraregional (intrametropolitan) level. Thus, we arrive at three main fields where we will give a literature review.

Section 4.3: productivity, costs; regional level
Section 4.4: industrial location; regional level
Section 4.5: location of activities; urban level

Finally, in section 4.6 special attention will be given to two case studies of major European projects: The Danish fixed links and the Channel tunnel.

4.2 Methods for Analyzing Transport Infrastructure Impacts

A summary of methods is given in Table 4.1. *Models* are an important tool in the analysis of economic impacts of infrastructure investments. An important line of model-based research addresses the productivity aspects of infrastructure by means of *production functions*. These functions allow one to carry out a systematic statistical analysis of the joint contribution of infrastructure and other production factors (labour, private capital) to production. Despite the level of sophistication of the statistical methods used, there is still substantial uncertainty about infrastructure impacts, however, as will be discussed in section 4.3. *Location models* concern locational patterns of private capital and/or employment as a function of infrastructure provision. These models entail a statistical analysis of the role of infrastructure vis-a-vis other locational factors such as labour market variables, urbanization economies and investment subsidies. These models will be discussed in section 4.4. A disadvantage of the model approaches mentioned here is that they deal with limited parts of the sphere of impact and hence are incomplete. A more ambitious way of analyzing the impacts of infrastructure improvement is to use *partial or general equilibrium models*. It appears, however, that most of the general equilibrium models built give a rather superficial treatment of transport infrastructure. A final group of models concerns land-use/transport interactions. These models usually address transport and land-use at the metropolitan level. As will be shown in section 4.5, they build upon the well-established four-stage transport models by taking into account feed-backs of transport flows and infrastructure on the location of economic activity, including residential location. In this respect these models are richer than the above models, where impacts on locational decisions of households are neglected.

A property of the models mentioned above is that they allow one to disentangle the impact of infrastructure from that of other factors with respect to economic performance of a zone or region.

The *quasi experimental approach* (Isserman, 1990) achieves the same result, but without using models to be estimated. This is done by using control regions which are (approximately) identical to the region where an infrastructure improvement took place. If infrastructure improvement did not occur in the control regions, and if these regions are indeed identical to the study region, the difference between the control regions and the study region may indeed be attributed to the infrastructure improvement. The difficulty with this approach is of course the selection of control regions, because regions are never perfectly identical. Isserman (1990) has developed statistical tools to determine an appropriate set of control regions and to test whether the difference between the economic development of the study region and the control regions is significant.

Table 4.1 Methods for the analysis of infrastructure impacts

Method	Aggregate data	Disaggregate data[1]
Modelling	Production functions Location models General equilibrium models with transport as component Land-use/transport models	Stated choice models Revealed choice models
Other	Quasi-experimental method Direct interviews with entrepreneurs	Quasi-experimental method Expert judgement Transport cost approach

1. Data at the individual firm level. 'Aggregate' implies some degree of aggregation to e.g., the urban, regional, sectoral, or national levels

Stated choice and *revealed choice* methods can both be used in the case of disaggregate data. These methods are used, for example, for the analysis of tendencies for firms and households to relocate, where the attractiveness of spatial alternatives depends on their accessibility. The usual advantages of stated choice methods (the choice set can be clearly defined, multicollinearity problems can be overcome) and its disadvantages (the formulated choice problem may not make sense to the respondent, and one never knows whether stated choices correspond to actual behaviour) apply here (see, for a review Kroes & Sheldon, 1988).

Among the non-modelling approaches with disaggregate data we mention the computation of transport infrastructure impacts on transport costs and productivity of firms. This entails a detailed analysis of time use in transport activities such as loading, unloading, waiting, riding, etc. In many policy oriented studies, use is made of direct interviews of entrepreneurs on items such as:

- intensity of present use of transport infrastructure;
- level of satisfaction about infrastructure;
- bottlenecks in infrastructure as a determinant in past relocation or expansion decisions;
- relative importance of infrastructure as a determinant in possible future relocation or expansion decisions;
- impacts on (employment in) firm, of existing bottlenecks in infrastructure supply;
- impacts on (employment in) firm if infrastructure would be improved;
- perceptions on the performance of alternative regions as a location of production.

These questions aim at getting information on the mental steps entrepreneurs make when they take locational decisions. Most of these queries are of a clearly subjective nature. They correspond to a great extent to questions raised in the stated choice experiments mentioned above, but the way they are formulated is usually less precise. Another difference is that they are analyzed in a rather simple and straightforward way.

There are several reasons why the results of this interview approach may be biased. For example, it may be difficult for entrepreneurs to admit that their location decisions in the past were wrong, and they may give an overly positive opinion on their present location, accordingly. On the other hand, entrepreneurs may overstate the importance of bottlenecks of infrastructure in the expectation that governments will improve them. Nevertheless, information obtained in this way is often an important complement to other types of information. This is especially true since entrepreneurs often take their decisions on the basis of perceptions which are not always corresponding to reality (cf. Pellenbarg, 1985). These perceptions are usually not taken into account in revealed preference based studies.

In the following sections we will review empirical infrastructure impact studies using the above classification of methods. We begin with productivity related studies (section 4.3). In section 4.4 we discuss studies concerning the impact of infrastructure on (regional) location. Finally, the impacts on land use in urban areas will be reviewed in section 4.5. In each of these fields we will start with model based approaches, followed by non-model approaches.

4.3 Transport Infrastructure and Regional/National Productivity

We begin with a survey of productivity effects of infrastructure improvement. Most production function studies use the following formulation (see section 3.2):

$$Q_r = g(K_r, L_r, IN_r) \tag{1}$$

where IN_r is the size of the infrastructure stock in region r. The impacts of infrastructure in surrounding areas on productivity in region r is usually ignored. A summary of modelling approaches is given in Table 4.2.

Table 4.2 Applications of the production function approach to infrastructure modelling with regional data

Author	Country	No. of sectors	No. of types infra.	Presence of: labour	private capital	Form of production function
Biehl et al. ('86)	E.U.	1	1	yes	no	Cobb-Douglas
Blum ('82)	F.R.G.	3	8	no	no	Cobb-Douglas
Andersson et al. ('89)	Sweden	1	7	yes	yes	Cobb-Douglas[1]
Snickars/Granh. ('81)	Sweden	21	5	yes	yes	Leontief
Nijkamp ('86)	Netherlands	1	3	yes	no	Cobb-Douglas
Fukuchi ('78)	Japan	3	3	yes	yes	Cobb-Douglas
Kawashima ('78)	Japan	8	1	yes	no	linear
Mera ('73)	Japan	3	4	yes	yes	Cobb-Douglas
Costa et al. ('87)	USA	3	1	yes	yes	translog
Deno ('88)	USA	1	3	yes	yes	translog
Sasaki et al. ('95)	Japan	1	1	yes	yes	Cobb-Douglas
Forslund & Joh. ('95)	Sweden	1	1	yes	yes	Cobb-Douglas
Seitz ('95a)	Germany	1	1	yes	yes	translog
Prud'homme ('96)	France	1	1	yes	yes	Cobb-Douglas

1 = modified

The output elasticity of infrastructure is often substantial in these studies; values around .40 as also reported by Aschauer (1989) are no exception. However, Munnell (1993), in a review of studies for the USA, notes that more recent estimates of the elasticity are considerably lower. Indeed, an output elasticity of about .40 would be difficult to believe. The marginal contribution of an investment in infrastructure to output would equal the product of the output elasticity and the output-infrastructure ratio Q/IN. The latter is the product of: a) the output-capital ratio Q/(K+IN) and b) the inverse of the share of infrastructure in total capital IN/(IN+K).

The values of a and b will be about .25 and 5 (cf. Chapter 2) so that the product will be about 1.25. Thus, the marginal contribution of infrastructure to output would be about (1.25)*(.40)=.50. This outcome can be considered as extreme: it would imply a very short pay-back period of only two years for investments in public capital.

A closer look at the literature reveals that estimation of production functions by means of time series data is plagued by technical difficulties. Some authors use level variables. This is no problem when the variables on production, capital, labour, etc.) are stationary. However, non-stationarity (implying that the variables drift over time, see Steward, 1994) can usually not be ruled out with these data (Sturm & de Haan, 1995). The problem is that with non-stationary data, an analysis with variables measured at levels may have spurious results (Granger & Newbold, 1974); ordinary test statistics would lead to the conclusion that a relationship exists between variables, which in reality does not exist. An analysis in terms of levels is, however, acceptable when the variables are co-integrated, i.e., they move together in time. Tests for this can be found for example, in Engle and Granger (1987).

If the co-integration test fails, one may proceed with estimating the equation in terms of *changes* in levels (first differences between subsequent periods) instead of levels. However, as indicated by Munnell (1993), an analysis in terms of first differences only addresses the *short-run* relationships between infrastructure and production, which is a pity because one would like to obtain the long-run effects. The estimation results based on estimations in first differences are often implausible, (Munnell, 1993, Sturm & de Haan, 1995); with this specification very extreme values may be found for one or more of the output elasticities of private capital, public capital or labour. Even though much research has already been carried out in this well-defined field, estimation problems remain pertinent. The choice between estimating a short-run relationship with often implausible outcomes and a long-run relationship which is probably spurious because of non-stationarity problems, is not attractive. More refined dynamic models may do a better job, but the robustness of the results with respect to the various choices of how to specify the relationships may remain a weak point.

An alternative approach is to conduct the analysis on the basis of cross-sectional data (multi-region, or multi-country). By doing so, the problems of temporal auto-correlation, which cause the problems mentioned above can be ruled out. However, instead of temporal auto-correlation, one now has to confront the problem of spatial auto-correlation, a topic that has received little systematic attention in this context (see, however, Kelejian & Robinson, 1997 and Rietveld & Wintershoven, 1998).

Sectoral detail is important in the analysis of productivity effects. This is shown by Fukuchi (1978) and Blum (1982), who found that the productivity increase due to infrastructure may be quite different among different economic sectors. It is also confirmed by Biehl et al. (1986), who found that an index of sectoral composition of regional economies explains much more of the variance in regional per capita income than does infrastructure.

Detailed treatments of infrastructure are given by Blum (1982), Andersson et al. (1989) and Forslund and Johansson (1995). As far as transport infrastructure is concerned, Blum identified (in a regional study of the former Federal Republic Germany):

- Long distance highways
- All other roads
- Railroads
- Ports

For both types of roads and for ports, significant results were obtained. For railroads, Blum found zero and even negative effects.

Andersson et al. (1989) distinguished the following aspects of transport infrastructure for the Swedish regions:

- Main roads
- Railroads
- Airport capacity
- Travel time to major metropolitan area
- Interregional travel time

For 1970, Andersson et al. found that the impact of railroads on regional production was stronger than that of main roads. In 1980, this situation had reversed. Airport capacity itself does not have an influence on regional production, according to the estimates. However, if taken in conjunction with Research & Development, it can be shown to have a positive effect.

Forslund and Johansson (1995) focus on the contribution of road infrastructure to production levels in Swedish municipalities. Their main contribution is that they do not directly include road infrastructure in the production function, but only indirectly via the concept of accessibility (see Chapter 2). The authors distinguish intramunicipality accessibility (a density index where capacities and speeds are taken into account), and 'international freight accessibility', the latter being an index of time distance via the road system to major ports. Both accessibility indices appear to be significant in the production function. A point of comment is that the intramunicipality index may also function as a proxy of various aspects of urbanization economies, so that it is not easy to say whether this result reflects the impact of accessibility only, or the joint effect of accessibility and urbanization economies (this problem will also appear in Chapter 10). Another interesting aspect of the Forslund and Johansson approach is the dynamic formulation of the production function. They find clear evidence of a long-run adjustment of production to changes in infrastructure: it might even take some 15 years before the adjustment process to changes in infrastructure has come to a halt.

As can be seen in Table 4.2, the specification of the production function is not always complete: sometimes private capital is lacking as a production factor and in the case of Blum (1982), infrastructure is even the only production factor taken into account. Biehl et al. (1986) use the term 'quasi production function' to represent this omitted variable problem. A quasi production function can be conceived of as a reduced form of a more elaborate model with a complete production function and with functions describing how labour and capital respond to changes in infrastructure. Thus, the quasi production function not only takes into account the productivity effect of infrastructure, but also includes the relocation of labour and capital as a consequence of the provision of infrastructure. A problem with this approach is that it is based on very simple response functions of labour and capital. If, for example, the regional labour force would not only depend on infrastructure, but also on wages, wages would have to figure in the reduced form quasi production function (cf. Rietveld, 1989). This casts some doubt on the appropriateness of the quasi production function approach: it could easily lead to an overestimate of the importance of infrastructure.

Another way of addressing the contribution of transport infrastructure to production processes is to consider transport cost shares of firms in various sectors. Table 4.3 shows transport cost shares for selected US industries.

Table 4.3 Transport cost shares for selected US industries (1982)

Industry	inbound and outbound transport costs (% of output)
	Low
Leather and leather products	3
Electrical and electronic machinery	4
Printing and publishing	4
Apparel and other textiles	4
Instruments	4
Machinery excluding electrical instruments	5
Tobacco manufacturers	5
	Medium
Rubber and plastics products	7
Transportation equipment	8
Miscellaneous manufacturing	8
Fabricated metal products	8
Textile mill products	8
Primary metal industries	9
Paper and allied products	11
	High
Furniture and fixtures	12
Food and kindred products	13
Chemicals	14
Lumber and wood products	18
Petroleum products	24
Stone, clay and glass products	27

Source: Anderson (1983)

The Table shows that cost shares range from a low 3 % for leather products to a high 27 % for construction materials such as stone, clay and glass. For the majority of the industries the share is lower than 10 %. These outcomes are confirmed in a survey of about 200 firms in the manufacturing and service sector in Great Britain. Diamond and Spence (1989) find that the share of transport costs in total costs is about 6 % with variations between 12 % for the wholesale distribution industry and 3 % for motor vehicle production. They also find that in the service sector transport cost shares tend to be higher than in manufacturing. This is partly related to the high costs of business travel in producer services.

The above figures are important for an assessment of the contribution of infrastructure to productivity or costs, but they do not give an explicit answer on how costs change when infrastructure improves. For this purpose *cost functions* can be used. The endogenous variable in the analysis is total cost with prices and the level of public capital as explanatory variables. This allows one to draw conclusions about the effect of infrastructure on cost levels of firms. In addition, when flexible functional forms are used, one may draw conclusions about the degree of substitution and complementarity between the various production factors. Seitz (1995a) concludes in a study of productivity in German cities by means of translog functions that the relationship between private and public capital is characterized by complementarity: an increase in public capital leads to an increase in private capital, and a decrease of labour (assuming that total output is constant).

Of course, as indicated in Chapter 3, when output will change as a result of the decrease in costs, employment may nevertheless increase, but this obviously depends on price elasticities.

A major difference between the cost function approach and the production function approach is that the cost function approach is based on an assumption of cost minimization, whereas in the production function approach such a behavioural assumption does not figure. This gives the cost function approach a more complex structure and also implies that the estimation results may depend on a larger number of choices of the specification.

A more micro-oriented approach is to study the potential effects of infrastructure improvement at the firm level. For example, NEA (1990) studied the effects of building a new highway in a Dutch region with a low quality of regional and interregional roads. The direct gains of the construction of a highway in this region for the average firm only led to a 1-2 % reduction of transport costs. A major reason for this result is that the total time gains of faster transport are small compared with the total time needed for loading, unloading, etc. In addition, the new highway can only be used for part of the trips. These results suggest that the overall supply of the (road) transport infrastructure is in many regions in Western Europe already so extensive that a further improvement only has limited effects on costs. A possible exception to this statement is the case where an infrastructure project is a missing link in a network, or where the project helps to remove high levels of congestion.

Nevertheless, one must be aware that there are several reasons why transport infrastructure remains important. First, even though the impact on total transport costs (and hence also total costs) is small, the impact on the profit margin may be substantial. Second, the above figures do not take into account the impacts on overall distribution costs which may occur for example by changes in the physical distribution network (number of distribution centres, size of stocks). Third, one should not only take into account the improvement of average travel time, but also the improvement of its variance. A smaller variance in travel times as a consequence of a decrease of congestion enables firms to save time because transport schedules can be planned with less slack. Fourth, good infrastructure not only influences transport costs, but also other types of costs as explained earlier in this section. Fifth, improved transport infrastructure also has impacts on the demand side: it makes new markets accessible and leads to economies of scale, accordingly. Finally, transport infrastructure may be important for firms and regions since it influences the status of locations. This also holds true in the reverse: congested areas often have a negative reputation. Perception plays an important role here. For example, in the NEA (1990) study it was found that perceived time losses due to congestion by firms were much larger than the actual time losses.

The above studies focused on the transport costs related to freight. In particular sectors, such as business services, and particular types of activities within larger firms, business travel is also an important cost component. In a recent study for the European Commission (E.C., 1993), it was found that the introduction of high-speed rail in the European Union would give rise to a considerable reduction in time costs of business travel between particular origin-destination pairs. Two components can be distinguished here: a direct reduction of travel time and an indirect effect owing to the higher quality of services provided by a high-speed rail system. The latter effect consists of the following components:

- in travel work capacity (better opportunities for productive work while being in the vehicle);
- access time saving (measured in terms of the share of travel time spent on

checking in, boarding, etc.);
- new opportunities for day-return trips (no need to make an overnight stay);
- higher service frequencies (smaller difference between preferred and scheduled travel time);
- greater service reliability (low risk of delays).

The economic valuation of these quality aspects of transport infrastructure improvements is estimated to be of a size comparable to the economic value of the direct travel time reductions, as considered in the conventional cost-benefit analysis. This result underlines the importance of taking into account the quality aspects of the services provided by infrastructure.

4.4 Transport Infrastructure and Relocation of Economic Activity

Above we focussed on the productivity effects of infrastructure. As has been explained in section 3.3, another effect of infrastructure is that it leads to a relocation of labour and capital when accessibility or productivity of regions changes. One approach to investigate these effects is to formulate a model for the location of economic activity, such as:

$$L_r = f(A_r, w_r, z_r) \qquad (2)$$

where L_r, A_r, w_r and z_r denote employment, accessibility (as defined in Chapter 2), wage rate, and other relevant location factors in region r, respectively. One way to arrive at a specification for equation (2) is to formulate it as a factor demand function based on the production function formulated in (1) (see also Seitz, 1995a, and Chapter 7). A similar model can be formulated for investments in private capital. Examples of such an approach can be found in Dodgson (1974), Botham (1983), Kau (1976), Mills and Carlino (1989), and Sasaki et al. (1995) for the impacts of roads and Evers et al. (1987), Plassard (1992) and Sasaki et al. (1997) for the impacts of high-speed rail. Other studies of this type relate to the impact of infrastructure investments at large (den Hartog et al., 1986), and the impact of the extension of the European Union on accessibility, and hence on economic development in European cities (Cheshire, 1990). These studies mainly reach to the conclusion that the effects of infrastructure change on employment are rather modest.

A problem with the specification of the above equation is that it may give a misleading impression on the overall impact of infrastructure improvement on regional development. If we estimate the equation on the basis of cross-sectional data and find a positive derivative of regional employment L_r with respect to accessibility A_r, it is tempting to infer that an increase in accessibility leads to a corresponding increase in regional activity. Since an infrastructure improvement in one region in general leads to an improvement in accessibility in all regions, the conclusion would be that employment expands in all regions. This is not plausible, however, since part of the employment change is due to relocation and has a distributional nature, accordingly. Such a lack of attention with respect to the appropriate specification of the equation may easily lead to an overestimate of the impacts of infrastructure improvement. The problem can be solved by formulating the equation in a dynamic form, or by using an accessibility measure in relative terms as is done by Evers et al. (1987).

The empirical analysis of the impacts of infrastructure investments on employment is hampered by causality problems. In addition to the usual assumption that infrastructure investments induce employment change, one may also argue that infrastructure investments take place where they are most needed, i.e. where employment growth occurs. Pierce and Haugh (1977) have developed statistical tests based on correlations between time series with different lag intervals to determine which of the two regimes actually applies.

Using this approach, den Hartog et al. (1986) found for the Netherlands that there is indeed a causal relationship between public and private investments, taking place with an interval of 3 to 4 years. For the reverse relationship (i.e., where public investments are caused by private investments), no statistical confirmation could be found.

A disadvantage of the approach above is that it is not possible to separate indirect, crowding out and spin-off effects. Since the lag interval is rather short (3 to 4 years), it is not clear whether the causal relationship refers to short-run (indirect or crowding out) effects or long-run spin-offs. In order to overcome this difficulty, one may change the spatial level of analysis. An important part of the short-term effects of an infrastructure project will take place outside the region where the project is undertaken. Long-run spin-off effects are likely to be concentrated in the project region, however. Therefore, den Hartog et al. (1986) also carried out an analysis at the provincial level. In each region, private investment IP as a share of the gross domestic product Q is explained by government investment IG as a share of the gross domestic product:

$$\Delta(IP/Q)_{r,t} = \Sigma_{i=0}^{k} \alpha_i \Delta(IG/Q)_{N,t-i} + \Sigma_{i=0}^{k} \beta_i \{ \Delta(IG/Q)_{r,t-i} - \Delta(IG/Q)_{N,t-i} \} \quad (3)$$

The subscripts *t* and *r* refer to time and region. The national level is represented by *N*. Thus, in equation (3) for each region, private investments are explained by government investments at both the national and regional level with certain lags.

Spin-off effects can be detected by means of the β_i coefficients in a cumulative way. Empirical results from equation (3) show that spin-off effects are indeed significant for an interval of 0-5 years. However, den Hartog et al. (1986) indicate that this positive result depends strongly on one particular province (Zeeland), which happened to attract huge levels of government investments during the period considered. If the province is deleted, spin-off effects are no longer statistically significant. Thus, with the given approach, only when investments are large is it possible to show that significant spin-offs occur in the regions. Another example of a causality test is given by Seitz (1995b). He finds that in many cases causality tests do not lead to clear answers. This obviously is a problem in impact research, because unclear causality implies that even when a significant impact is found based on an assumed direction of causality, it is not permissible to use this impact for predictions of effects of future projects.

In causality tests, causality is interpreted as a sequence of events. With infrastructure, the problem is that *anticipation* may take place. For example, when it is announced that an expressway will be built between A and B, firms and households may anticipate the completion of the new road by relocating before the project has been completed. This phenomenon is especially relevant in the context of infrastructure projects, because the period of construction is often very lengthy. The appropriate causality structure would be to relate locational decisions of firms to the decision to build the road, not to the completion of the road itself. However, this is usually not done in analyses of this type.

The modelling approaches reviewed above are partial. As a consequence they are not entirely adequate for investigating the full range of impacts of infrastructure improvement on the spatial economy. For example, the impacts of infrastructure on trade flows do not receive the attention they deserve. Such feed-backs are covered by a range of multi/interregional economic models such as developed by Amano and Fujita (1970), Daluwatte and Ando (1992), Sasaki et al. (1995, 1997) for Japan, Madsen and Jensen-Butler (1991) for Denmark, Treyz (1980) and Liew and Liew (1985) for the USA, and Elbers (1992) for Nepal. The two last mentioned models are of the computable general equilibrium type. The advantage is that they take into account all relevant feed-backs of infrastructure investments, including the possible crowding out effects on private investments and employment due to higher interest rates and wages.

Given the weak way in which infrastructure is usually included in national and (inter)regional economic models, another approach is to use in a first phase transport models (complemented with other methods) to determine the direct impacts on transport flows, and costs (and sometimes a first order employment effect). In a second phase these results are fed into an economic model (see, for example Bradley, 1997). This allows the use of large existing general purpose models for the analysis of infrastructure impacts. Note that specific assumptions have to be made to enable the bridging of the gap between the concepts and definitions used in both phases. Of course this two-stage approach only makes sense when large infrastructure projects are concerned, because otherwise the feed-back effects via the national economy are negligible.

This two-stage approach allows one to incorporate both short-term multiplier effects (related to the construction phases; see Chapter 3), and long-term effects on the factor markets. Especially when factor markets are tight, crowding out effects may substantially reduce the initial positive employment effects of infrastructure investments. Much depends on the spatial and sectoral aspects of the projects: the market conditions in the regions and sectors that will be the main beneficiaries will influence the feed-back effects.

Non-modelling approaches to the impact of infrastructure on location decisions of firms usually relate to direct statements of entrepreneurs. This approach is followed among others by Morita (1992) for Japan, Diamond and Spence (1989) for the UK, Healey & Baker (1996) for European metropolitan areas, and Chapters 8 and 9 give examples for the Netherlands. In most of these studies, entrepreneurs assign an important role to infrastructure as a location factor. A clear example is given by Healey & Baker (1996), who find that infrastructure related criteria are very important location factors for large European firms (Table 4.4). The three most important factors relate to market accessibility, interregional and international transport links and telecommunications.

Table 4.4 Percentage of firms which consider the locational factor concerned as absolutely essential for locating their business (1996)

Factor	%
Easy access to markets, clients or customers	63
Transport links with other cities and international centres	52
Quality of telecommunications	46
Cost and availability of staff	43
Business climate and availability of financial incentives	36
Value for money of office space	26
Availability of office space	22
Ease of travelling around within the city	22
Languages spoken	18
High quality of environment	11
Quality of life for employees	10

Source: Healey & Baker (1996)

One should not forget, however, that the results of Table 4.4 with their high importance for accessibility have been obtained for firms with an international orientation. Other types of firms give higher priority to labour market and tax variables as can be inferred from Table 4.5, where the results are presented of a survey conducted among some 9000 firms in the European Union (E.C., 1990a). The importance attached to infrastructure is highest in lagging regions where infrastructure quality will be low, but also here it receives only the 7th rank. In the other types of regions infrastructure receives even lower ranks.

It is interesting to compare these results with those of Morita (1992) for Japanese firms, who finds as the four most important location factors: availability of labour force, wages, land prices, and preferential treatment in the context of regional policies. Accessibility of inputs, markets, seaports and

Table 4.5 Firms' priorities for improving regional competitiveness (ranked by determinant)

Determinants of competitiveness	Lagging regions	Regions industrial decline	Favoured regions
1 Labour market			
Indirect labour costs	4	1	1
Regulation of labour market	5	8	6
Wages and salaries	13	4	5
2 Macro-economic outlook			
Economic growth	5	3	4
Sectoral outlook	12	10	9
3 Financial markets			
Cost of credit	1	6	6
Corporate taxes	2	5	3
Exchange rates	10	8	8
Availability of risk capital	20	19	23
4 Infrastructure			
Transport network	7	11	10
Supply & cost of energy	8	12	18
Industrial sites	14	17	25
Communication system	17	23	14
Waste disposal	26	21	14
5 Educational system			
Supply of qualified labour	3	1	2
School facilities	15	26	26
Training facilities	28	33	33
Supply of unskilled labour	30	15	11
Proximity of higher education	34	33	29
6 National policies			
Industrial policy	9	18	12
Administrative procedures	16	25	20
Other national factors	25	28	32
Legal regulations	29	22	19
7 Regional policies			
Regional incentives	11	14	20
Cooperation with local authorities	24	20	24
Other regional factors	31	32	31
Local taxes	33	7	12

Table 4.5 continued

Determinants of competitiveness	Lagging regions	Regions in industrial decline	Favoured regions
8 Regional economic structure			
Servicing of machinery	18	31	27
Proximity to suppliers	19	23	28
Proximity to customers	21	15	22
Banks, insurance lawyers	22	30	33
Business culture	26	26	30
Advertising & consultancy	36	36	35
9 Social facilities			
Social climate	23	12	17
Cost of housing	31	29	16
Cultural & social facilities	35	35	36
Leisure facilities	37	37	37

Notes:

1 The ranks were obtained from the frequency of company replies to the request to list the 3 determinants of competitiveness with the highest priority for improvement.

2 Lagging regions are those with a GDP per capita which is generally 25 percent below EU average. These 'Objective 1' regions are at the time of the survey located in the Mediterranean, except for Ireland.

3 Regions in industrial decline usually have a relatively high unemployment rate. These 'Objective 2' regions are usually older industrial areas and 'totally eligible' areas have a high population density as well as high rates of unemployment.

4 Favoured regions are a sample of regions which did not receive regional assistance.

Source: E.C. (1990a)

airports are considered as much less important. These results clearly reflect the scarcity of land and labour in Japan.

We conclude that interviews of entrepreneurs on the relative importance of locational factors may lead to rather largely varying outcomes, depending on the types of firms and the types of regions considered. Therefore, one should be careful with transferring the outcomes from a particular study to other ranges of application. Much depends on the types of firms involved, the present quality of the infrastructure, and the spatial level of the analysis. In addition, as will also be explained in Chapter 11, the result of a direct rating or ranking of locational factors may lead to some ambiguity of the outcomes.

4.5 Transport Infrastructure and Intra-Urban Development

In this section we give a short review of approaches to the impact of transport infrastructure on intra-urban development. These approaches tie in with what has been written in section 3.4 about labour and housing markets. After the first integrated transport/land-use model developed by Lowry (1964), a considerable number of models have been developed and applied to a variety of metropolitan areas. A detailed review of such models is given by Webster et al. (1988). Well-known examples of models included in the review are MEP (Marcial Echenique and Partners, Echenique, 1986), LILT (Leeds Integrated Land Use Transport Model, Mackett, 1981) and the Dortmund model developed by Wegener (1986). A major part of the models consists of standard transport models with trip generation, trip distribution, mode choice, and route choice as common elements. These transport models are linked to models describing feedbacks of transport related variables on location decisions of firms and households (Berechman & Gordon, 1986). The feed-back structure is presented in Figure 4.1. The Figure indicates that 'accessibility' is a core concept in the interface between transport and the locational patterns of activities. The Figure also indicates that two steps have to be made to connect accessibility with activity patterns: location decisions of investors and location decisions of users. In addition to the models mentioned above, important monographs in this field are Anas (1982) and de la Barra (1989). More recent contributions can be found among others in Anderstig and Mattson (1991), Mackett (1996) and Seo et al. (1996).

A systematic comparison of modelling results of the spatial impacts of infrastructure improvement is given in Webster et al. (1988). It appears that the impact of a uniform improvement of infrastructure on urban form is limited; most models surveyed predict that the effect of an increase of 20 % in speeds of motorized transport modes on decentralization (suburbanization) of population is very small compared with the autonomous development in the spatial distribution of population. Slightly larger figures are found for the impact of uniform speed change on the spatial distribution of employment. In an update (Dasgupta & Webster, 1992), again small impacts on the decentralization process in urban areas are observed.

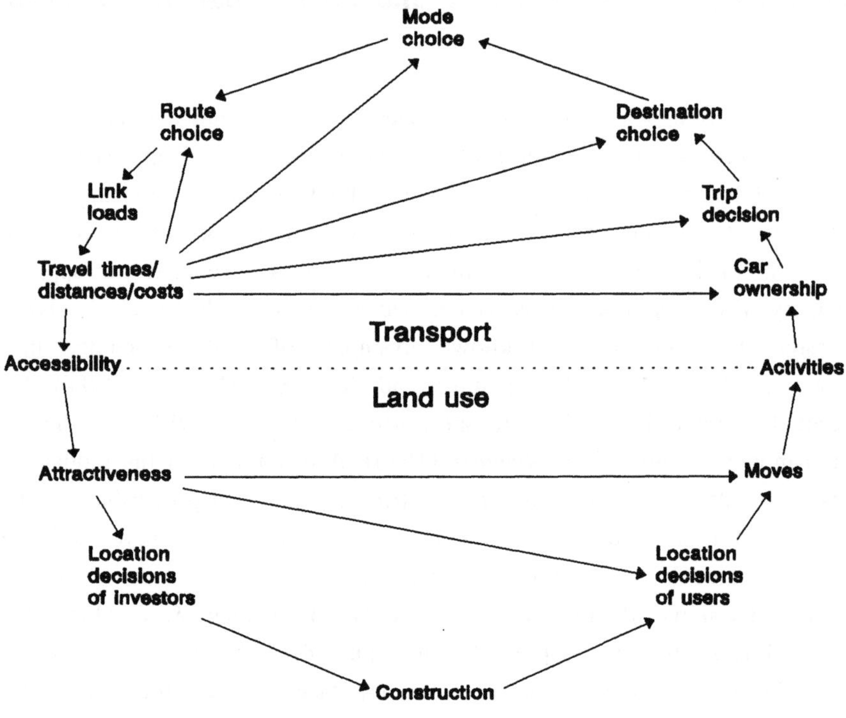

Source: Wegener (1996)

Figure 4.1 The land-use/transport feedback cycle

The above results hold true for the distribution of total employment and population. Of course for certain subgroups (e.g., low-income households, or service sector employment), somewhat more marked developments may occur. Another remark is that the above results are based on general infrastructure improvements in an entire urban area. More drastic changes in urban form may be expected when one studies the effects of improvement in infrastructure at certain areas of the city. However, as indicated by Dasgupta and

Webster (1992), the impact of specific infrastructure improvement such as the construction of a ringroad or a cross-town metroline on the overall structure of cities is again small.

The conclusion is that in most cities surveyed, the quality of transport infrastructure is already high. Accessibility of most zones in urban areas is already reasonable so that one may not expect a considerable change in urban form when infrastructure is improved. Other comparative studies in this field can be found in Mackett (1993) and Wegener (1994).

It must be noted that the rather indecisive results found with the models surveyed here only apply when infrastructure impacts are considered with respect to urban form. Impacts on the functioning of the transport system are of a more immediate nature and are often numerous.

An important dimension of the transport/land-use interaction is that it is important to study the spatial location of activities as well as their density. Density is a key factor in transport demand (Newman & Kenworthy, 1989), since it determines average distances between origins and destinations. In addition, density is an important determinant of the success of public transport. However, it is not only density that counts, but also the mixture of activities. For example, with strongly mixed land use, the balances of workers and workplaces per zone are close to one; this implies that commuting may be a mainly short distance phenomenon. However, since jobs and residences are heterogeneous, and information in these markets is incomplete, the potential of short distance commuting can often not be realized. Especially when transport costs are low and transport networks offer high quality, the gap between the potential reduction and the actual reduction in tranport costs by mixed land uses will be large. Transport/land-use models can be used to compare the results of market driven developments in land use with more land use planning driven developments and can investigate the implications for transport demand and energy consumption (Wegener, 1996).

A property of urban land-use/transport models is that they are based on the assumption that total population and employment are a given. This means that these models can only deal with the spatial distribution of activities, not with the total level of these activities. These models only address distributive effects within an urban area. Generative effects on employment or investments are excluded from consideration. For example, the question whether an improvement in transport infrastructure in a metropolitan area improves competitiveness so that it attracts more investments (in basic sectors) than it would otherwise do cannot be studied within the structure of these models. The reason is of course that for this purpose one needs insight into the locational behaviour of firms with an interregional orientation. These firms make a comparison between different (metropolitan) regions. Therefore, in order to address questions of the type raised above, one must compare the infrastructure quality in several competing metropolitan regions. This would entail the development of linkages between urban transport/land-use models and interregional trade and location models. This is indeed a promising direction of future model development.

The locational decisions of firms in the above models are modelled mainly on the basis of an accessibility measure for the zones in the urban system, where land rents may have an equilibrating impact on supply and demand of land for different purposes. The way accessibility is modelled is inevitably somewhat superficial in such models if one considers the large number of inputs and services needed and the complexity of the distribution channels used. Direct interviews with entrepreneurs to obtain the relevant factors determining the attractiveness of a location are a useful complement in this respect. A good example is a study by NSS (1991), which is based on interviews with 1250 entrepreneurs in the Randstad (the highly urbanized area around Amsterdam, Rotterdam, The Hague and Utrecht) in the Netherlands. The importance of 25 location factors is indicated on a scale from 0 to 10. It seems that there is a group of 11 factors which have clearly higher scores than the rest (Table 4.6). Among these, four refer to the quality of the building and its immediate environment, one relates to the educational level of staff and six relate to the quality of infrastructure: roads, parking, loading/unloading possibilities; telecommunications facilities and services; public transport. The dominant position of road transport is clear.

Remarkable are the low ratings of distance to both airport and seaport and the infrastructure facilities for freight transport (combined transport, waterways, rail). The spatial level at which the analysis takes place is important here. Schiphol airport and Rotterdam seaport are both located in the Randstad, and any point in the Randstad is located within a distance of approximately 80 km from these ports. Therefore, the cause of the low rating for both ports is probably not that they are unimportant as a location factor, but rather that they are easily accessible from anywhere in the examined area and do not influence the decision of where to locate within the region.

Table 4.6 Average importance attached to business environment by companies in Randstad Holland (n=1250)

Location factor	Score	Standard deviation
Accessibility by road	9.0	1.5
Parking facilities	8.4	2.0
Quality of staff	8.1	2.5
Availability of telecommunications facilities	7.9	2.9
Representativity of building	7.3	2.6
Accessibility by public transport	6.7	3.1
Rent / Lot price	6.7	3.2
Opportunities for charging / discharging	6.1	3.2
Opportunities to expand	5.8	3.3
Representativity of direct environment	5.8	2.9
Availability of telecommunications services	5.3	3.5
Presence of suppliers	3.8	3.6
Availability of logistic services	3.4	3.6
Quality of landscape	3.2	3.2
Proximity to airport	3.1	3.5
Availability of educational facilities	3.0	3.5
Presence of international companies	2.9	3.5
Proximity to seaport	2.6	3.2
Proximity to distribution centre	2.4	3.5
Presence of similar companies	2.4	3.2
Proximity to customs warehouse	2.2	3.3
Presence of knowledge centres	2.1	3.1
Opportunities for combined transport	2.0	3.3
Proximity to waterway	1.6	3.1
Opportunities for rail transport	1.1	2.6

Source: NSS (1991)

Other examples of studies of this type can be found in Wilson et al. (1982), Pellenbarg (1985), Korteweg (1992) and Chapter 8. Although less extreme, here again the importance of infrastructure (especially road infrastructure) is considerable.

4.6 Major European Infrastructure Projects: Two Case Studies

4.6.1 The Danish Fixed Links[1]

In Denmark at the end of the 1990s three major fixed links across sea straits are under construction or in the planning stage (see Figure 4.2). These fixed links are each 15-20 km. in length and represent major investments on the European scale.

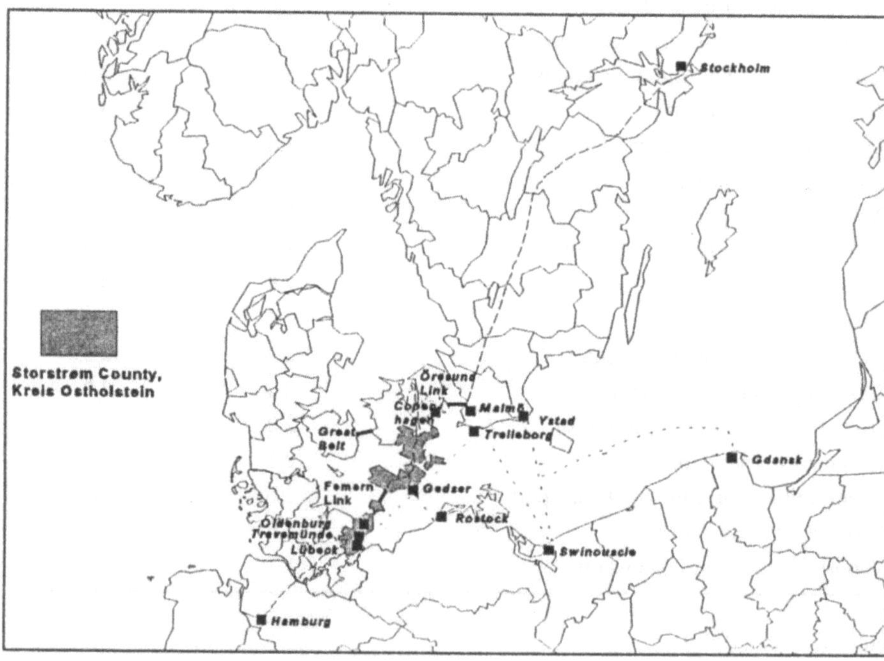

Source: Jensen-Butler and Madsen (1998)

Figure 4.2 The three fixed links (the broken line is the proposed high-speed rail link Hamburg-Copenhagen-Stockholm, the dotted lines are ferry routes)

[1] For this section extensive use has been made of Jensen-Butler and Madsen (1998).

The Great Belt link - connecting east and west Denmark - has opened for rail traffic in 1997 and for road traffic a year later. The Great Belt link is a national link and the other two fixed links are international. The Øresund link between Denmark and Sweden (Copenhagen - Malmø) will open in 1999. The proposal for the last fixed link - between Denmark and Germany (across the Femern Belt) - has not yet been finalized. There are two primary alternatives: either a rail tunnel (vehicles on a rail shuttle) or a multi modal bridge. The option of building a high-speed rail link from Hamburg to Copenhagen and farther into Sweden is still under discussion.

The Great Belt, Øresund and Femern Belt are real barriers in - national and international - traffic flows. Cars and trains need to be transferred by ferry-boat, which is rather time consuming. Consequently, the fixed links lead to large travel time gains (Great Belt; down from 75 to 17 min., Øresund; down from 75 to 20 min., and Femern; down from 90 to 20 min.). The tolls on these links are planned to equal the price of the ferry tickets today, so the users need not pay for the travel time gains.

The three projects cannot be examined in isolation, since synergy effects arise from interaction between the projects. The links constructed / under construction (Great Belt and Øresund) will have major impacts on the international transport flows on the Femern crossing. Therefore, in order to isolate the pure effects of the Femern link, two analyses have been undertaken: the effects of the Great Belt and Øresund link *with* and *without* the Femern link. The effect of the Femern link is considered to be the difference between the two.

To measure the medium term effects of the opening of the links on production, employment and incomes by region (including changes in patterns of regional trade), three regional economic effects of investment in transport infrastructure are identified:

- changes in transport technology (opening fixed link and closure of the ferry routes),
- changes in patterns of regional competitiveness (caused by changes in costs and prices),
- changes in traffic flows along different corridors

Changes in transport technology. In case of the Femern link a clear change in transport technology appears. A ferry route is replaced by either a rail tunnel or a multi-modal bridge. The direct effects on employment in the transport sector is a concrete assessment of the changes in employment related to each transport alternative. The indirect and induced effects are modelled using a one-region macro-economic model: EMIL (see, for a description Jensen-Butler & Madsen, 1998).

The *direct* employment effect due to the opening of the Great Belt and Øresund fixed links on the Danish region Storstrøm - adjacent to the Femern link - is a loss of 850 ferry service related jobs in case of a rail shuttle crossing, or 950 jobs in case of a multi-modal bridge (the difference of 100 jobs is caused by the people needed to load and unload the rail shuttle). Those losses are caused by reduced traffic flows using the Femern corridor. The construction of the Femern link will lead to an additional loss of 200 jobs, due to the closure of the ferry.

The direct, indirect and induced effects on employment in Storstrøm county for the three fixed links is a loss of 1.330 jobs in case of a rail shuttle or 1.460 jobs in case of a multi-modal bridge (about 1.7 per cent of the regional labour force, and about 1 per cent of regional GDP). The isolated effects of the Femern link are 1.080 and 1.200 jobs, respectively.

Changes in regional competitiveness. As transport costs change after the construction of the fixed links, regional accessibility also changes. In turn this will cause changes in patterns of regional competitiveness. A simple potential (gravity) model involving 100 zones in Europe was used to calculate potential values for four situations: a) before the opening of any fixed link, b) after the opening of the Great Belt link, c) after the opening of the Øresund link and d) after the opening of the Femern link. Figure 4.3 shows the changes in regional economic potential as each of the three fixed links is added to the transport system. Whereas the effects of the Great Belt link are confined to Denmark, the effects of the other links are international, often in regions at great distance from the links. Jensen-Butler and Madsen argue that the strong effects in southern Scandinavia can be explained by the improved access of this area to the large German market, whereas the northern part of Germany is hardly affected by the improved access to the rather small Scandinavian market. The total employment effects are estimated by the changes in the potential values and the changes in economic activity as calculated by the interregional model. In Table 4.7 the estimates of total employment change are given. In Denmark, it appears that Greater Copenhagen benefits most from the improved network.

The long-term effects on the labour market strongly depend on infrastructure improvements additional to the Femern fixed link, in particular the exten-

Great Belt

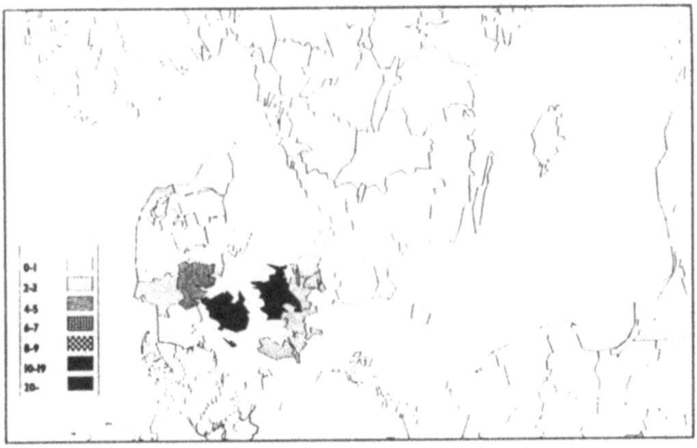

Great Belt and Øresund

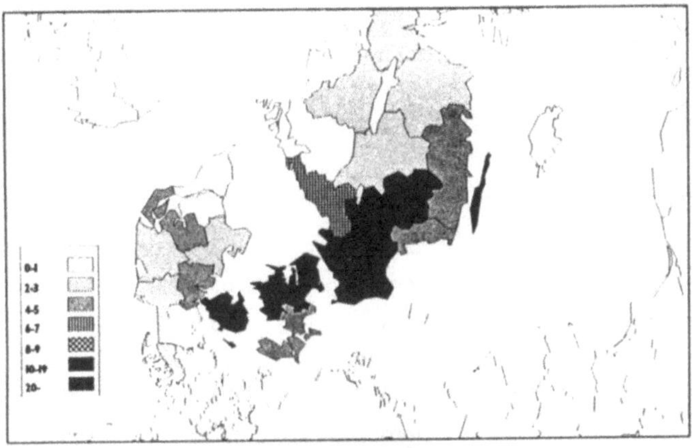

Great Belt, Øresund and Femern Belt

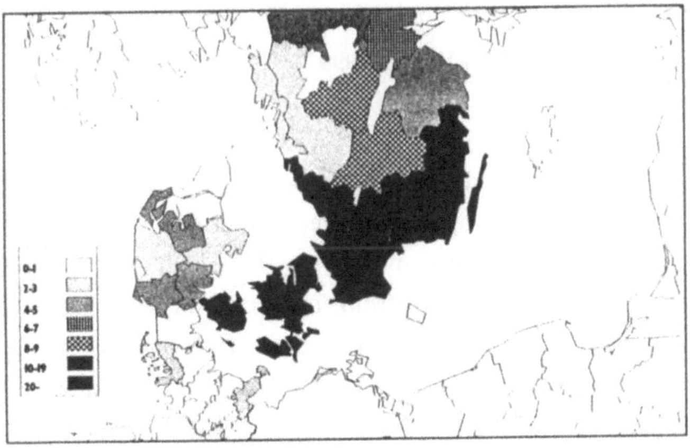

Source: Jensen-Butler & Madsen, 1998

Figure 4.3 Cumulative changes in economic potential (per thousand) after the opening of the Great Belt, Øresund and Femern fixed link, respectively.

Table 4.7 Successive total employment effects (number of jobs) of increased competitivity as a consequence of the three fixed links. The results are cumulative down columns.

	Storstrøm county	Denmark	Slesvig Holstein
Great Belt link	50	2.000	0
Øresund link	100	4.000	100
Femern Belt link	300	5.500	500

Source: Jensen-Butler & Madsen, 1998

Table 4.8 Market shares for transport corridors between the Nordic countries and the rest of Europe

		Today	2 fixed links	3 fixed links
Land border	cars	38.3	40.4	38.3
Jutland - Slesvig Holstein	lorries	65.6	68.0	64.1
Femern corridor	cars	33.0	30.8	41.9
	lorries	12.2	11.2	19.7
Other crossings	cars	28.7	28.8	19.8
	lorries	22.2	20.8	16.2

Source: Jensen-Butler & Madsen, 1998

sion of the high-speed rail line between Hamburg and Copenhagen. In the event that there is an upgraded intercity service, more significant effects will result in Denmark, since commuters will be able to buy cheap housing, enjoy a green environment in Storstrøm, and continue working in Copenhagen. This qualified labour force might attract firms to locate in this green area centrally located between Copenhagen, Hamburg and Berlin. As an illustration, Jensen-Butler and Madsen calculated that a relocation of 1.000 inmigrants in the Storstrøm county, of which 500 are working in Greater Copenhagen, will induce a local employment effect of 270 jobs (0.2 % increase in regional GDP).

Changes in traffic flows. The Femern link will have major effects on the choice of traffic corridor between Scandinavia and the rest of Europe. A trip distribution model, a modal split model and a route choice model have been used to study the changes in traffic flows by using data for Europe and the western Baltic area. As shown in Table 4.8, the Femern fixed link will more than compensate for the initial loss of its market share caused by the opening of the Great Belt and Øresund links. These changes in traffic flows create changes in the employment in the transport sector. Using ad hoc estimates, the Storstrøm county will gain 200 jobs, including indirect and induced effects. However, the decrease in traffic on the other ferry link in Storstrøm (Gedser-Rostock) is estimated to provoke the loss of 500 jobs. So the overall employment impact for Storstrøm county will be 300 lost jobs.

The above described three types of regional economic effects show that the overall employment effect of the Femern Belt - excluding the effects which may arise from new patterns of commuting - is negative for the regions adjacent to this link (Storstrøm county). The negative employment effect is primarily a result of the opening of the Femern link (ferry closures), rather than the effects of the two other fixed links. The positive medium term effect on regional competitiveness is rather small for Storstrøm and cannot compensate for the loss of jobs arising from ferry closures. The more remote regions of Greater Copenhagen and southern Sweden will benefit most from the fixed links; however, the overall regional economic effects are expected to be modest.

4.6.2 The Channel Tunnel[2]

An interesting example of a combined qualitative and quantitative research approach to the impact of a large infrastructure investment was the Channel Tunnel study commissioned by the European Commission.

In this project both a qualitative regional analysis and a quantitative transport and regional economic model are applied. In the qualitative approach, 13 regions within the European Union which have representative problems or characteristics regarding the impact of the Tunnel, are selected. The location of the regions is spread throughout the European Union. Some regions are located nearby the Channel Tunnel, for instance Kent and Nord

[2] For this section extensive use has been made of Fayman et al. (1995).

Pas de Calais, but others like Ireland, Piemonte (Italy) and Norte (Portugal) are located at quite a distance. The main purpose of the qualitative analysis is to include 'soft' economic factors - attitudinal responses and subjective judgements - which may influence the way regions adjust to the Channel Tunnel. The regional monographs contain basic indicators for the region and the findings of in-depth interviews with regional experts and decision makers.

For the quantitative approach, the MEPLAN transport and regional economic model (see section 4.4) was used (33 regions have been distinguished). The model consists of three modules: 1) the *regional economic module* estimates the changes in the regional patterns of production and consumption in each of the 33 economic sectors; 2) the *interface module* converts trade from units of annual value to units of daily tonnes or passengers moved along the transport system; and 3) the *transport module* takes the origin-destination matrices of daily flows of freight and passengers, divides them into the main modes of transport and assigns them to the appropriate transport network. Beginning with 1986, the model is run at five-year intervals until the year 2001. Five scenarios were calibrated: A) the current infrastructure network without the Tunnel; B) (without Tunnel) and B1 (with Tunnel) limited network development with substantial motorway construction and a medium level of rail upgrading; and C) (without Tunnel) and C1 (with Tunnel) assuming an extended network with a substantial number of new high-speed rail lines.

Because of the already well-developed motorway network in the European Union, car travel times are reduced by about two hours in 2001 for most regions (exactly the time savings of the Channel Tunnel compared with the ferries). However, the Channel Tunnel has major impacts on the travel time by rail, because it is integrated into the high-speed rail network. When the high-speed network is extended (scenario C1), most cross-Channel journeys by train will by 2001 be halved in time. The train will become the fastest surface transport mode in Europe and is for many relations even competitive with air travel.

The model predicts a rise of cross-Channel passengers from 67 million in 1986 to 135 million for 2001. Without the Channel Tunnel, both the number of ferry passengers and air passengers would nearly double over this period. With the Tunnel, the number of air passengers increases by 'only' 80 %, whereas the number of ferry passengers reduces by 20 %. About four million cross-Channel passages per year would be Tunnel-instigated (would not be made without the Tunnel). In the period 1986-2001, the number of lorries will increase from 2.1 million to 5.5 million, of which 1.6 million will go through the Tunnel. The forecasts for rail freight strongly depend on the introduction of high-quality through rail freight services. Without such a service the rail freight will be tripled; with this service the model predicts a sixteen-fold increase. These results are in line with the results of in-depth interviews with ferry companies and port authorities in maritime regions.

The effects of the Channel Tunnel on European transport flows are rather complex. In general, the ferries in its vicinity will face a major reduction of transport volumes in the years after the Tunnel starts operating. However, in the longer term there will be a secure future, because surface cross-Channel transport will grow significantly. The ferry Channel-crossings at some distance from the Tunnel only have to deal with temporally small decreases and a slight reduction in growth potential. There are three regions benefitting most from the Channel Tunnel: Nord-Pas-de-Calais for lorry traffic, West-Vlaanderen for unaccompanied RoRo traffic and - on the English side - Kent. The regions that prefer short ferry crossings today will prefer the Tunnel. They are located in a central corridor along the extended Tunnel axis on both sides of the Channel. The large time gains caused by the future high-speed rail network leads to a shift from car to train passengers along the high-speed lines as far as Germany, Northern Italy and Southern France. Transport flows from outside this region are rather indifferent to the construction of the Channel Tunnel and depend on other infrastructure development.

Concerning regional development: the model forecasts the change in total value added in the year 2001. The conclusion is that although positive effects of the Tunnel were found in the regions near the Tunnel and negative in the remote areas, the isolated Tunnel effect is small, compared with the autonomous growth. The regional analysis forecasts positive impacts in study regions at a distance up to 400 km. from the Tunnel and none or negative impacts beyond that distance. The regional monographs allows us to distinguish between the subjective expectations of the regional actors and the model based results. Taking the model based results as a reference point, it was found that regional actors tend to overestimate the likely impacts of the Channel Tunnel in regions near the Tunnel and underestimate them in remote regions.

Two interesting interpretations of the differences in the results between the model and the qualitative regional analysis are given, which - in our opinion -

emphasize the importance of combining qualitative and quantitative approaches to reach maximal knowledge of the spatial economic impacts of infrastructure investments. First, the model focusses on changes in transport costs assuming a uniform impact structure, whereas the regional analysis may take into account factors such as strategic considerations of carriers and specific regional features. Second, the results of the model show that even remote areas may be - negatively - affected by the Tunnel, whereas regional actors are not aware of the risks they may encounter.

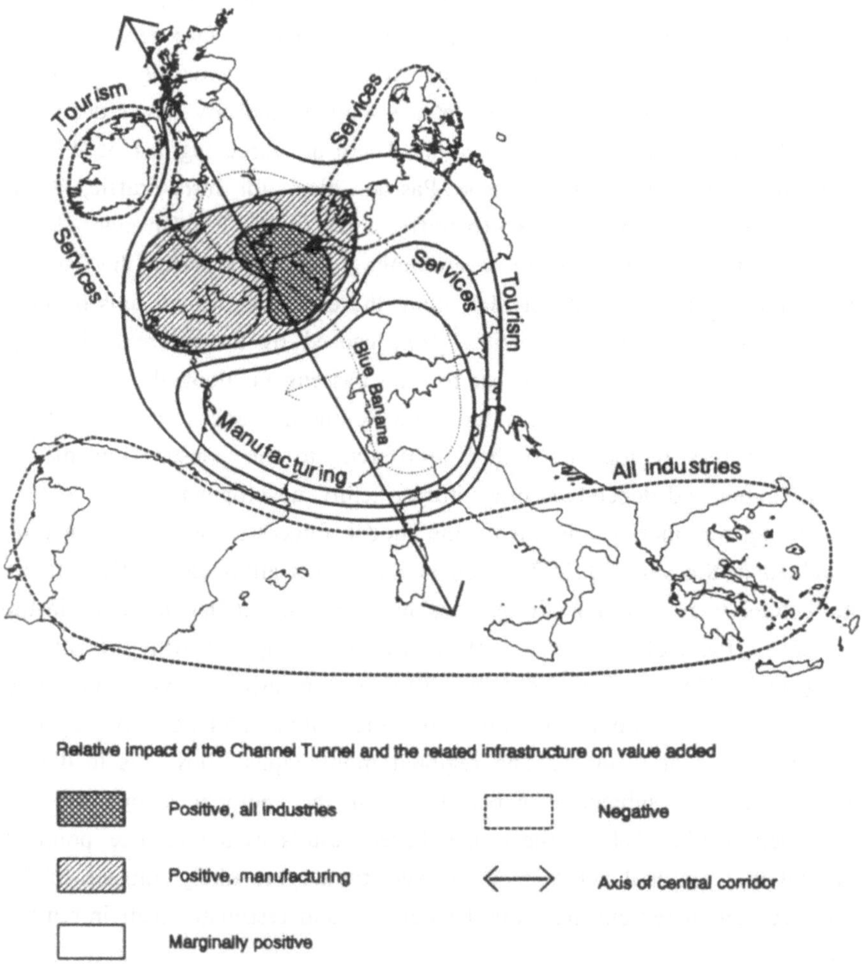

Source: Fayman et al. (1995)

Figure 4.4 Impacts of the Channel Tunnel on economic development

The final result of such a qualitative and quantitative study can be generalized in a map for all regions of the European Union (see Figure 4.4). In Figure 4.4 the main areas of relative growth and decline of value added induced by the Channel Tunnel and the related transport infrastructure are shown for manufacturing, services and tourism. In general, the triangle London-Brussels-Paris will benefit most. Furthermore, expansion will be found around connecting motorways and in particular high-speed lines. However, it is important to note that because of polarization effects there will be deprived regions next to regions with positive impacts (interstitial spaces on both sides of the Tunnel along the continental sea shore).

4.7 Concluding Remarks

A broad variety of methods is available to study economic impacts of infrastructure. Given the limitations inherent to both modelling and non-modelling approaches, they can often be considered (and used) as complements rather than as substitutes.

The direct contribution of infrastructure improvement to a reduction in transport costs is in general small in industrialized countries. The overall supply of transport infrastructure is already so extensive in many regions in Western Europe that a further improvement only has limited effects on costs. A possible exception to this statement is the case where an infrastructure project is a missing link in a network, or where the project helps to remove high levels of congestion.

Production function models often indicate substantial impacts of infrastructure on regional productivity, but it must be mentioned that this may be (partly) due to an improper specification of the production functions; problems of non-stationarity have not always been accommodated in a proper way. An interesting result of the use of cost functions is that infrastructure seems to be a complement to private capital and a substitute for labour. This would imply that - if output effects are ignored - infrastructure investment induces adjustments in production processes such that more private capital and less labour are used. In most production function models infrastructure is dealt with in a very aggregate way. The number of studies where a disaggregate approach is used is small. The results suggest that roads have stronger effects than the other infrastructure elements. A promising approach to productivity analysis is where the contribution of infrastructure is not measured via the stock of infrastructure, but via the services provided by that stock (accessibility).

Model based studies on the employment impacts of infrastructure at the (inter)regional level indicate that the effects are usually rather modest. This is no surprise in view of the theoretical results presented in Chapter 3, and given the substitution effect between infrastructure and labour mentioned above.

Non-modelling approaches point to the fact that infrastructure and accessibility are important in the locational patterns of firms with an international orientation. For firms in general, infrastructure appears to achieve a much less important position. It also appears that in lagging regions with a low current level of infrastructure, the importance attached to infrastructure as a location factor is higher than in more developed regions. Clearly, the present level of infrastructure supply appears to play a large role. The spatial level of analysis is also important. Infrastructure investments may lead to both generative effects and distributive effects. Part of the relocations related to the distributive effects will occur within the region itself and hence remain unobserved when a regional level of analysis is adopted. They will certainly have to be taken into account when the analysis has a more local orientation. For both spatial levels of analysis proper attention should be given to the distinction between generative and distributive effects.

At the intra-urban level, transport infrastructure impacts are usually studied by means of integrated land-use/transport models. Most models surveyed indicate rather modest effects of transport infrastructure on urban form. For example, the effect of an increase of 20 % in speeds of motorized transport modes on decentralization (suburbanization) of population is very small compared with the autonomous development in the spatial distribution of population. Slightly larger figures are found for the impact of uniform speed change on the spatial distribution of employment. It must be added, however, that effects for specific subgroups and in specific zones may be much larger. A property of these models is that they are based on the assumption that total population and employment are given. This means that the models can only deal with the *spatial distribution* of activities, not with the total *level* of these activities. These models only address distributive effects within an urban area.

Non-modelling approaches to the importance of infrastructure and accessibility at the local level in locational behaviour point to the conclusion that this is indeed a prime factor. For example, the two most important location factors mentioned (among a list of some 25 items) in a Dutch survey concern 1. accessibility by road, and 2. parking facilities. It is at this spatial level that spatial distributive effects are large so that one may expect clearly visible effects of infrastructure improvement.

The above conclusions also appeared to be quite relevant in the studies of the effects of large infrastructure projects (the Channel tunnel and three Danish fixed links). Both studies showed that the impacts of infrastructure improvements are relatively small compared with the autonomous growth. However, the spatial distribution of the impacts deserves attention. In case of the Danish fixed links, it is not the regions adjacent to the fixed links that profit most; they suffer due to a loss of jobs in the ferry sector. On the other hand, the greater Copenhagen region located somewhat farther away, profits most. The Channel Tunnel study clearly shows the importance of combining qualitative and quantitative approaches to achieve a full understanding of the impacts of infrastructure improvements. A qualitative approach can adjust aspects like strategic considerations of carriers, region specific aspects, and the impact of entrepreneurial perceptions; whereas the model approach shows that even remote areas will be - negatively - affected by major infrastructure improvements which is not foreseen by experts in those regions.

A basic weakness of most model studies is that causality problems are ignored: many decisions on infrastructure investments in the past were made as a response to observed or expected economic growth. Therefore, these decisions are not exogenous, which is usually assumed in the causality structures of models used.

Data sets on transport infrastructure are relatively weak, especially at lower spatial levels. Lack of coherence of data has hindered research, especially at the international level. This is a pity, since the forces leading to economic integration of countries induce a strong need for infrastructure studies with an international orientation.

It is dangerous to speak in general terms about impacts of infrastructure investments on employment. The different types of transport infrastructure (roads, rail, canals, airports) are in different stages of their life cycle and have different degrees of spatial penetration. Most of the studies reviewed in this Chapter refer to road infrastructure. This is an infrastructure type with a relatively high density in many parts of the higher income countries. In such a situation the extension of the highway system usually has a less dramatic impact on employment compared with low highway densities. It may also be expected to have smaller impacts compared with transport systems which are in a more dynamic stage of development (high-speed rail, the airline system). The fact that many studies reviewed here focus on road infrastructure in high income countries may be an explanation of the result that in several cases a rather limited impact of infrastructure on the economy was found.

PART B

European Accessibility Studies

Introduction to Part B

In this part of the book case studies are presented on the accessibility of major European agglomerations in the European transport infrastructure network. As already mentioned in Chapter 2, accessibility is one of the main concepts of the book. In this part we concentrate on the European level. In Part C the accessibility concept is dealt with in infrastructure impact and locational studies at the regional and urban level of spatial aggregation. Finally, in Part D the accessibility concept is used in infrastructure supply studies.

In Chapter 5 we present the results of a study on the relative position of 42 major urban agglomerations in three European infrastructure networks: air, rail and road. Furthermore, the effects of planned or possible future improvements in those infrastructure networks on the relative position of these cities are analyzed. Next to this, the implications of national borders as barriers in the international transport flows are studied for the relative position of the agglomerations in the road network. In addition, reductions in the barriers caused for instance by the European unification and improved relations with former communist countries are analyzed. Finally, the relative position of the agglomerations in the European infrastructure networks is calculated, when the fastest travel mode between the cities is chosen. For short distances this will be the car or train; for longer distances the airplane will be the main transport mode.

A last topic of interest in this Chapter is the equity in accessibility among cities. For each network, the equity in accessibility is studied in a base year as well as after the implementation of improvements.

We used a gravity type model in the analyses described above (see Chapter 2). However, in the recent past, a number of other studies using partly different approaches, have been undertaken on the relative position of agglomerations in European infrastructure networks. Therefore, we carry out a comparative analysis among seven of those studies in Chapter 6. In this Chapter we focus on the sensitivity of the results for alternative methodologies and operationalizations. The different operationalizations of the accessibility concept are analyzed on the basis of the notions given in Chapter 2. To allow a comparison of the results achieved in seven studies, standardized indices are constructed. Those indices are analyzed in a quantitative and a qualitative way for multi-modal, air and rail (including future high-speed lines) transport networks. Next, the results of different transport modes are compared by correlation analysis. The conclusion will be that the differences in the accessibility rankings of cities depend more on the choice of the transport network than on the operationalization of the accessibility concept, whereas the opposite holds true when one is interested in the equity in accessibility among the European agglomerations.

5 The Accessibility of Cities in European Infrastructure Networks

5.1 Introduction

The ongoing process of European integration causes an increasing competition among major urban agglomerations in Western Europe. The target of integration is the removal of all barriers to international trade, which includes among others, the harmonisation of fiscal policies. An implication is that several of the policy instruments which national governments could use in the past to promote development of their major urban agglomerations are no longer applicable. The development of urban infrastructure is seen as one of the final opportunities of the national government to advance their cities in the international competition. It is no surprise then, to see that in a number of recent studies, urban infrastructure plays a role as a determinant of competitiveness of urban regions (Biehl et al., 1986, DATAR, 1989, Cheshire, 1990, NEI, 1991, Healey & Baker, 1996). In most of these studies, attention is focused on the intra-metropolitan infrastructure. This includes the supply of transport infrastructure such as highways and (light) rail in metropolitan areas. In some studies a broader range of infrastructure types is also taken into account, so that education, culture and environmental amenities are also included.

In addition to intra-metropolitan infrastructure, inter-metropolitan infrastructure will also be important for the urban areas concerned. The free market compels cities to be outward oriented. Good connections in the international infrastructure networks will be a critical success factor in the distribution of economic activity in Europe. This raises the issue of the appropriateness of European infrastructure networks. The road and railway network have been planned in the past with a clear national orientation. This is no surprise, since the domestic component was dominant in trade and communication flows. During the last decades however, one observes a tendency that international flows grow faster than domestic flow. This implies an increase in demand for international infrastructure. The supply response can be observed in the examples of the development of the Channel Tunnel, a connection between Sweden and Denmark, and of tunnels through the Alps. Another example is the design of an international network of high-speed rail connections. In air transport, physical constraints do not play an important role in international communication. Nevertheless, it can be observed that also here certain non-physical barriers to international communication exist.

This Chapter describes the relative position of urban agglomerations in three European infrastructure networks: road, rail and air (see Bruinsma & Rietveld, 1993). The effects of planned or possible future major improvements in these networks on these accessibility patterns are studied. In addition, for the road network an analysis is given of the effects of national borders - which are seen as barriers in international interaction patterns - and the decline of these border effects through major political changes.

5.2 Methodology

The accessibility of each agglomeration is measured by the following simple gravity type formulation in which travel time is the main indicator (cf Chapter 2):

$$A_i = \Sigma_j \; 1/T_{ij}^{\,c} \qquad (1)$$

where A_i is the accessibility of agglomeration i and T_{ij} the travel time from i to agglomeration j

The gravity parameter c is assumed to equal 1. The total travel time T is measured in minutes and consists of three elements:

$$T = V + RT + I \qquad (2)$$

where:

V = penalty because one cannot depart at the desired moment (rail, ferries and air)

RT = real travel time while moving

I = check in and check out time (ferries and air)

The penalty V is estimated as follows:

$$V = \frac{1}{4}\, E/F \tag{3}$$

where E is the effective travel period during which one can depart - for instance between 06.00 hours and 18.00 hours - and F is the frequency of the connection. For example, if one would go at an arbitrary moment to the airport, average waiting time would be half of the average time in between two departures (which equals ½ E/F). We suppose most travellers know their departure time, but we still give a penalty because travellers cannot leave at the moment they most desire. Therefore, we reduce the penalty of ½ (as would occur in the case of an arbitrary arrival on the airport) to ¼ to express this inconvenience (see Bruinsma & Rietveld, 1991).

If interaction is supposed to depend on the size of the agglomerations with which an agglomeration interacts, then weighting can take place by use of the population size of those agglomerations. This leads to the next formula, where P_j is the population of agglomeration j:

$$A_i = \sum_j P_j/T_{ij}^{\,c} \tag{4}$$

The weighting by population size makes it necessary to include the internal interaction in agglomeration i. This leads to the final gravity formula:

$$A_i = P_i/t_i^{\,c} + \sum_j P_j/T_{ij}^{\,c} \tag{5}$$

The impact of the internal interaction is relatively large, since the intra-urban travel time is relatively small, that - in the gravity formula - implies a relatively large value. Therefore, it is important to pay some attention to the measurement of the internal interaction. When an agglomeration would be circular and all zones would be used in an equally intensive way, the average distance between two arbitrary zones can be shown to equal:

$$d = r(\pi-1)/\pi \tag{6}$$

However, most agglomerations are not circular. Since the surface O equals πr^2, the formula can be rewriten as:

$$d = \sqrt{(O/\pi)}(\pi-1)/\pi \tag{7}$$

This implies that d equals $.68\sqrt{(O/\pi)}$. Based on somewhat different assumptions, Rich (1980) arrives at a similar result:

$$d = \sqrt{(O/\pi)}/2 \tag{8}$$

A problem in the present study is that we do not have standardized data on the surface O of the cities. Therefore O has to be approximated. This is done by assuming a given population density in all cities involved of 2000 persons per square kilometre. An exception is made for a limited number of polycentric cities where we assume a density of 1000 persons per square kilometre, implying an increase in average distance[1].

The value of the share of the internal interaction in the total accessibility score of agglomeration i depends on two factors. The share is higher the larger the population size of agglomeration i, and the share is lower the larger the number of connections with the agglomerations j located nearby.

To measure the interaction pattern for air, rail and road networks, we selected the 42 agglomerations in Europe excluding the former U.S.S.R. with a population size of over 1 million. Polycentric urban areas such as for instance, the Randstad, the Ruhr-area and Mid-England, are divided into areas with a maximum size of 50 kilometres. This implies that the Randstad is divided into two, the Ruhr-area into three, and Mid-England into four urban agglomerations.

Data on travel time and frequencies between those agglomerations are obtained from ABC World Airways Guide (1991) (air), Thomas Cook European Timetable (1991) (rail, ferries), and the Michelin Wegenatlas van Europa (1988). The data of the road network are converted from distance into

---

[1] This issue of the internal share in total accessibility deserves more systematic attention than is usually given, especially since the data on average internal distance and speed are often weak. See also Bröcker (1989) and Frost and Spence (1995). This issue also receives attention in Chapters 2, 6 and 10.

travel time. For highways we used an average speed of 90 kilometres per hour, for roads of a lower quality, this Figure is 60. Within the urban agglomerations we used an average speed of 30 kilometres per hour.

5.3 Air Traffic

To measure the accessibility of the 42 European agglomerations, we start with the study of direct air connections. The index which is not weighted for population size reflects only the spatial dimension of the location in the network, whereas the weighted index in which the internal interaction is included reflects a combination of the mass of the agglomeration itself and its external contacts.

According to the unweighted index, London and Paris have the best accessibility in air traffic. Their position is not very dominant, however; the difference with other European cities such as Frankfurt and Zurich being rather small (see Table 5.1, column 1). However, when the weighted index is used, their position is much more dominant (see Table 5.1, column 2 and Figure 5.1). This rising dominance is completely explained by the share of the internal interaction. For London and Paris, the share of the internal interaction is (both in a relative and an absolute sense) rather large compared with its smaller competitors like Frankfurt, Zurich, Brussels and Milan. The scores of Essen and Lodz, which have no airport, completely depend on their internal interaction. As a consequence, the spread of the index is relatively large (100-17), compared with the other modes of transport, in which all agglomerations are linked.

A problem with airports is that their home markets often exceed the urban agglomeration. For instance, Schiphol does not only serve Amsterdam, but also Rotterdam, and the same could be said for the relation between Frankfurt and the Ruhr-area. For this reason we have included the indirect connections in the gravity model. As can be seen in Table 5.1, column 3, the scores of agglomerations with no or small airports rise sharply. In the ranked scores Rotterdam and Essen rise respectively, 18 and 14 places. Important to note is that the third position of Frankfurt becomes less clear. In the situation with only direct flights, only 7 competitors are within the 10 percentage points distance of Frankfurt. When the indirect flights are included, the number of competitors has risen to 13. We conclude that in this context the mainport position of Frankfurt is not very evident.

Table 5.1 Accessibility of European cities by air traffic

	1	2	3	4	5		1	2	3	4	5
London	100	99	99	99	98	Dublin	43	50	66	51	47
Paris	96	100	100	100	100	Athens	41	57	64	59	54
Frankfurt	90	75	77	76	77	Birmingham	41	48	62	50	47
Zurich	87	73	76	75	69	Stockholm	41	50	60	51	47
Brussels	83	70	74	72	66	Belgrade	39	47	55	51	46
Amsterdam	83	70	73	72	66	Lyon	39	48	65	50	46
Milan	75	69	72	70	65	Bucharest	36	48	56	52	47
Munich	69	65	70	67	62	Cologne	36	42	60	43	41
Copenhagen	68	64	70	66	60	Lisbon	35	49	59	50	47
Rome	64	66	73	68	63	Sofia	35	43	51	47	43
Berlin	64	69	75	72	66	Zagreb	30	36	51	39	37
Düsseldorf	63	60	69	62	58	Marseille	24	39	61	41	38
Vienna	62	61	70	64	58	Turin	22	35	60	36	35
Madrid	57	67	73	67	63	Newcastle	21	32	55	33	32
Hamburg	56	60	71	61	57	Naples	20	33	54	34	33
Barcelona	55	60	67	62	57	Leeds	13	31	56	31	30
Budapest	47	55	63	59	53	Genoa	13	28	54	29	29
Prague	47	48	59	53	47	Rotterdam	11	32	66	32	32
Warsaw	46	52	58	56	50	Liverpool	6	21	54	21	20
Manchester	44	51	63	52	49	Essen	0	18	60	17	16
Istanbul	43	65	70	67	62	Lodz	0	17	49	16	16

1 = unweighted 2 = weighted 3 = indirect connections
4 = even growth 5 = mainport

We have formulated two scenarios for future developments. The first scenario is based on an even growth assumption: the frequencies of all flights are doubled. In the second scenario it is assumed that the total volume of flights is again doubled, but that the growth is uniquely concentrated on the mainports London, Paris and Frankfurt; the frequencies of all flights to and from these airports are multiplied by four; the frequencies of the other flights remain unchanged. These scenarios describe the two extremes between which the future of the airline system may be expected to be: growth without structural change, and the development of a limited number of large mainports.

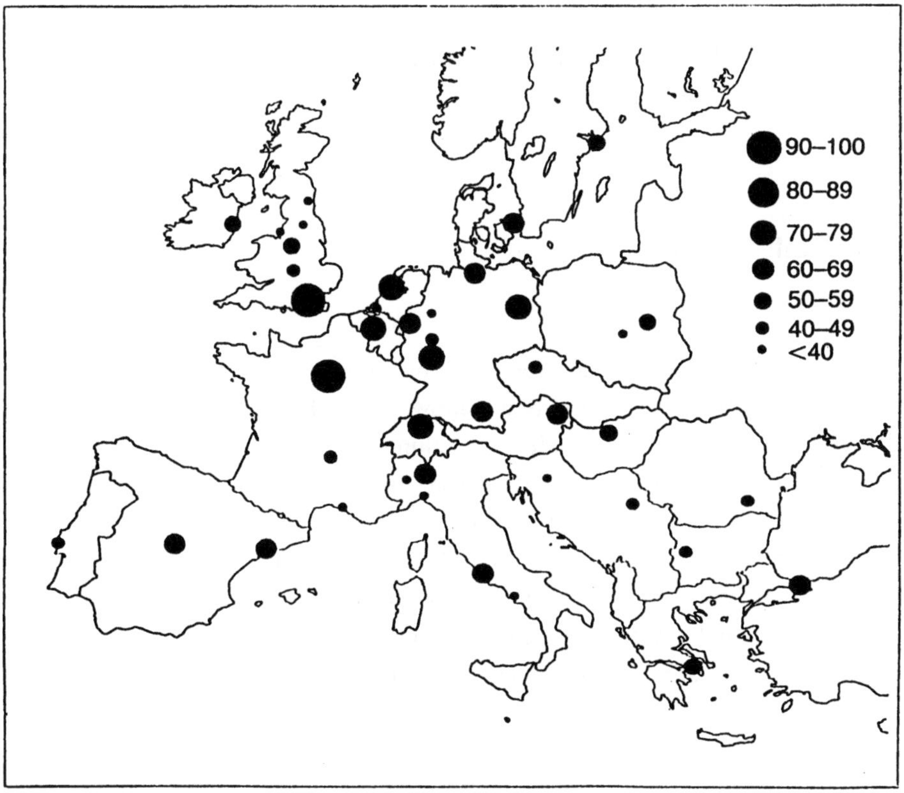

Figure 5.1 Accessibility of European cities by air traffic, weighted by population size

In the even-growth scenario, especially the East-European agglomerations improve their scores (Table 5.1, column 4). Those cities have relatively large numbers of connections, but have low frequencies on those connections. A doubling of the frequencies leads to a relatively sharp fall in the penalty V, which means a substantial shortening of the total travel time T. As a consequence, the accessibility of these cities rises.

In the mainport scenario the mainports do not become as dominant as one might expect (Table 5.1, column 5). The explanation is the reverse of the former one. The connections of the mainports are already flown with high frequencies, so that the penalty V does not change much (this is the consequence of putting the frequency F in the denominator of the formula of the penalty). The relative scores of the nearest competitors of the mainports tend to decrease by about four points, which means a strengthening of the mainport positions. Interesting is the fact that the individual index of small airports is relatively stable. An explanation is that those airports have relatively many connections with mainports.

5.4 Rail Traffic

In the situation where no weighting by population size has occurred, the Central German cities score best followed by the Mid-England cities, the Benelux cities and Paris (Table 5.2, column 1). The Mid-England cities are not centrally located in the European railway system. They score high because here a relatively large number of cities are located near to each other. In the gravity model, interaction over short distances is relatively high.

When weighting for population size takes place, Paris and London become dominant again by their large internal interaction (Table 5.2, column 2 and Figure 5.2). The scores of other large agglomerations also improves considerably.

For rail traffic only one future scenario is formulated. The high-speed links (HSL) as proposed by the European Commission (E.C., 1990b) are integrated in the model, although the measures already taken in the various countries make the E.C. proposal plausible. One must of course always consider that at the level of individual links, differences will emerge between proposal and realization. The high-speed rail network is planned in Northwestern Europe, the area which currently is already best accessible by rail traffic. Before explaining the results, we must indicate how travel times are computed in the HSL case. We have not changed the frequencies of the existing connections. We only reduced the travel time on the sections of the network where high speed services are planned.

Table 5.2 Accessibility of European cities by rail traffic

	1	2	3		1	2	3
Düsseldorf	100	90	83	Vienna	37	51	54
Cologne	92	85	84	Berlin	35	62	55
Essen	86	81	74	Marseille	35	51	56
Manchester	66	77	68	Rome	34	57	52
Leeds	63	77	69	Budapest	31	51	43
Rotterdam	63	67	68	Prague	30	45	38
Brussels	63	71	82	Copenhagen	28	44	42
Amsterdam	62	66	67	Zagreb	27	42	35
Paris	61	100	100	Naples	26	42	44
Liverpool	59	71	63	Warsaw	26	44	37
Frankfurt	58	64	71	Belgrade	24	42	35
Birmingham	57	76	69	Lodz	24	39	32
London	54	96	90	Barcelona	23	46	49
Milan	50	57	62	Dublin	22	39	33
Hamburg	48	61	56	Madrid	20	53	50
Turin	47	52	54	Stockholm	18	40	34
Munich	45	54	54	Sofia	17	35	29
Lyon	45	63	68	Bucharest	16	42	34
Zurich	44	54	59	Lisbon	13	42	36
Genoa	43	49	47	Istanbul	12	58	47
Newcastle	43	55	50	Athens	11	46	37

1 = unweighted   2 = weighted   3 = HSL

Interesting is that it is Brussels and not Paris which gains most by the introduction of high-speed links (Table 5.2, column 3). Brussels gains most both in a relative and an absolute sense. The central location of Brussels in between the large population concentrations of London, Paris and the Ruhr-area explains this favourable development.

The development of the high-speed rail network is of course most favourable for the cities located on this network, especially Northwestern Europe. Interesting is the relatively large impact of the high-speed rail network on the urban axis of Northern Italy towards Barcelona. For those cities not only do their links with Northwestern Europe improve, but so do their mutual links, which are currently rather bad.

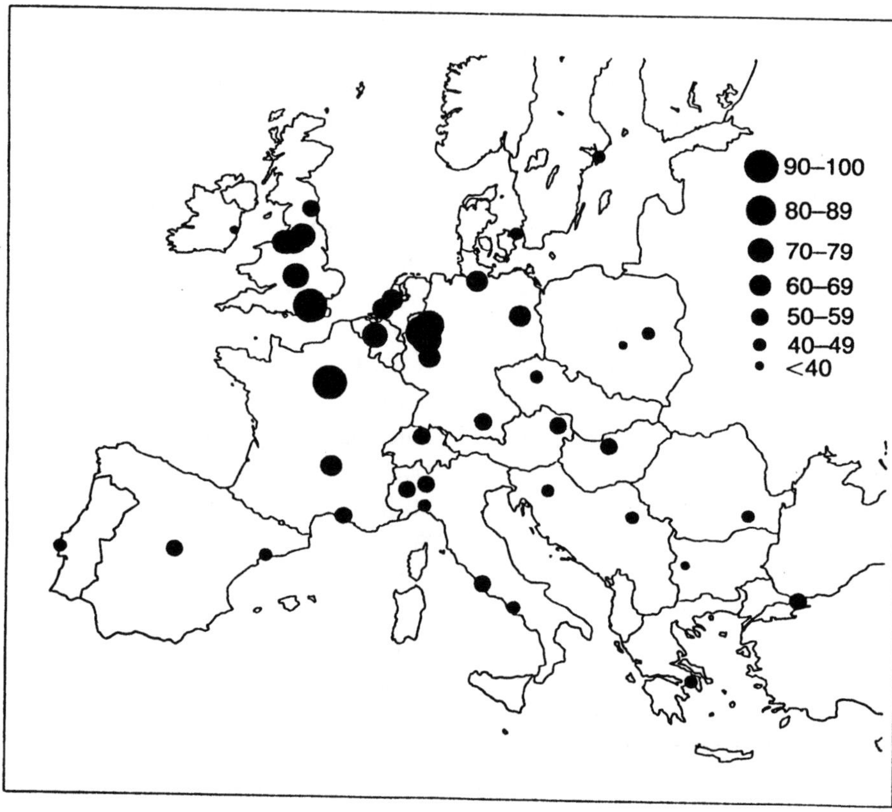

Figure 5.2 Accessibility of European cities by rail traffic, weighted by population size

5.5 Road Traffic

The results of the road network are to a great extent similar to the results of the rail network. Again in the unweighted situation, the cities in the Ruhr-area score best and in the situation weighted for population size, Paris and London become dominant (Table 5.3, column 1 and 2 and Figure 5.3).

Table 5.3 Accessibility of European cities by road traffic

	1	2	3	4		1	2	3	4
Düsseldorf	100	78	79	78	London	72	94	94	95
Essen	98	77	78	77	Vienna	71	60	62	61
Cologne	97	75	75	74	Prague	71	58	61	60
Brussels	87	70	70	70	Marseille	69	56	57	56
Rotterdam	85	69	69	69	Zagreb	67	54	57	55
Leeds	85	74	74	74	Budapest	66	61	64	62
Amsterdam	85	67	67	67	Rome	64	63	64	62
Frankfurt	84	70	70	69	Copenhagen	60	52	52	56
Manchester	84	71	71	71	Lodz	60	49	55	54
Liverpool	81	68	68	68	Belgrade	60	52	55	54
Milan	80	65	65	64	Barcelona	59	54	54	54
Turin	78	61	61	60	Naples	59	49	50	49
Munich	78	63	64	63	Warsaw	58	51	57	56
Zurich	78	63	63	62	Sofia	55	45	49	48
Birmingham	77	70	70	71	Dublin	52	43	43	43
Genoa	77	59	60	59	Madrid	51	58	60	59
Paris	76	100	100	100	Bucharest	51	50	54	53
Lyon	75	62	62	62	Stockholm	50	45	46	47
Newcastle	75	60	60	61	Istanbul	47	67	70	68
Hamburg	74	66	66	66	Lisbon	46	48	50	49
Berlin	72	74	75	74	Athens	44	52	55	54

1 = unweighted
2 = weighted
3 = road improvement
4 = road improvement/bridges/tunnels

We have formulated two future scenarios. In the first scenario all connections considered achieve highway quality (Table 5.3, column 3); in the second scenario in addition to the improvement of the road network itself, the ferries are replaced by bridges and tunnels (Table 5.3, column 4). Both scenarios are plausible. The first entails the upgrading of road quality in the Northern, Eastern and Southern part of Europe; this is what one may expect when economic conditions improve in these areas. Agreement on most of the bridges and tunnels mentioned in the second scenario has already been reached among the governments concerned.

The improvement of the road network leads to better scores for the East and South European cities and Stockholm. At the moment most of the roads in these regions are of low quality. The relatively strong rise in the scores of the East European cities compared with the South European cities and Stockholm can be explained by the shorter distance of the East European cities to the centrally located agglomerations. In the gravity model a gain of 10 minutes travel time on a trip of one hour has a greater effect on accessibility than the same gain in travel time on a two hour trip.

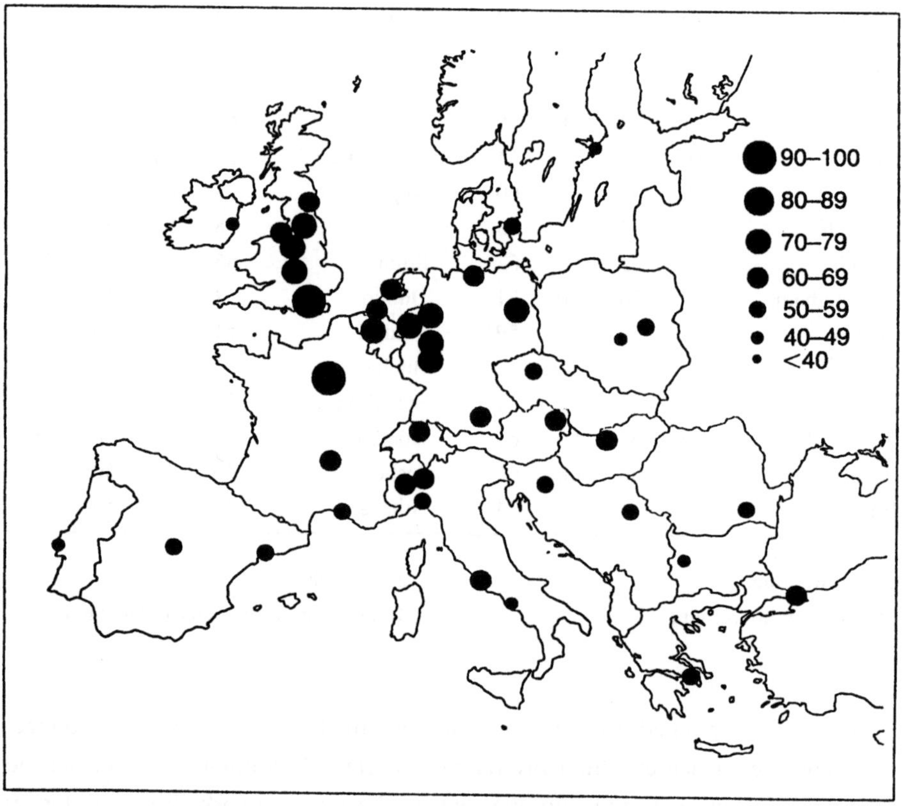

Figure 5.3 Accessibility for road traffic, weighted by population size

When, in addition to the improvement of the road network itself, ferries are replaced by bridges and tunnels (the Channel-tunnel, links between Sweden - Denmark and Denmark - Germany), the impact on the accessibility index seems to be marginal. However, the individual index of, for instance, the

English cities shows a relatively sharp rise. This rise in the individual score does not appear in the final index because Paris (the reference city) also profits from the improvement of the road network as well as from the Channel-tunnel.

5.6 National Borders as Barriers in the Road Network

It is generally recognized that national borders function as barriers in international interaction patterns. Research has proved that the crossing of national borders is more than merely the physical crossing of an administrative barrier, which could lead to extra travel time, for instance, as a consequence of customs formalities. A national border can often be understood as a non-physical barrier with an economic, political, cultural, or language dimension (see Nijkamp et al., 1990).

In most of the research on the effects of borders on international interaction patterns, it is attempted to trace the difference in interaction patterns of two cities located in the same country compared with two cities located in different countries. The results show a reduction in interaction of in between 70 or 80 percent in the case when a border has to be crossed within Northwest Europe (see Bröcker, 1984 and Nuesser, 1985). The interaction is even more disrupted when borders between other countries are crossed (see Rietveld & Jansen, 1991).

Another approach would be to analyze the density of infrastructure networks (both road and rail) in border regions, national border zones (contains all regions of a country bordering on another country), and on the border line itself, and to compare these figures with the national average. One might expect that network densities in border areas are in general lower than their national average. We have tested this for a number of Western and central European regions (for the E.C. NUTS-II regions have been used (EUROSTAT, 1990)). It appears that our expectation is not confirmed: the density of the highway network in 44 % of the border regions and 49 % of the national border zones is above the national average. Only 26 % of the national border zones score below the national average. In 25 % of the cases an equal score is found. This high score can be explained by the fact that some countries (Luxembourg, Switzerland and Austria) could not be divided into border zones.

The explanation of these unexpected results is that at the level of spatial aggregation used, many border regions have large centres within their area. For example, NorthRhein Westfalia, which includes the Ruhr-area, is a border region in Germany. Therefore, we have checked to which extent population density can explain the observed results (see also Chapter 13). Regression results show that infrastructure density is closely related to population density. Corrected for population density, border regions still have slightly higher infrastructure densities however. Thus, at this level of spatial aggregation, no clear sign can be found of a disadvantageous position of border regions in terms of infrastructure densities.

Completely different results are obtained when one focuses on infrastructure densities on the border line itself. The density on the border line is in nearly all cases below the regional average (on average only 29 % of the regional average). Exceptions have to be made for areas in which a highway is located on an extremely short border line as, for instance, the region Pais Vasco in Spain.

Similar results are found if the density on border lines is compared with the density in national border zones. Although still clearly below the national average, the density on the border line is relatively high on the mutual Benelux borders, on the borders between the Benelux and Germany, and on the borders between the Benelux and France.

We conclude that, although the border regions are relatively well-equipped with highway infrastructure, the density on the border line itself remains far below the national average. Thus, it appears that the orientation of the well-equipped border regions is focused on the national instead of the international economy.

Our conclusion is that national borders exert a barrier effect on international communication. In the case of the road transport, this barrier effect is due to both non-physical factors such as language differentials, and physical factors such as the low density of roads on border lines. In the case of air traffic, the physical network infrastructure is not biased against border crossing flights. However, here too barrier affects will occur, since the frequencies of flights between cities tend to be lower when these cities are located in different countries compared with only one country. With rail both elements play a role: network densities on borders are relatively low, and the frequencies of international trains are also relatively low.

It is interesting to investigate the impact of barrier effects on the accessibility measures. We will focus on road transport here. Barrier effects are taken into account by reduction factors. The reduction factors for the different combinations of countries are given in Table 5.4. The domestic interaction flows are not reduced.

Table 5.4 Border related reduction factors for road transport

	E.U. country	E.F.T.A. country	E.E. country
E.U. country	.250	.167	.125
E.F.T.A. country	.167	.167	.125
East European country	.125	.125	.167

The results are striking compared with the situation without barriers (Table 5.5, column 1 and 2, Figure 5.3 and 5.4). London takes over the first position from Paris.

The explanation is rather simple. Reduction factors are only used for international connections. In our sample six English and seven German agglomerations are included. They are all ranked within the first 20. Only large agglomerations such as Paris, Istanbul, Rome, Madrid, and Athens are ranked in between them. The share of the internal interaction in their total score can be as high as 96 %, as is the case for Istanbul.

Major losers are the agglomerations in smaller countries which are rather centrally located, such as Brussels, Amsterdam, Rotterdam, Zurich, and Prague: their score largely depends on connections with foreign agglomerations.

We have formulated four scenarios of future political developments which could decrease the barrier effects of national borders. In the first scenario we assumed that in Europe after 1992, the cross-national interaction would be less disturbed than at present. So the reduction factor declines from .250 to .333. In the second scenario, above this development within the E.U., we expect that the political transition in Eastern Europe leads to an easier access of those countries. So all relations with East European countries receive a reduction factor of .167. In the third scenario we analyze the changes when the E.F.T.A. countries are welcomed into the E.U., and in the final scenario Hungary also becomes an E.U. member.

Table 5.5 Impact of borders as barriers in the road network on the accessibility of European cities

	1	2	3	4	5	6		1	2	3	4	5	6
Paris	100	97	1	-	-	-	Budapest	61	48	-1	1	-	4
London	94	100	-	-	-	-	Turin	61	49	1	1	-	-
Düsseldorf	78	69	1	-	1	-	Newcastle	60	61	-	-	-	-
Essen	77	68	1	-	-	1	Vienna	60	43	-	-	5	-
Cologne	75	64	1	-	1	-	Genoa	59	50	1	-	-	-
Berlin	74	71	1	-	-	-	Madrid	58	59	-	-	-	-
Leeds	74	77	-	-	-	-	Prague	58	38	-	1	-	-
Manchester	71	73	1	-	-	-	Marseille	56	47	2	-	-	-
Brussels	70	41	4	-	1	-	Barcelona	54	48	1	-	-	-
Birmingham	70	70	1	-	-	-	Zagreb	54	39	-1	1	-	-
Frankfurt	70	59	2	-	-	-	Copenhagen	52	42	1	-	1	-
Rotterdam	69	49	3	-	-	-	Athens	52	55	-	-	-	-
Liverpool	68	69	1	-	-	-	Belgrade	52	41	-	1	-1	-
Istanbul	67	73	-2	-	2	-	Warsaw	51	44	-	-	-	-
Amsterdam	67	47	3	-	-	-	Bucharest	50	45	-1	1	-	-
Hamburg	66	60	1	-	-	-	Naples	49	46	1	-	-	-
Milan	65	54	2	-	-	-	Lodz	49	39	-	1	-	-1
Rome	63	62	-	-	-	-	Lisbon	48	48	1	-	-	-
Munich	63	62	1	1	-	-	Sofia	45	35	-	-	-	-
Zurich	63	40	-1	-	7	-	Stockholm	45	41	-	-	3	-
Lyon	62	51	1	-	1	-	Dublin	43	36	1	-	-	-

1 = without border effect
2 = with border effect
3 = change due to scenario 1
4 = difference between scenario 2 to 1
5 = difference between 3 to 2
6 = difference between 4 to 3

One might expect some major changes in the accessibility index, but as shown in Table 5.5, columns 3-6, the changes are moderate. The smaller agglomerations in relatively small countries gain most from Europe 1992. In the second scenario, the rise in accessibility of Eastern European countries is marginal, Warsaw and Sofia do not even gain one percentage point. The acceptance of the E.F.T.A. countries in the E.U. leads to the first rise of real importance. However, notice that here the possibility to cross the border is doubled. When Hungary is accepted as a member of the E.U., a similar rise would occur for Budapest.

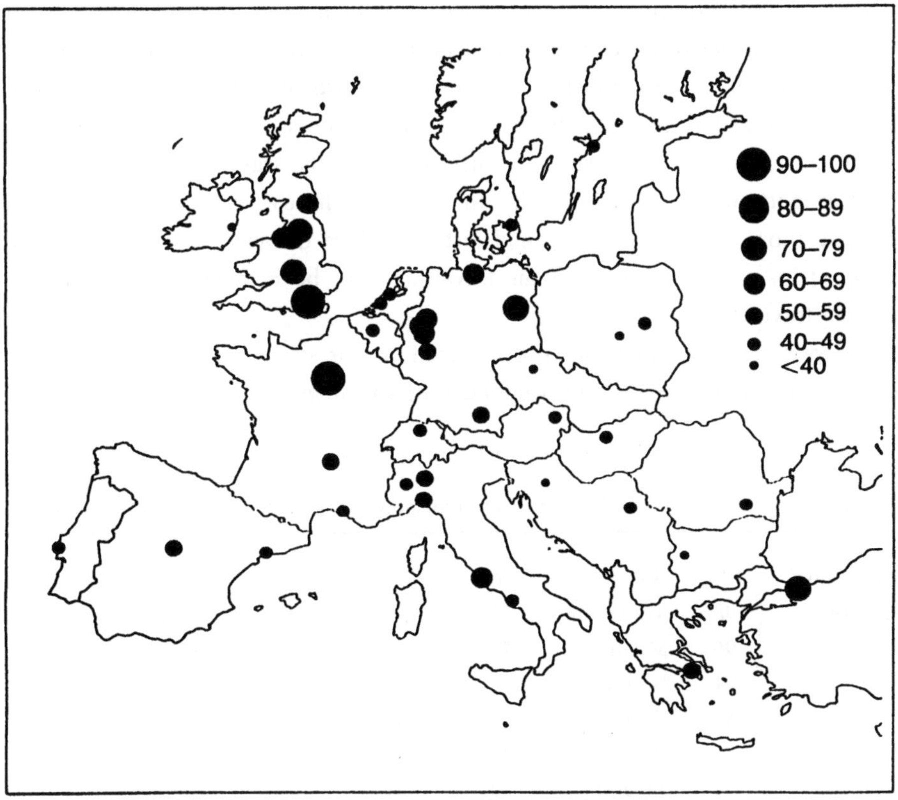

Figure 5.4 Impact of borders as barriers in the road network on the accessibility index of European cities

The conclusion is that there has to be a relatively large decline in the barrier related reduction factor before a substantial rise in the accessibility score of an agglomeration occurs. It would be rather short-sighted to be only concerned about the extension of physical infrastructure networks, however. The non-physical - organizational and political - barriers seem to be as important as those of physical barriers for an improvement of accessibility.

5.7 Integration of the Transport Modes: the Shortest Travel Time

In the preceding sections we have studied the accessibility for the individual infrastructure networks. An important reason to proceed thusly is that changes in accessibility as a consequence of certain major improvements in the individual networks can be shown most clearly by this way. However, the real interaction pattern consists of a mix of those infrastructure modes. The choice of transport mode depends on the preferences of individual travellers. Their preferences will be strongly influenced by the travel time and price of the transport modes. On short trips the share of the car and train will be high because those modes are cheap and fast over short distances. However, on the longer trips the high price of air travelling will be compensated for by the gain in travel time.

The prices of transport are not included in our study, so we only used the shortest travel time for the measurement of an integrated accessibility index.

Table 5.6 Accessibility by use of the transport mode with the shortest travel time

Düsseldorf	100	Birmingham	76	Lyon	67	Prague	57
London	98	Rotterdam	74	Copenhagen	67	Lisbon	57
Paris	96	Milan	73	Istanbul	67	Genoa	56
Manchester	91	Berlin	73	Barcelona	64	Bucharest	54
Essen	89	Zurich	73	Dublin	63	Belgrade	53
Leeds	87	Rome	70	Turin	62	Naples	53
Cologne	87	Madrid	70	Athens	61	Zagreb	50
Liverpool	86	Hamburg	70	Budapest	61	Sofia	49
Amsterdam	81	Munich	68	Marseille	59	Lodz	49
Brussels	78	Vienna	68	Stockholm	58		
Frankfurt	77	Newcastle	67	Warsaw	57		

When the indirect flights are included, the airplane is the fastest transport mode for 93 % of the connections (1604). The car and train are fastest for respectively, 5 (92) and 2 (26) percent of the connections. The 118 non-airplane connections consist of 36 % of connections for which no direct air connection is available (mainly domestic connections).

The results, weighted for population size, are given in Table 5.6 and Figure 5.5. Düsseldorf, an agglomeration with a small population (low share of the internal interaction) scores best. Düsseldorf has fast road and rail links with other cities in the Ruhr-area, which means a high interaction between those cities. Furthermore, Düsseldorf can use the airport of Frankfurt for missing air connections.

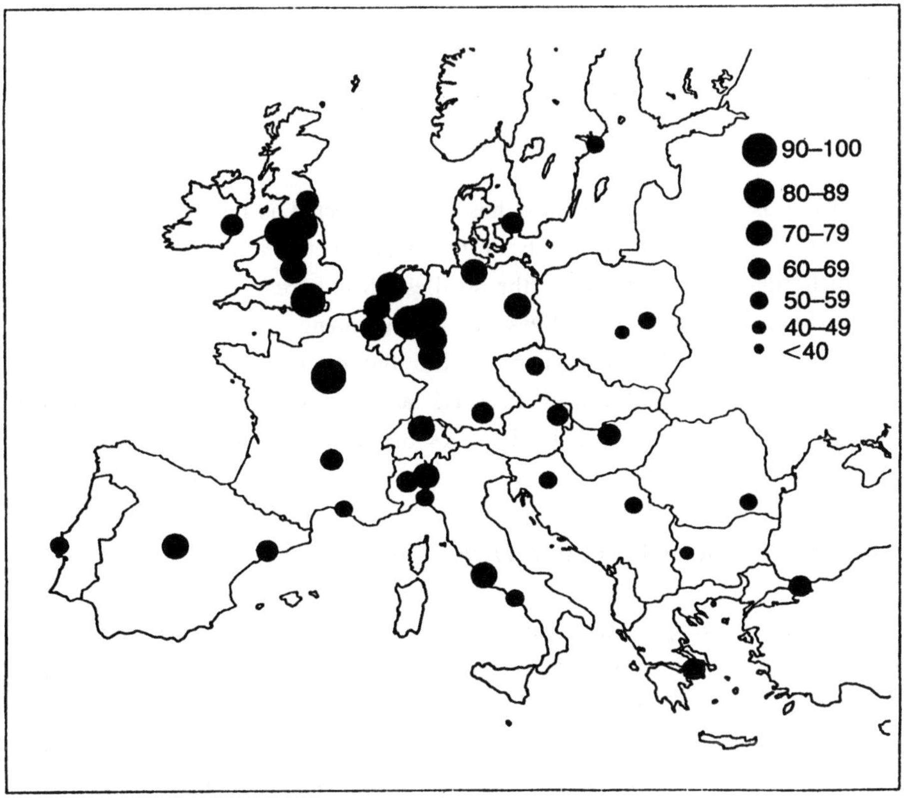

Figure 5.5 Accessibility of European cities by use of the transport mode with the shortest travel time

Although the air traffic deals with 93 percent of the connections, the interaction pattern is quite different from the pattern of the air traffic (compare Table 5.1, column 3). This can be explained by analyzing the car and train connections. The car and train connections all concern connections over short distances, which results in high interaction flows in the gravity model. Another interesting result is the relatively small spread in the integrated index (100-49) compared with the spread in the index of road traffic (100-43) and rail traffic (100-35). This can be explained by the reduction of the large travel time differences which exists for car and train traffic between short and long connections by introducing the air traffic for the long distances. Thus, the integrated index shows a greater equity in accessibility compared with the individual transport modes.

5.8 Equity in Accessibility

A major issue which until now has been neglected is the equity in accessibility of the agglomerations in the infrastructure networks.

Which transport modes show the smallest differences in accessibility? Does this difference increase or decrease as a consequence of the improvements in the network? Which transport mode has the highest average accessibility? Does this average increase or decrease as a consequence of the improvements? What are the mutual relationships between these improvements?

5.8.1 Air Traffic and Shortest Travel Time

In Table 5.7 we present some key summary indicators on accessibility for air traffic, weighted by population size. The coefficient of variation is rather high, which means that there are large differences in the accessibility of the agglomerations. This could be expected because, for instance, Essen and Lodz have no airport and as a consequence have very low scores. This also explains why the impact of the future scenarios on the coefficient of variation is rather small. The situation of Lodz and Essen does not change.

The fact that the doubling of all frequencies results in larger effects than the mainport development is shown in the coefficient of variation and in the average accessibility. In both situations the development of the doubling of all frequencies is favourable.

Major changes are recognizable when indirect flights are included. Compared with the situation without indirect flights, the inequity in accessibility is halved and the average accessibility rises by one third. It might be clear that in this situation the accessibility of agglomerations with no or bad air traffic connections rises sharply, while the accessibility of the best accessible agglomerations rises only marginally by adding the indirect flights.

Table 5.7 Development average accessibility by air traffic and shortest travel time

	Actual	Doubled frequency	Mainport orientation	Indirect connections	Shortest time
Average score	264671	285942	274660	340625	379371
Standard deviation	94164	99410	97520	55933	74206
Coefficient of variation	.356	.348	.355	.164	.196
Average accessibility	100	108.0	103.8	128.7	143.3

If we compare the results of air traffic, in which indirect flights are included, with the results based on the fastest transport mode, some interesting changes occur. Although the average accessibility rises sharply, there is a slight decrease in the equity of accessibility. This can be explained as follows. The average accessibility rises because for the lacking inland connections or for unfavourable air connections, one can choose to take a trip by car or by train. In all cases this concerns connections over short distances, which according to the gravity model, generate large interaction flows. The agglomerations for which those connections are included, already had a relatively high accessibility in the reference case (cities in the Ruhr-area and Mid-England). Their accessibility tends to rise sharply while the accessibility of the low scoring agglomerations hardly rises. So the equity in accessibility decreases.

5.8.2 Rail traffic

As is the case for air traffic without indirect flights, the equity in accessibility for rail traffic is rather low (Table 5.8). Especially the connections in East and South Europe are rather insufficient. The development of high-speed links in Northwest Europe only worsens the situation. The rise in inequity is evident.

Table 5.8 Development average accessibility by rail traffic

	Actual	High-speed links
Average score	211304	205695
Standard deviation	59321	79673
Coefficient of variation	.281	.318
Average accessibility	100	118.6

Notable, however, is the sharp rise in the average accessibility as a consequence of the high-speed links. Of the formulated future scenarios, for all transport modes, the impact of the high-speed links on the average accessibility is by far the largest. However, the measures needed are also far-reaching (both in network changes and in investment volumes).

5.8.3 Road Traffic

The inequity in accessibility for the road network is relatively small (Table 5.9). The inequity further decreases when all roads become of the highway quality. This is not surprising because this refers to road improvements in areas already peripherally located: South Europe, East Europe and Sweden.

Table 5.9 Development average accessibility by road traffic

	Actual	Highway	Bridge/tunnel	Barriers
Average score	212914	218698	222930	145251
Standard deviation	40843	38534	40150	39867
Coefficient of variation	.192	.176	.180	.274
Average accessibility	100	102.7	104.7	68.2

So the score of the lowest scoring agglomerations tends to rise. The construction of tunnels and bridges leads to a rising inequity. Paris, London - the highest scoring cities - and the English cities are favoured by the Channel-tunnel, whereas the score of the low scoring agglomerations does not substantially rise. The investment program foreseen has only a marginal impact on average accessibility in Europe. However, the rise in inequity cannot compensate for the decrease in inequity achieved by the construction of highways.

The impact of national borders as barriers on the international interaction patterns is rather drastic. Compared with the reference situation, the inequity of accessibility rises by over 40 %, and the average accessibility decreases by over 30 %. This shows once more the impact of non-physical barriers on the accessibility of urban agglomerations.

Table 5.10 Development average accessibility by road traffic, barriers included

	Barriers	Scen. 1	Scen. 2	Scen. 3	Scen. 4
Average score	145251	149356	150237	150227	152829
Standard deviation	39867	40911	40467	40052	39977
Coefficient of variation	.274	.274	.269	.263	.262
Average accessibility	100	102.8	103.4	104.8	105.2

In Table 5.10 the results are shown when the barriers are reduced by political changes as formulated in section 6. A decrease of the barriers within the E.U. by one third (scenario 1) as a consequence of the common market, has only a small impact on the average accessibility. The impact on the equity issue is negligible. The impact is rather meagre if one bears in mind that here the barriers of 33 out of the 42 agglomerations are concerned. The same could be said for scenario 2, where the barriers for the eight East-European agglomerations also decreases by one third. The coefficient of variation decreases because here the relative score of the lowest scoring agglomerations rises. Scenario 3 concerns only the four E.F.T.A. agglomerations of which barriers decrease from .167 to .333 (that means a doubling of the possibility to cross a border). The equity in accessibility increases in a relatively sharp way, as does the average accessibility. The same pattern occurs when Budapest undergoes a similar decrease of barrier effect (scenario 4).

Keeping in mind the coefficient of variation before introducing the barrier effects (Table 5.9), one may conclude that the existence of non-physical barriers has strong impacts on the accessibility of cities.

5.8.4 The Transport Modes Compared

In Table 5.11 the average accessibility of each transport mode is given as a percentage of the average accessibility of the air traffic in which indirect connections are excluded. The sequence in which the transport modes are ranked may not cause surprise. However, the difference in the values is notable. Once more the effect of borders as barriers in interaction patterns stands out.

Table 5.11 Average accessibility of the transport modes in 1991 (air traffic = 100)

Road traffic, barriers included	54.9
Rail traffic	79.8
Road traffic, barriers excluded	80.4
Air traffic, indirect flights excluded	100
Air traffic, indirect flight included	128.7
Shortest travel time	143.3

Although the value for road traffic (barriers excluded) is higher than the value for rail traffic, the difference is marginal. This can be explained by the fast and frequent train connections between city centres within countries by the national railway companies. Those connections are often faster than car connections. On those short distances a small difference in the total travel time leads to rather large differences in accessibility. It is because of this effect that the average accessibility of rail traffic becomes close to that of the average accessibility of road traffic.

5.9 Concluding Remarks

The above numerical results depend on various assumptions about parameters, conventions used in measuring travel times, and on the ways the scenarios have been formulated. We believe that the main patterns emerging are fairly robust.

Inequalities in accessibility are least pronounced in the road network. In the rail network inequalities are clearly higher. In the air system inequalities depend strongly on whether or not indirect flights are considered: with direct flights only, inequality is high, whereas when indirect flights are also considered, inequality is low.

The scenario studies reveal that the impacts of changes in the road and air system on average accessibility will be rather small. In the rail system larger impacts on average accessibility may be expected.

Existing inequalities in accessibility are expected to remain rather constant in the air system. For roads, the improvement of the system in Eastern and Southern Europe is expected to contribute to improvements of relative accessibility of the cities there, thus leading to a decrease in inequalities. For rail the reverse is expected to occur: cities in Northwestern Europe will benefit most from the high-speed links, thus leading to an increase of existing inequalities.

In our study we also investigated non-physical aspects of borders. Their effects on accessibility of cities in smaller countries is considerable. Therefore, non-physical aspects of networks should receive due attention in future studies and policies on infrastructure networks in Europe.

This Chapter is based on one particular accessibility concept. As shown in Chapter 2, there are several other ways to conceptualize accessibility. In the next Chapter we will conduct a comparative analysis of European accessibility studies to discover the extent to which results depend on the specific ways of measurement used.

6 A Comparison of European Accessibility Studies[1]

6.1 Introduction

During most of the 20th century the main attention of national governments in infrastructure policy was focused on the - further - extension of the national infrastructure networks to improve the internal accessibility. However, the opening of the European market and the incentives given by the European Commission to construct 'Trans European Networks' has led to a recent emphasis on international infrastructure networks. To create a competitive market all major urban regions must have good access to this common European market, so they will all be connected to highway networks, high-speed rail networks and so on. Considering the resulting maps made by the European Commission, Europe will be covered by a dense pattern of infrastructure networks. What would be the effects of such an investment policy on the accessibility of European cities?

Several studies on this important topic have been produced in the last decade. It is striking that in these studies, accessibility was defined and/or operationalized in different ways. In this Chapter the results of seven studies on the accessibility of cities in Europe are compared (see Bruinsma & Rietveld, 1998a). In section 2 the different operationalizations of the seven studies are discussed, based on the notions given in Chapter 2. For each

---

[1] The authors thank Nadine Cattan, DATAR, Ulf Erlandsson, Javier Gutiérres, Klaus Spiekermann and Michael Wegener for making their data available for this comparison.

study, accessibility indices have been constructed which are analyzed in a qualitative and a quantitative way in section 3. The conclusion will be that the differences in the city rankings resulting from the seven studies that can be explained by different operationalizations are often smaller than the differences that can be explained by the type of infrastructure network analyzed. In other words, the differences in the accessibility rankings of cities depend more on the choice of the network (air, rail, road, or a combination of those types) than on the operationalization of the accessibility concept. However, considering the equity in accessibility of the European agglomerations, the operationalization of accessibility seems to be crucial.

6.2 A Comparison of Accessibility Measures

In the above section it is mentioned that the accessibility of cities can be measured in various ways. But what are the consequences of these different approaches for the ranking of cities on accessibility characteristics within a European context? Does it result in completely different rankings of cities, or are the differences in rankings rather small, apart from the measure used? In the present section we will discuss and compare the results of seven studies, using partly different, partly the same approaches. The seven studies are in chronological order:

- Datar (1989)
- Cattan (1992)
- Erlandsson and Lindell (1993)
- Bruinsma and Rietveld (1993, see also Chapter 5)
- Healey and Baker (1994)
- Spiekermann and Wegener (1996)
- Gutiérrez, González and Gómez (1996)

In section 6.2.1 the different measures used in those studies are described and their consequences for the ranking of the cities from a methodological viewpoint are analyzed. In section 6.2.2 the impacts of the approaches are analyzed in line with the discussion on accessibility as given in Chapter 2. In section 6.3 - in which the results are compared - the focus of attention will shift to the rankings for each modality, amongst others by studying correlations.

6.2.1 Methodological Aspects of the Seven Studies

An overview of the main characteristics of the approaches of the seven studies is given in Table 6.1. For an overview of the various operationalizations of the accessibility concept we refer to Chapter 2.

Erlandsson and Lindell (EL) distinguish for 97 cities inbound and outbound contact potentials (acc9). We will use the term 'accessibility' in this context. Inbound accessibility of a city is measured as the total number of people living in urban areas who can travel to that city remain there for at least four hours, and travel back on the same day. Outbound accessibility of a city is defined in a similar way as the total number of people living in urban areas who can be paid a visit from that city, again with the restriction that the duration of the stay is at least four hours, and that the return trip takes place on the same day. This is indeed a relevant concept for business travel, since generalized costs of communication increase considerably when one has to stay overnight. A disadvantage of the definition is that 'population' is not always an appropriate measure of the economic importance of a city. One would prefer to use the number of workers in particular economic sectors. The EL measure is based on a joint analysis of all relevant transport modes. Differences in departure and arrival times of trains and airplanes cause the differences between inbound and outbound accessibility of cities. If accessibility only depended upon the road system, inbound and outbound accessibility would be identical.

Bruinsma and Rietveld (BR) define accessibility in the context of a gravity type model (acc8). For a complete description of this approach, we refer to Chapter 5; here we give only a brief sketch. For 42 agglomerations with over 1 million inhabitants this measure has been computed by BR for various transport modes: airlines, railways, road transport, and combinations of these. For rail and air total travel times also depend on waiting times, which are related to frequencies. For air two indicators are computed: one includes only direct flights, the other also includes transfer flights. The BR approach shares the disadvantage with the EL approach in that the importance of cities is

Table 6.1 Overview of the main characteristics

	Erlandsson & Lindell	Bruinsma & Rietveld	Spiekermann & Wegener	Gutiérrez et al.	Cattan	Datar	Healey & Baker
type of indicator	acc9	acc8	acc9/acc8	acc6	acc6	acc1/acc2	acc11
demarcation	whole of Europe	Europe, except former USSR	Europe except eastern part former USSR	European Union	European Union, Austria and Switzerland	European Union, Austria and Switzerland	Europe
spatial unit	cities > 100.000 inhabitants	urban areas > 1 million inhabitants	grids (raster cells 10 km width)	4.000 nodes	cities > 200.000 inhabitants + airport	cities > 200.000 inhabitants	major cities and possible newcomers as location for multinationals
modalities	fastest travel mode	fastest travel mode, car, train, airplane	train	train	train, airplane	airport, port, telecommunications	all modes
orientation	passengers	passengers	passengers	passengers	passengers	passengers, freight, information	passengers, freight
mass indicator	population	population	population	GDP	no. of planes, trains leaving to destination		attractiveness as location
transport cost indicator	travel time	travel time	travel time	travel time	travel time		

measured in terms of population rather than a more relevant economic variable.

Spiekerman and Wegener (SW) measure the accessibility of cities only through the rail network. They compute the accessibility conform acc9 (daily accessibility) as well as conform acc8 (potential accessibility). SW try to incorporate the fact that accessibility is continuous in space. To achieve this SW disaggregated the European territory into some 70.000 square raster cells of 10 km width. For each country first the population of large cities was allocated to cells, the remaining population of the country was distributed equally across the rest of the country. So besides the large cities, a homogenous population density over the country is assumed[2]. The access time from each cell to the nearest node of the rail network was calculated assuming a travel speed of 30 km/h where distance is measured as the crow flies. The total travel time consists of: the access time, the travel time on the network, and the terminal time to the destination cell. If the travel time between two cells is shorter than travel time over the network, the first is used.

In line with the continuous space approach of SW, Gutiérrez et al. (GU) develop a - potential - accessibility indicator by the rail network. Unlike SW, who covered Europe with a raster grid, GU defined 4.000 nodes and 7.000 arcs linking these nodes. However, GU used a model in which the indicator is specified as a - by gross domestic product - weighted average of travel times (acc6). So in this approach, there is no distance decay as is the case in the above mentioned gravity type models.

Cattan (CA) used acc6 to compute the accessibility indicator for a sample of 90 cities with over 200.000 inhabitants with an airport. In this study accessibility measures are computed for rail and air traffic. The air traffic indicator - by direct links - includes flight time and the time needed to travel from the city centre to the airport. The rail accessibility is calculated by the travel time between stations. Weighing has taken place by the number of planes (trains) leaving to certain destinations. The problem of using the number of trains or planes to weigh the interaction is that the number of trains and planes is not perfectly correlated with the size of the traffic flows. The numbers of seats varies among different types of airplanes. Most interna-

[2] This can be refined by using a regional subdivision. Spiekermann and Wegener informed us that they are presently improving their approach on the basis of data on cities larger than 50,000 inhabitants and the NUTS-3 population figures.

tional trains are blocktrains of which the blocks consist of different numbers of coaches having different destinations, and changes of train several times during the journey. Therefore, the actual traffic flows are not perfectly correlated with the number of trains and airplanes. CA had tested for this problem by computing the air attractivity of the cities defined by the number of passengers instead of the number of planes. The results show about the same hierarchy of cities (correlation coefficient 0.9).

In the remaining two studies to discuss - the Datar (DA) and Healey & Baker (HB) study - the accessibility of European cities is measured in a more or less qualitative manner, instead of the above described approaches in which accessibility is measured in quantitative terms.

In the DA study the socio-economic performance of 165 European cities with a population of more than 200.000 inhabitants is given. The performance of 'functional urban regions' is measured by means of 16 indicators consisting of three infrastructure indicators: airports, ports and telecommunication. This can be conceived of as a variant of acc1 or acc2. The cities have been rated on a scale from 0 (most attractive) to 5 (least attractive). The data set has been checked by experts from different European countries. Although this study brings together interesting information, it can be criticized for various reasons. First, for several of the underlying variables, quantitative data are readily available, so rating means an unnecessary loss of information. Second, no basis is given for the assumption of equal weights of the indicators, although DA mentions that sensitivity analysis reveals that other assumptions lead to approximately the same result.

An alternative approach is followed by HB, who measure perceptions of accessibility, rather than accessibility itself (acc11). They study the attractiveness of European cities as locations for large companies. A stated preference approach is followed by interviewing 500 senior managers of large companies in industry, trade and services from nine European countries. The respondents are asked to rate the three cities which are the best locations in terms of various location factors. The two accessibility variables are: 'easy access to markets, customers or clients' and 'transport links with other cities and internationally'. Thus, the responses relate to perceptions of the attractiveness. Although this is an interesting approach, some shortcomings should be mentioned. First, respondents only had to mention three cities, which has the effect that the dominant position of London and Paris as attractive locations for company headquarters in Europe tend to be overrated when average scores are computed (see section 6.3.1). Second, the set of countries from which respondents are interviewed is limited to nine, whereas cities from 18 countries are taken into account. This may produce a bias against cities in non-represented countries. Third, a more general problem is that familiarity with a certain location may lead to a bias for that location. Fourth, a problem that is inherent to stated preference is that there is no guarantee that it is followed by actual location behaviour. Nevertheless, the perceptions observed in this way are a relevant piece of information since they play a role in company location decisions.

6.2.2 A Comparison of Approaches

The seven studies are different in various respects, so that it is not surprising to find that they yield different results. One source of difference concerns the demarcation of the research area and the choice of the set of cities (see Table 6.1). The boundaries of the area under research are rather arbitrary; EL include cities in the former USSR, which are excluded in BR, the area under study of SW is somewhere in between. GU restrict the area to the European Union; CA and DA include Austria and Switzerland. HB confine themselves to a rather small set of major cities and possible newcomers. From a methodological point of view, the spatial demarcation issue is especially of importance in acc6 type of models where the impact of distance decay is not included; as with the studies of GU and CA. In the other models the impact of cities outside the research area becomes less important by an increasing distance to the research area. Nevertheless, the scores of cities in the fringe of the chosen study area will be influenced negatively by the exclusion of cities nearby, but which are located on the 'wrong' side of the demarcation line.

Not only the set of cities involved, but also the way cities are configured and the way their total population size is measured, is arbitrary. There is no standard data base for this purpose. EL, BR, CA and DA all use their own definition of urban areas. An extra complication is formed by the delimitation of urban areas in large city regions such as the German Ruhr area, the Dutch Randstad, and the English Midlands. GU is confronted with the same difficulties by computing the gross domestic product of the 4.000 nodes. Even more detailed is the distribution of population in a 70.000 cells raster as used by SW.

In the HB study, the problem of delimitation is avoided, but the disadvantage is of course that one does not know exactly to what spatial unit the answers of the experts relate: to the core of the city, or to the larger metropolitan area.

Another problem worthy of mention is the handling of the cities' own mass in computing the accessibility indicator. The economic strength of a city not only depends on its external relations, but also on its internal relations. If interaction is supposed to depend on the size of the agglomerations with which an agglomeration interacts, then not only the external interaction to other cities, but also the internal interaction within the city need to be included. This holds especially for the gravity type models in which interaction over short distances is relatively strong.

A difference between EL and BR is that in the former, attention is paid to a-symmetries in rail and air connections, which is not taken into account in the latter. Indeed it makes a difference when the first flight Copenhagen-Vienna leaves from Copenhagen at 7.30 a.m., or when it only leaves at 10.00 a.m. Another reason why the measures may differ is that the considered transport modes are different. EL consider all transport modes jointly, but in principle, it would not be a problem were their approach to be repeated separately for each travel mode, as is done by BR.

6.3 Comparison of the Rankings of Each Transport Mode

In this section we will analyze the results achieved from the approaches used in the seven studies. In order to make the results comparable, the rankings of cities in those studies are converted into indices in which the most accessible city receives the value 100 (see Appendix 6.I at the end of this Chapter). The comparisons will be presented per modality, starting with the studies which give rankings for multi-modal approaches (section 6.3.1), followed by the air mode (6.3.2), the rail mode (6.3.3), and the rail mode after completion of the high-speed rail links (6.3.4). In the comparison a qualitative description of the differences in results is first given, followed by a quantitative analysis. In the qualitative description, the three most accessible cities of each study are compared (the three least accessible cities cannot be discussed, since those cities are not involved in all studies). Beyond this, for each transport mode a picture is drawn of the most accessible area as a summation of the ten most accessible cities of each ranking. Finally, it is investigated to what extent the differences in outcomes are related to differences in the methodologies used. In the quantitative analysis, the rankings of the studies are compared by correlation analysis and the equity in accessibility is compared by the coefficient of variation (see Appendix 6.I and 6.II). This section will be concluded by a short comparison of the results between the modes.

6.3.1 Multi-modal

Four out of the seven studies give a ranking of cities for the accessibility by multi-modal means. EL give a ranking for inbound and outbound accessibility and HB for access to markets, clients and customers, and access by external transport links. This brings the total number of rankings to six (Table 6.2).

When we compare the three most scoring cities in those rankings, only Paris is present in all rankings. However, Paris is only once ranked as the most accessible city (EL; inbound). London is listed in five rankings, three of which rate it as the most accessible citty (DA and both HB studies). The other top three ranked cities are Frankfurt (4 times, most accessible according to EL; outbound), Brussels (2 times), Düsseldorf (most accessible according to BR), Amsterdam and Rotterdam (both once).

The most accessible area in Europe - as a summation of the ten most accessible cities of each ranking - is a rather circular area containing London, the Benelux cities, the Ruhr area, and Paris (see Figure 6.1). There is one rather large spur containing Munich, Switzerland, Milan, Marseille, and Spain and two small spurs; one towards the Midlands and the second in the direction of Berlin, Hamburg and Copenhagen.

It is remarkable that London scores best in the non-model approaches in which accessibility is measured in subjective terms. In the case of HB, this might partly be explained by a British bias. The fact that Paris scores best in the inbound accessibility measure, but not in the outbound accessibility measure might be explained by the asymmetry in rail and air connections. In general, Paris is each morning first feedered by the other cities before the departures towards other European cities take place (this also holds true for

London and, to a less extent, Frankfurt). The high score of Brussels, Amsterdam and Düsseldorf - all in model approaches - can be explained by their central position in between the densely populated areas of Paris, London and the Ruhr area. The high score of Rotterdam in the ranking of DA is explained by the rather different criteria of this study on which the accessibility measure is based (airports, ports and telecommunications). The fact that Rotterdam is the world's largest harbour seems of decisive importance here.

Table 6.2 Indices of the accessibility of cities (multi modal)

		1)	2)	3)	4)	5)	6)			1)	2)	3)	4)	5)	6)
A	Vienna	68	61	58	27	9	3	GB	Liverpool	86	57	57	27	--	--
B	Brussels	78	93	92	33	50	33	GB	London	98	93	85	100	100	100
BG	Sofia	49	1	1	--	--	--	GB	Manchester	91	62	60	43	10	6
CH	Geneva	--	61	74	33	14	11	GB	Newcastle	67	47	42	27	--	--
CH	Zurich	73	83	84	30	11	20	GR	Athens	61	23	4	43	--	--
CR	Zagreb	50	1	4	--	--	--	H	Budapest	61	14	24	--	7	--
CZ	Prague	57	26	43	--	5	2	I	Genoa	56	17	13	30	--	--
D	Berlin	73	86	53	38	19	6	I	Milan	73	60	71	43	29	10
D	Cologne	87	74	49	38	--	--	I	Naples	53	22	15	33	--	--
D	Düsseldorf	100	77	67	38	30	13	I	Rome	70	60	50	38	5	3
D	Essen	89	--	--	25	--	--	I	Turin	62	49	40	27	4	2
D	Frankfurt	77	92	100	60	78	80	IR	Dublin	63	54	46	27	--	--
D	Hamburg	70	73	58	50	15	8	NL	Amsterdam	81	96	84	50	26	35
D	Munich	68	73	63	38	16	9	NL	Rotterdam	74	85	64	60	--	--
DK	Copenhag.	67	58	58	50	3	4	P	Lisbon	57	34	24	38	3	--
E	Barcelona	64	53	39	43	15	6	PL	Lodz	49	4	1	--	--	--
E	Madrid	70	57	50	38	15	5	PL	Warsaw	57	12	18	--	8	--
F	Lyon	67	49	47	30	10	8	R	Bucharest	54	3	2	--	--	--
F	Marseille	59	31	5	50	--	--	RU	Moscow	--	24	35	--	3	--
F	Paris	96	100	94	60	86	79	S	Stockholm	58	58	30	--	4	1
GB	Birmingh.	76	63	52	27	--	--	SB	Belgrade	53	2	8	--	--	--
GB	Glasgow	--	51	38	38	3	3	TR	Istanbul	67	9	6	--	--	--
GB	Leeds	87	49	32	23	--	--								

1) BR; fastest travel mode
2) EL; inbound accessibility
3) EL; outbound accessibility
4) DA; airports, ports and telecommunications
5) HB; access to markets
6) HB; access external transport links

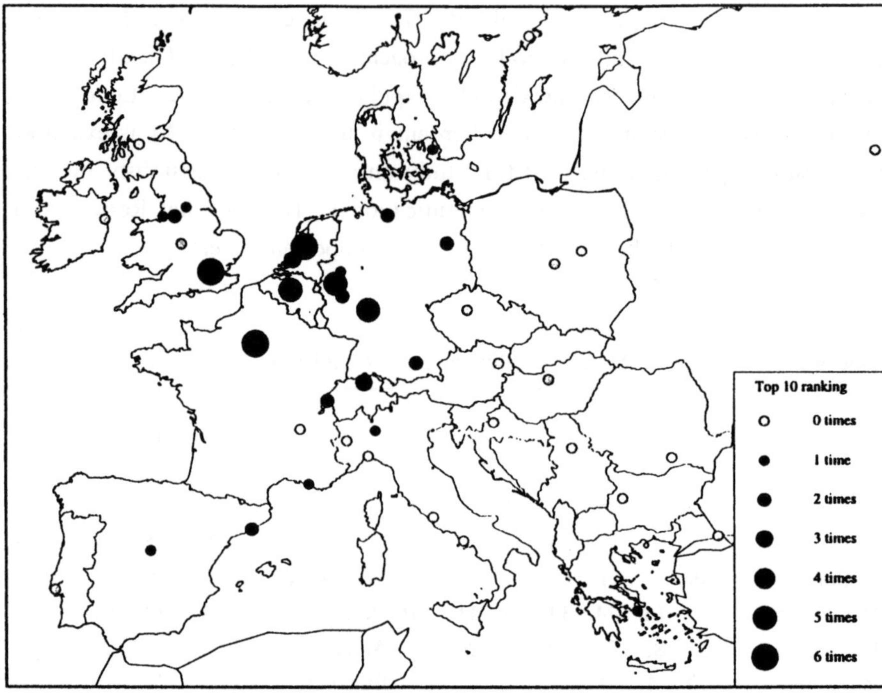

Figure 6.1 The best accessible area by the fastest travel mode (as a summation of the 10 best accessible cities of each ranking)

The different operationalization of accessibility by DA also becomes clear when we compare the rankings by using - Pearson - correlation coefficients (see Table 6.3). Especially the correlations with the quantitative approaches (BR and EL) are low. However, the correlations with the other more qualitative rankings of HB are rather good. Most closely correlated are the rankings based on the same approach. The mutual correlations between the results of the quantitative approaches seem to be higher than the correlations between the results of the qualitative and quantitative approaches.

The last comparison concerns the equity in accessibility as given by the coefficient of variation (see Table 6.3, last column). The lower this coefficient is, the more evenly accessible are the cities within the European infrastructure network. The approach of BR results in the most even distribution of accessibility over Europe. The range of outcomes in this study is relatively low (from 100 till about 50). This can partly be explained by the impact of the travel time decay as discussed in Chapter 2.

Table 6.3 Correlation coefficients for multi-modal accessibility indices

		1)	2)	3)	4)	5)	6)	cov
Bruinsma & Rietveld	1)	---	.785	.758	.326	.706	.640	.195
Erlandsson & Lindell	2)	n=41	---	.926	.459	.671	.722	.584
Erlandsson & Lindell	3)	n=41	n=44	---	.378	.737	.767	.628
Datar	4)	n=32	n=33	n=33	---	.781	.813	.366
Healey & Baker	5)	n=24	n=27	n=27	n=22	---	.966	1.211
Healey & Baker	6)	n=21	n=23	n=23	n=21	n=23	---	1.417

2) inbound accessibility 3) outbound accessibility 5) access to markets
6) access by external links cov = coefficient of variation

The worst equity in accessibility is found in the HB measures. In this study, the gap between the top three: London, Paris and Frankfurt and the rest is very large. A possible explanation is that the respondents only had the opportunity to rank the three most accessible destinations. In a second stage the accessibility scores were computed on the basis of a weighted summation of rank numbers. A probable consequence is that the differences between the accessibility scores of the cities are overestimated. A city with a reasonable degree of perceived accessibility will not easily enter the top three of respondents, so that it will receive a very low score[3].

6.3.2 Air Traffic

The scores of the cities by air traffic in the various studies is given in Table 6.4. The indicators unanimously show that Paris, London and - at some distance - Frankfurt are the three most accessible cities. Other cities with relatively high scores in all studies are: Brussels, Zurich, Amsterdam, Milan, Madrid and - to a smaller extent - Rome and Berlin.

The most accessible area is most dispersed for the air transport mode; it

[3] This might be an indication of the sensitivity of the equity measure (as measured by the coefficient of variation) for the set of cities selected and the operationalization of accessibility.

contains nearly all major international airports (see Figure 6.2). In the DA ranking, which is based on expert opinion, more cities receive a high ranking. However, the experts had only a limited range (a six point scale) to rank the cities.

Table 6.4 Indices of the accessibility of cities by air traffic

		1)	2)	3)	4)			1)	2)	3)	4)
A	Vienna	61	70	60	15	GB	Liverpool	21	54	20	--
B	Brussels	70	74	80	30	GB	London	99	99	100	100
BG	Sofia	43	51	--	--	GB	Manchester	51	63	80	17
CH	Geneva	--	--	60	25	GB	Newcastle	32	55	20	4
CH	Zurich	73	76	80	34	GR	Athens	57	64	80	9
CR	Zagreb	36	51	--	--	H	Budapest	55	63	--	--
CZ	Prague	48	59	--	--	I	Genoa	28	54	0	2
D	Berlin	69	75	80	17	I	Milan	69	72	80	28
D	Cologne	42	60	40	16	I	Naples	33	54	40	5
D	Düsseldorf	60	69	80	27	I	Rome	66	73	80	25
D	Essen	18	60	0	--	I	Turin	35	60	40	4
D	Frankfurt	75	77	100	53	IRL	Dublin	50	66	40	12
D	Hamburg	60	71	60	21	NL	Amsterdam	70	73	80	37
D	Munich	65	70	80	33	NL	Rotterdam	32	66	60	5
DK	Copenhagen	64	70	80	17	P	Lisbon	49	59	60	8
E	Barcelona	60	67	80	18	PL	Lodz	17	49	--	--
E	Madrid	67	73	80	27	PL	Warsaw	52	58	--	--
F	Lyon	48	65	40	13	R	Bucharest	48	56	--	--
F	Marseille	39	61	60	11	RU	Moscow	--	--	--	--
F	Paris	100	100	100	71	S	Stockholm	50	60	--	--
GB	Birmingham	48	62	40	12	SB	Belgrade	47	55	--	--
GB	Glasgow	--	--	60	12	TR	Istanbul	65	70	--	--
GB	Leeds	31	56	0	4						

1) BR; transfer flights excluded
2) BR; transfer flights included
3) DA; expert opinion
4) CA; transfer flights excluded

At first glance, there seems to be a strong similarity in results achieved by these rather different approaches. However, it is of great importance if transfer flights are included in or excluded from the measuring procedure. The difference is measured in the BR study. Cities without or with only a very small airport score very low when only direct flights are used (following acc8: the score of, for instance, Essen and Lodz which have no airport is explained by their own mass). When transfer flights are allowed or passengers are permitted to travel via nearby airports, those cities and cities such as Liverpool, Leeds and Newcastle - all well-connected with London - show a sharp increase in accessibility. The same holds true for most Italian and French cities which are missing an international airport.

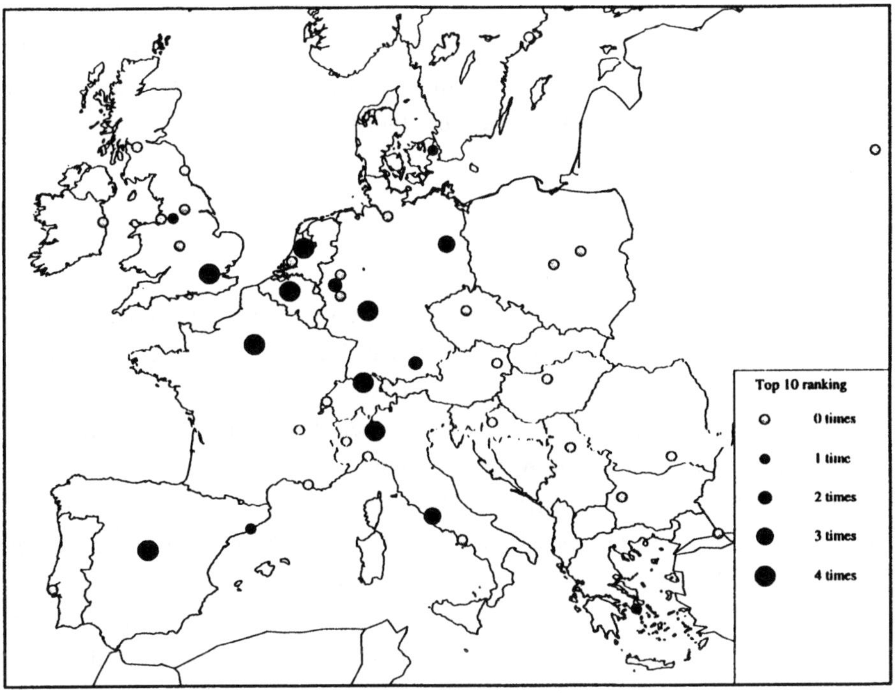

Figure 6.2 The best accessible area via the air network (as a summation of the 10 best accessible cities of each ranking)

When we analyze the correlations between the rankings (Table 6.5), then again it is clear that the rankings show strong similarities. Even the somewhat crude ranking of DA (based on a 6 point scale) correlates rather well to the other rankings.

Table 6.5 Correlation coefficients for air traffic accessibility indices

		1)	2)	3)	4)	cov
Bruinsma & Rietveld	1)	---	.899	.843	.880	.353
Bruinsma & Rietveld	2)	n=42	---	.767	.920	.165
Datar	3)	n=32	n=32	---	.686	.464
Cattan	4)	n=30	n=30	n=32	---	.909

1) transfer flights excluded 2) transfer flights included
cov = coefficient of variation

The importance of transfer flights is again obvious if attention is paid to the equity in accessibility. The ranking without those flights is clearly less evenly distributed than the ranking with transfer flights.

6.3.3 Rail Traffic

Accessibility via rail traffic is interesting because it is undergoing a strong improvement by the construction of high-speed rail lines. On this last topic we will return in section 6.3.4. Here we will concentrate on the existing network (including the high-speed link between Paris and Lyon, which was already operational at the time of these studies).

The three most accessible cities are Paris (4 times present in the top three), cities located in the Ruhr-area (Düsseldorf, Cologne (both 3 times) and Essen (once)), Brussels, London (both 2 times) and once Lyon (see Table 6.6). The most accessible area by rail in Europe (see Figure 6.3) is, to a large extent, similar to that for the multi-modal case. Only the spurs are slightly different. The spur towards Switzerland and Spain is missing, Lyon appears to be important and also the spur into the Midlands is more strongly accentuated.

A closer look into the scores of the rankings by the different approaches shows that compared to the range of the potential accessibility rankings, (BR and SW), the range of scores in the daily accessibility ranking is larger. The same holds true for the ranking of CA, for which approach the number of trains that leave to a destination is used as a weighing factor.

According to Table 6.7, the rankings correlate rather well; only the correlations between GU on the one hand and BR and CA on the other hand are rather low. The broader range in scores of the rankings of the daily accessibility and of the CA ranking - compared with the rankings of the potential accessibility - are reflected in the coefficient of variation of the scores. The potential accessibility rankings result in a more equally distributed accessibility throughout Europe.

Table 6.6 Indices of the accessibility of cities by rail traffic

		1)	2)	3)	4)	5)			1)	2)	3)	4)	5)
A	Vienna	51	25	64	27	--	GB	Liverpool	71	--	58	44	51
B	Brussels	71	33	87	100	90	GB	London	96	100	83	52	62
BG	Sofia	35	--	42	14	--	GB	Manchester	77	33	63	62	51
CH	Geneva	--	17	47	19	--	GB	Newcastle	55	22	51	22	52
CH	Zurich	54	36	71	36	--	GR	Athens	46	2	42	11	19
CR	Zagreb	42	--	47	8	--	H	Budapest	51	--	62	25	--
CZ	Prague	45	--	60	24	--	I	Genoa	49	20	46	21	70
D	Berlin	62	7	78	40	54	I	Milan	57	48	68	45	70
D	Cologne	85	65	92	90	82	I	Naples	42	29	54	32	49
D	Düsseldorf	90	47	92	74	81	I	Rome	57	51	71	39	55
D	Essen	81	--	93	63	79	I	Turin	52	13	62	32	68
D	Frankfurt	64	39	71	69	79	IR	Dublin	39	1	44	10	24
D	Hamburg	61	39	76	48	67	NL	Amsterdam	66	41	79	61	77
D	Munich	54	48	73	39	71	NL	Rotterdam	67	31	71	40	81
DK	Copenhagen	44	5	39	5	46	P	Lisbon	42	3	42	13	29
E	Barcelona	46	9	47	10	57	PL	Lodz	39	--	47	22	--
E	Madrid	53	11	57	20	41	PL	Warsaw	44	--	60	24	--
F	Lyon	63	29	76	56	91	R	Bucharest	42	--	49	17	--
F	Marseille	51	32	50	15	74	RU	Moscow	--	--	51	23	--
F	Paris	100	65	100	73	100	S	Stockholm	40	--	41	7	--
GB	Birmingham	76	47	67	47	55	SB	Belgrade	42	--	51	16	--
GB	Glasgow	--	8	47	18	45	TR	Istanbul	58	--	52	34	--
GB	Leeds	77	25	67	44	54							

1) BR 2) CA 3) SW; potential accessibility
4) SW; daily accessibility 5) GU

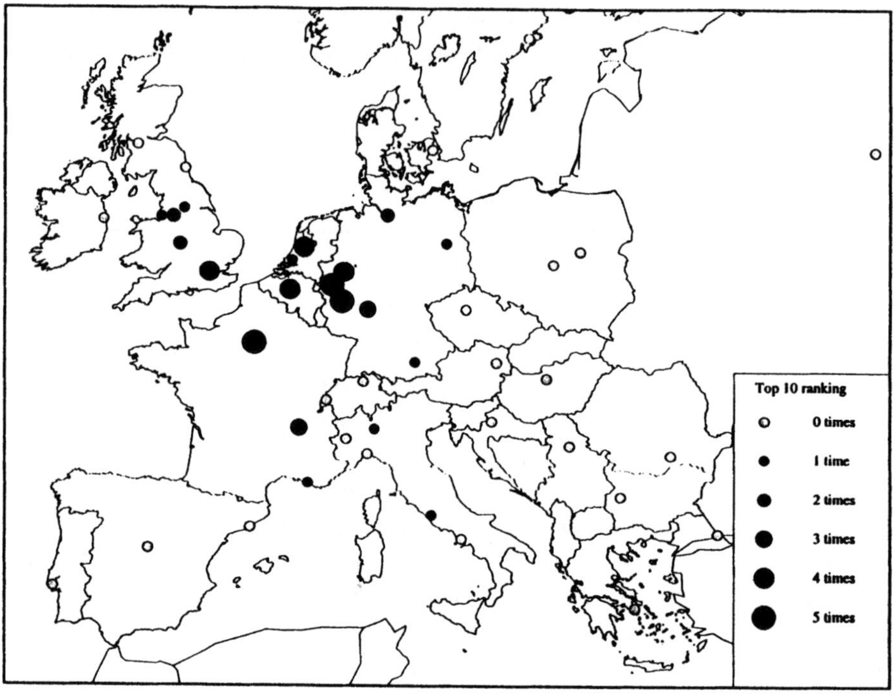

Figure 6.3 The best accessible area via the actual rail network (as a summation of the 10 best accessible cities of each ranking)

Table 6.7 Correlation coefficients for rail traffic accessibility indices

		1)	2)	3)	4)	5)	cov
Bruinsma & Rietveld	1)	---	.765	.845	.826	.580	.281
Cattan	2)	n=30	---	.747	.657	.584	.702
Spiekermann & Wegener	3)	n=42	n=32	---	.894	.766	.258
Spiekermann & Wegener	4)	n=42	n=32	n=45	---	.729	.640
Gutiérrez et al.	5)	n=30	n=29	n=31	n=31	---	.311

3) potential accessibility 4) daily accessibility cov = coefficient of variation

6.3.4 High-Speed Trains Network

The construction of the high-speed rail network in Northwestern Europe leads to an improvement of the accessibility of the cities which were already best accessible by the rail network (see Table 6.8). The three most accessible cities are nodes in the high-speed line network: Paris, Cologne (both in all 4 top three rankings), Brussels (3 times) and London (once). The most accessible area is narrowed to London, the Benelux cities, the Ruhr-area, Frankfurt,

Table 6.8 Indices of the accessibility of cities by rail traffic 2010

		1)	2)	3)	4)			1)	2)	3)	4)
A	Vienna	46	69	33	--	GB	Liverpool	63	55	32	61
B	Brussels	82	90	100	95	GB	London	90	86	76	82
BG	Sofia	29	51	25	--	GB	Manchester	68	63	40	61
CH	Geneva	--	44	12	--	GB	Newcastle	50	43	14	58
CH	Zurich	59	71	51	--	GR	Athens	37	42	7	12
CR	Zagreb	35	53	19	--	H	Budapest	43	67	30	--
CZ	Prague	38	67	38	--	I	Genoa	47	44	14	66
D	Berlin	55	78	46	64	I	Milan	62	74	56	71
D	Cologne	84	89	92	92	I	Naples	44	55	26	46
D	Düsseldorf	83	84	60	89	I	Rome	52	69	32	51
D	Essen	74	83	49	89	I	Turin	54	72	55	72
D	Frankfurt	71	76	74	90	IRL	Dublin	33	38	6	27
D	Hamburg	56	72	43	69	NL	Amsterdam	67	79	76	76
D	Munich	54	74	39	72	NL	Rotterdam	68	66	32	82
DK	Copenhagen	42	48	21	40	P	Lisbon	36	47	14	32
E	Barcelona	49	56	21	60	PL	Lodz	32	44	14	--
E	Madrid	50	59	21	45	PL	Warsaw	37	63	30	--
F	Lyon	68	79	67	88	R	Bucharest	34	51	15	--
F	Marseille	56	57	32	66	RU	Moscow	--	39	15	--
F	Paris	100	100	99	100	S	Stockholm	34	45	7	--
GB	Birmingham	69	58	30	65	SB	Belgrade	35	59	26	--
GB	Glasgow	--	46	13	52	TR	Istanbul	47	53	25	--
GB	Leeds	69	61	32	61						

1) BR 2) SW; potential accessibility 3) SW; daily accessibility 4) GU

Lyon and Paris, with only small spurs to Northern Italy and the Midlands (see Figure 6.4).

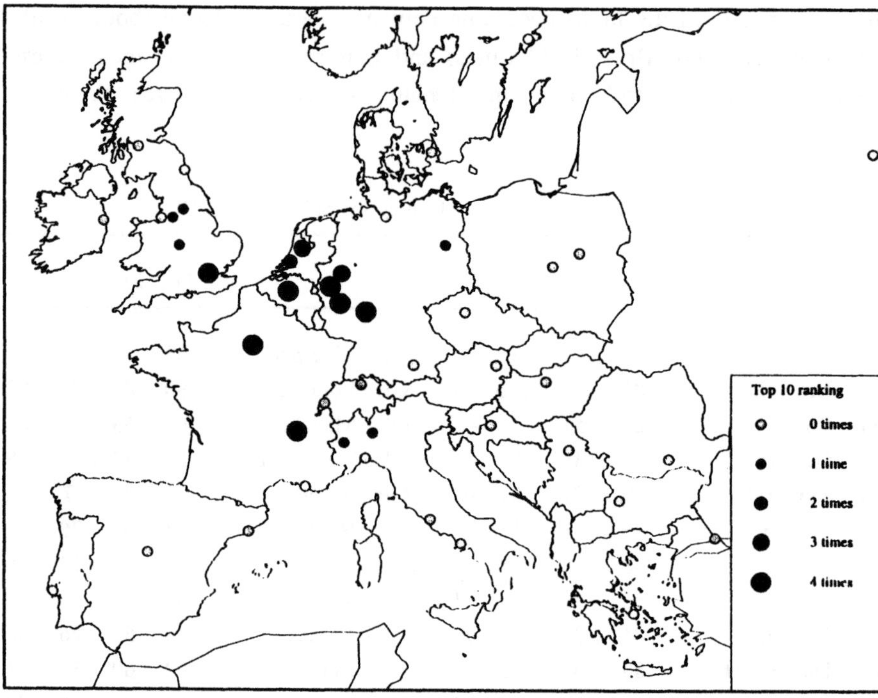

Figure 6.4 The best accessible area via the rail network after the construction of the high-speed lines (as a summation of the 10 best accessible cities of each ranking)

The accessibility rankings for the future rail network are all closely correlated (see Table 6.9). However, more important is the change in the dispersion of accessibility over Europe. As might be expected, a tendency is observed of a decrease in equity by improving the accessibility of the already best accessible cities[4].

---

[4] Note that we found in section 2.5 that an improvement of a link between a large and a small city leads to an accessibility advantage for the small city. An important difference with the proposed HST network is that it often connects

Table 6.9 Correlation coefficients and coefficients of variation for rail traffic accessibility indices in 2010

		1)	2)	3)	4)	cov	+/-	cor.
Bruinsma & Rietveld	1)	---	.812	.840	.867	.319	+.038	.955
Spiekermann & Wegener	2)	n=42	---	.929	.851	.245	-.013	.941
Spiekermann & Wegener	3)	n=42	n=45	---	.840	.662	+.022	.929
Gutiérrez et al.	4)	n=30	n=31	n=31	---	.315	+.004	.953

2) potential
3) daily

cov = coefficient of variation 1993
+/- = change in cov 1993-2010
cor = correlation indices 1993-2010

A last question might be if much really changes as a consequence of the high-speed lines. The answer seems to be no. The changes in the coefficient of variation are relatively low and the accessibility rankings of the existing network correlate highly with the future network. So from both an equity point of view and from a competing accessibility viewpoint, the average consequences of the construction of the high-speed rail network in the North-western part of Europe are limited.

Two points must be emphasized here. First, for individual cities, the consequences may be considerable. Second, in the Tables we only presented relative positions. The absolute change in accessibility (and travel times) owing to high-speed lines of European cities is substantial. The average accessibility is increasing by 18.6 % in the BR study; by 34.5 % (potential) and 63.8 % (daily) in the SW study and - depending on how one measures the increase - by 53.2 % (in travel time), or even 113.8 % (in accessibility scores) in the GU study as a consequence of the construction of the high-speed network.

---

cities which *both* already have a high level of accessibility by rail.

6.3.5 Comparison between Modes

Until now the comparison of the accessibility indices concerned the same type of infrastructure. Before we compare the intermodal indices, a hypotheses can be formulated. One might expect that air traffic indices (especially when transfer flights are excluded) will lead to low correlations with rail traffic indices, since there are relatively few short distance air connections. On those short distances, rail performance is high because of the fast heart-to-heart connections between cities. The same holds true for road infrastructure. The only study in which the accessibility of rail and road is measured (BR) shows a relatively high correlation between rail and road accessibility (.953). In Table 6.10 this hypothesis is tested. Indeed, the correlations between the air and the rail modes are low. With the exception of the correlation between CA rail and CA air, all correlation coefficients listed below are under the overall average of .635.

Table 6.10 Correlation coefficients of air mode indices with rail indices

	BR rail	CA rail	SW pot.	SW daily	GU rail
Bruinsma & Rietveld, transfer flights excluded	.320	.467	.381	.277	.158
Bruinsma & Rietveld, transfer flights included	.570	.563	.581	.437	.319
Datar, air traffic	.089	.298	.184	.130	.084
Cattan, air traffic	.586	.720	.542	.417	.352

Following the above line of thought, the fastest travel mode will contain train or car modes for short distance connections and the air mode for the longer distances. For instance, in the fastest travel mode index of BR, the airplane is the fastest travel mode for 93 % of all connections, the car for 5 %, and the train for 2 %. The impact of those 7 % non-airplane connections is shown in Appendix 6.II. The correlation between the fastest travel mode index and both air traffic indices is relatively low (.377 transfer flights excluded, .603 transfer flights included) compared with the correlation with the index of rail traffic (.946). This strong impact of only 7 % of the connections by car and train is to a great extent the effect of the distance decay factor which is incorporated in the BR gravity model. Short distance connections receive heavy weights. Without such a distance decay factor the impact of modes for the short distance connections might be quite different. In Table 6.11 the correlations of the fastest travel mode indices are given with the air traffic indices on the one hand and the rail traffic indices on the other.

Table 6.11 Correlation coefficients of fastest travel mode indices with air and rail indices

fastest	air traffic indices				rail traffic indices				
travel mode	BR[1)]	BR[2)]	DA	CA	BR	CA	SW[3)]	SW[4)]	GU
BR	.377	.603	.179	.619	.946	.678	.784	.706	.640
EL; inbound	.576	.738	.584	.549	.693	.549	.710	.655	.551
EL; outbound	.630	.749	.560	.512	.656	.512	.686	.566	.541
DA	.618	.713	.638	.569	.354	.569	.228	.148	.195
HB; markets	.829	.851	.702	.915	.773	.733	.640	.636	.530
HB; links	.806	.840	.665	.933	.700	.718	.544	.531	.546

1) transfer flights excluded 2) transfer flights included 3) potential 4) daily

This Table clearly shows higher correlations between all the fastest travel mode indices with the BR index, in which transfer flights are included, than with the BR index without transfer flights. A similar conclusion can be drawn concerning the higher correlations between the fastest travel mode indices and the SW potential accessibility rail index compared with the SW daily accessibility rail index.

Most striking in this Table is that in the studies in which the fastest travel mode indices are measured by quantitative approaches (BR and both EL indices), the correlations tend to be higher with rail traffic indices. However, in studies in which the fastest travel mode is measured in more qualitative ways (DA and both HB indices), the correlations with the air traffic indices tend to be higher. One might conclude that the models show the objective impact of rail transport on the short and medium distance connections. However, in the perception of experts and senior executives of companies, the accessibility of European cities can, to a large extent, be explained by air traffic. The impact of rail traffic on short and medium distance connections seems to be neglected.

A last step is to compare the intermodal correlations with the intramodal correlations. In Appendix 6.II the correlations between indices of the same type of infrastructure (intramodal) are printed bold. It appears that in general the intramodal correlations are higher than the intermodal correlations. The average of the intramodal correlations is .773 compared with the intermodal average of .558 (the overall average is .635).

Table 6.12 shows that all the average correlations of the intramodal comparisons are above the overall average. Only one intermodal average correlation is above this overall average.

Table 6.12 The average correlation of intramodal and intermodal comparisons

intramodal	*.773*	*intermodal*	*.558*
fastest travel mode	.696	rail - air	.374
air travel	.833	rail - fastest mode	.656
rail travel	.739	air - fastest mode	.591
rail travel 2010	.857		
rail-rail 2010 travel	.945	*overall average*	*.635*

A final comparison concerns the equity aspect of accessibility measured by the coefficient of variation. Table 6.13 shows all the equity measures given the type of infrastructure and the operationalization concept. Not much can be said concerning the type of infrastructure; one finds relatively high inequality indicators as well as low inequality indicators within the same type of infrastructure.

However, the type of operationalization seems to be of importance. 'Simple' types of operationalization tend to lead to high inequality indicators. For instance, the daily accessibility concept (acc9) leads to higher inequality in accessibility than the potential accessibility concept (acc 8). Another factor which might lead to low equity indicators is the fine tuning within an operationalization concept. For instance, the inequality indicator of GU, who used 4.000 nodes and 7.000 arcs linking these nodes in his concept, is lower compared with the indicator of CA, who used a sample of 90 cities. A similar impact might be expected if HB had asked the managers to mention more than the three best accessible cities. If they had asked to mention five cities, one may expect that the coefficient of variation will decrease. The difference in equity between GU and CA might be explained not only by the fine tuning in the choice of cities and links, but also by different weighing procedures. GU weighs by GDP while CA weighs by the number of trains leaving to each destination.

Table 6.13 Coefficient of variation for various types of infrastructure and operationalization concepts

	fastest travel mode	air	rail	HSL
Acc1/Acc2				
- Datar	.366	.464		
Acc6				
- Gutiérrez et al.			.311	.315
- Cattan		.702	.909	
Acc8				
- Bruinsma & Rietveld	.195	.353 / .165	.281	.319
- Spiekermann & Wegener			.258	.245
Acc9				
- Spiekermann & Wegener			.640	.662
- Erlandsson & Lindell	.584 / .628			
Acc11				
- Healey & Baker	1.211 / 1.417			

6.4 Conclusion

This Chapter gives an analysis of the results of seven studies in which the accessibility of cities is measured. It appears that a rather broad range of approaches (qualitative, quantitative) is used.

Four choices are of major importance in the measurement of accessibility:

* the demarcation of the area under research
* the selection of cities

* the operationalization concept
* the choice of the type of infrastructure

Considering the operationalization of accessibility, additional choices have to be made concerning weighing procedures (mass of cities, travel time/costs), the frequencies of flights/trains and the parameter in the spatial interaction models, the hour tolerance in the case of the daily accessibility concept, etc.

In this Chapter the results of seven studies on the accessibility of cities within Europe are compared. The comparison is focused on two aspects. First is the comparison of the rankings of the cities as a result of the different conceptualizations and the type of infrastructure involved. Second is the equity in accessibility given the conceptualization and the type of infrastructure.

Through a correlation analysis the similarity of rankings can be addressed. It appears that although different conceptualizations indeed lead to differences in rankings of cities, a clear tendency can be observed that, given a certain mode, the correlations are rather high (see Table 6.12 for the average intra-modal versus intermodal correlation coefficients). When one is interested in a *ranking* of cities in terms of accessibility, the choice of the accessibility concept tends to be less important than the choice of the type(s) of infrastructure to be considered. An interesting result is that in the qualitative approach, incorporating the perception of managers and experts, the accessibility by air appears to be the most important element in the overall accessibility of cities, whereas quantitative approaches emphasize short distance connections in which rail and road infrastructure are the most important modes.

When one is interested in *inequalities* in accessibility among cities, the operationalization appears to have a much larger impact compared with the studies of orders in rankings per se (see for instance Table 6.13). Thus when issues of 'cohesion' in Europe are studied by means of accessibility indicators, analysts must be aware that their results depend to a considerable extent on the specific concepts used.

Appendices

Appendix 6.I Accessibility indices

Bruinsma & Rietveld (1993)
1) fastest travel mode
2) air traffic, transfer flights excluded
3) air traffic, transfer flights included
4) rail traffic
5) road traffic
6) rail traffic, future HST network included

Erlandsson & Lindell (1993)
7) inbound accessibility
8) outbound accessibility

Datar (1989)
9) airports, sea/inlandports & telecommunications
10) airports

Cattan (1992)
11) rail traffic
12) air traffic

Spiekermann & Wegener (1996)
13) rail traffic, potential accessibility 1993
14) rail traffic, potential accessibility 2010
15) rail traffic, daily accessibility 1993
16) rail traffic, daily accessibility 2010

Gutiérrez, Gonzáles & Gómez (1996)
17) rail traffic 1993
18) rail traffic 2010

Healey & Baker (1994)
19) access to markets
20) accessibility by external transport links

Appendix 6.I Accessibility indices (continued)

		1)	2)	3)	4)	5)	6)	7)	8)	9)	10)
A	Vienna	68	61	70	51	60	46	61	58	27	60
B	Brussels	78	70	74	71	70	82	93	92	33	80
BG	Sofia	49	43	51	35	45	29	1	1	--	--
CH	Geneva	--	--	--	--	--	--	61	74	33	60
CH	Zurich	73	73	76	54	63	59	83	84	30	80
CR	Zagreb	50	36	51	42	54	35	1	4	--	--
CZ	Praque	57	48	59	45	58	38	26	43	--	--
D	Berlin	73	69	75	62	74	55	86	53	38	80
D	Cologne	87	42	60	85	75	84	74	49	38	40
D	Düsseldorf	100	60	69	90	78	83	77	67	38	80
D	Essen	89	18	60	81	77	74	--	--	25	0
D	Frankfurt	77	75	77	64	70	71	92	100	60	100
D	Hamburg	70	60	71	61	66	56	73	58	50	60
D	Munich	68	65	70	54	63	54	73	63	38	80
DK	Copenhagen	67	64	70	44	52	42	58	58	50	80
E	Barcelona	64	60	67	46	54	49	53	39	43	80
E	Madrid	70	67	73	53	58	50	57	50	38	80
F	Lyon	67	48	65	63	62	68	49	47	30	40
F	Marseille	59	39	61	51	56	56	31	5	50	60
F	Paris	96	100	100	100	100	100	100	94	60	100
GB	Birmingham	76	48	62	76	70	69	63	52	27	40
GB	Glasgow	--	--	--	--	--	--	51	38	38	60
GB	Leeds	87	31	56	77	74	69	49	32	23	0
GB	Liverpool	86	21	54	71	68	63	57	57	27	20
GB	London	98	99	99	96	94	90	93	85	100	100
GB	Manchester	91	51	63	77	71	68	62	60	43	80
GB	Newcastle	67	32	55	55	60	50	47	42	27	20
GR	Athens	61	57	64	46	52	37	23	4	43	80
H	Budapest	61	55	63	51	61	43	14	24	--	--
I	Genoa	56	28	54	49	59	47	17	13	30	0
I	Milan	73	69	72	57	65	62	60	71	43	80
I	Naples	53	33	54	42	49	44	22	15	33	40
I	Rome	70	66	73	57	63	52	60	50	38	80
I	Turin	62	35	60	52	61	54	49	40	27	40
IRL	Dublin	63	50	66	39	43	33	54	46	27	40
NL	Amsterdam	81	70	73	66	67	67	96	84	50	80
NL	Rotterdam	74	32	66	67	69	68	85	64	60	60
P	Lisbon	57	49	59	42	48	36	34	24	38	60
PL	Lodz	49	17	49	39	49	32	4	1	--	--
PL	Warsaw	57	52	58	44	51	37	12	18	--	--
R	Bucharest	54	48	56	42	50	34	3	2	--	--
RU	Moscow	--	--	--	--	--	--	24	35	--	--
S	Stockholm	58	50	60	40	45	34	58	30	--	--
SB	Belgrade	53	47	55	42	52	35	2	8	--	--
TR	Istanbul	67	65	70	58	67	47	9	6	--	--
coefficient of variation		.195	.353	.165	.281	.193	.319	.584	.628	.366	.464

Appendix 6.I Accessibility indices (continued)

		11)	12)	13)	14)	15)	16)	17)	18)	19)	20)
A	Vienna	25	15	64	69	27	33	--	--	9	3
B	Brussels	33	30	87	90	100	100	90	95	50	33
BG	Sofia	--	--	42	51	14	25	--	--	--	--
CH	Geneva	17	25	47	44	19	12	--	--	14	11
CH	Zurich	36	34	71	71	36	51	--	--	11	20
CR	Zagreb	--	--	47	53	8	19	--	--	--	--
CZ	Praque	--	--	60	67	24	38	--	--	5	2
D	Berlin	7	17	78	78	40	46	54	64	19	6
D	Cologne	65	16	92	89	90	92	82	92	--	--
D	Düsseldorf	47	27	92	84	74	60	81	89	30	13
D	Essen	--	--	93	83	63	49	79	89	--	--
D	Frankfurt	39	53	71	76	69	74	79	90	78	80
D	Hamburg	39	21	76	72	48	43	67	69	15	8
D	Munich	48	33	73	74	39	39	71	72	16	9
DK	Copenhagen	5	17	39	48	5	21	46	40	3	4
E	Barcelona	9	18	47	56	10	21	57	60	15	6
E	Madrid	11	27	57	59	20	21	41	45	15	5
F	Lyon	29	13	76	79	56	67	91	88	10	8
F	Marseille	32	11	50	57	15	32	74	66	--	--
F	Paris	65	71	100	100	73	99	100	100	86	79
GB	Birmingham	47	12	67	58	47	30	55	65	--	--
GB	Glasgow	8	12	47	46	18	13	45	52	3	3
GB	Leeds	25	4	67	61	44	32	54	61	--	--
GB	Liverpool	--	--	58	55	44	32	51	61	--	--
GB	London	100	100	83	86	52	76	62	82	100	100
GB	Manchester	33	17	63	63	62	40	51	61	10	6
GB	Newcastle	22	4	51	43	22	14	52	58	--	--
GR	Athens	2	9	42	42	11	7	19	12	--	--
H	Budapest	--	--	62	67	25	30	--	--	7	--
I	Genoa	20	2	46	44	21	14	70	66	--	--
I	Milan	48	28	68	74	45	56	70	71	29	10
I	Naples	29	5	54	55	32	26	49	46	--	--
I	Rome	51	25	71	69	39	32	55	51	5	3
I	Turin	13	4	62	72	32	55	68	72	4	2
IRL	Dublin	1	12	44	38	10	6	24	27	--	--
NL	Amsterdam	41	37	79	79	61	76	77	76	26	35
NL	Rotterdam	31	5	71	66	40	32	81	82	--	--
P	Lisbon	3	8	42	47	13	14	29	32	3	--
PL	Lodz	--	--	47	44	22	14	--	--	--	--
PL	Warsaw	--	--	60	63	24	30	--	--	8	--
R	Bucharest	--	--	49	51	17	15	--	--	--	--
RU	Moscow	--	--	51	39	23	15	--	--	3	--
S	Stockholm	--	--	41	45	7	7	--	--	4	1
SB	Belgrade	--	--	51	59	16	26	--	--	--	--
TR	Istanbul	--	--	52	53	34	25	--	--	--	--
coefficient of variation		.702	.909	.258	.245	.640	.662	.311	.315	1.211	1.417

Appendix 6.II Overview of correlation coefficients (intramodal correlation coefficients are printed bold)

	1)	2)	3)	4)	5)	6)	7)	8)	9)	10)	11)	12)	13)	14)	15)	16)	17)	18)	19)
1)	---	.377	.603	.946	---	**.785**	**.758**	**.326**	.179	.678	.619	.784	---	.773	---	.428	---	**.706**	**.640**
2)	n=42	---	**.899**	.322	---	.576	.630	.618	**.843**	.467	**.880**	.381	---	.277	---	.158	---	.829	.806
3)	n=42	n=42	---	.570	---	.738	.749	.713	**.767**	.563	**.920**	.581	---	.437	---	.319	---	.851	.840
4)	n=42	n=42	n=42	---	**.955**	.693	.656	.354	.089	**.765**	.586	**.845**	---	**.826**	---	**.580**	---	.773	.700
5)	---	---	---	n=42	---	---	---	---	---	---	---	---	**.812**	---	**.840**	---	**.867**	---	---
6)	n=41	n=41	n=41	n=41	---	---	**.926**	**.459**	.584	.549	.680	.710	---	.655	---	.551	---	**.671**	**.722**
7)	n=41	n=41	n=41	n=41	---	n=44	---	**.378**	.560	.512	.709	.686	---	.566	---	.541	---	**.737**	**.767**
8)	n=32	n=32	n=32	n=32	---	n=33	n=33	---	.638	.569	.764	.228	---	.148	---	.195	---	**.781**	**.813**
9)	n=32	n=32	n=32	n=32	---	n=33	n=33	n=34	---	.298	**.686**	.184	---	.130	---	.084	---	.702	.665
10)	n=30	n=30	n=30	n=30	---	n=32	n=32	n=32	n=32	---	.720	**.747**	---	**.657**	---	**.584**	---	.733	.718
11)	n=30	n=30	n=30	n=30	---	n=32	n=32	n=32	n=32	n=32	---	.542	---	.417	---	.352	---	.915	.933
12)	n=42	n=42	n=42	n=42	---	n=44	n=44	n=34	n=34	n=32	n=32	---	**.941**	**.894**	---	**.766**	---	.640	.544
13)	---	---	---	---	n=42	---	---	---	---	---	---	n=45	---	---	**.929**	---	**.851**	---	---
14)	n=42	n=42	n=42	n=42	---	n=44	n=44	n=34	n=34	n=32	n=32	n=45	---	---	**.886**	**.729**	---	.636	.531
15)	---	---	---	---	n=42	---	---	---	---	---	---	---	n=45	n=45	---	---	**.840**	---	---
16)	n=30	n=30	n=30	n=30	---	n=30	n=30	n=31	n=31	n=29	n=29	n=31	---	n=31	---	---	**.953**	.530	.546
17)	---	---	---	---	n=30	---	---	---	---	---	---	---	n=31	---	n=31	n=31	---	---	---
18)	n=24	n=24	n=24	n=24	---	n=27	n=27	n=22	n=22	n=22	n=22	n=27	---	n=27	---	n=19	---	---	**.966**
19)	n=21	n=21	n=21	n=21	---	n=23	n=23	n=21	n=21	n=21	n=21	n=23	---	n=27	---	n=18	---	n=23	---

1) Bruinsma & Rietveld; fastest travel mode 1992
2) Bruinsma & Rietveld; air traffic, transfer flights excluded 1992
3) Bruinsma & Rietveld; air traffic, transfer flights included 1992
4) Bruinsma & Rietveld; rail traffic 1992
5) Bruinsma & Rietveld; rail traffic, future HSL network included
6) Erlandsson & Lindell; inbound accessibility 1992
7) Erlandsson & Lindell; outbound accessibility 1992
8) Datar; airports, sea/inlandports & telecommunications 1989
9) Datar; airports 1989
10) Cattan; rail traffic 1991
11) Cattan; air traffic 1991
12) Spiekermann & Wegener; rail, potential accessibility 1993
13) Spiekermann & Wegener; rail, potential accessibility 2010
14) Spiekermann & Wegener; rail, daily accessibility 1993
15) Spiekermann & Wegener; rail, daily accessibility 2010
16) Gutiérrez et al.; rail traffic 1993
17) Gutiérrez et al.; rail traffic 2010
18) Healey & Baker; access to markets
19) Healey & Baker; accessibility by external transport links

PART C

Regional and Urban Impact Studies

Introduction to Part C

In this part of the book case studies are presented concerning the impact of infrastructure on economic development at the regional and urban level of spatial aggregation. In the case studies a strong emphasis is laid upon the relation between infrastructure construction and the performance of firms. Along with model approaches, which are based on quantitative data in terms of changes in regional employment, etc., attention is paid to entrepreneurial surveys of a more qualitative character. In our opinion, the perceptions of entrepreneurs about the opportunities that new transport infrastructure offers them in the region is a valuable source of information for research in this field. For a good understanding of the complex interrelationship between the need to construct transport infrastructure and the ongoing economic development at various levels of spatial aggregation, one has to apply both quantitative and qualitative research approaches.

Chapters 7 and 8 deal with the impacts of the construction of an international transport corridor - the A1 highway in the eastern part of the Netherlands - on regional economic development. In Chapter 7 attention is paid to the application of regional models, whereas Chapter 8 focuses on entrepreneurial surveys.

In Chapter 7 a brief introduction is given of the development of regional employment and regional accessibility in the Netherlands in general and the area of research - the eastern part of the Netherlands - in particular, for the period 1970-1990. To measure the impact of the newly constructed A1 highway in the eastern part of the Netherlands, a reference region approach and a regional labour market model are applied.

In Chapter 8 the floor is given to the entrepreneurs. After all, it depends on the business community whether the expectations about the impacts of infrastructure projects (which may play a large role in the political decision making process) will be realized. After a few critical remarks on the application and interpretation of entrepreneurial surveys, some general results achieved by those approaches are presented. Then we focus on the entrepreneurs in the eastern part of the Netherlands and their perceptions of the impact that the A1 international transport corridor has on their firms. The impact of the transport corridor is placed in a regional context by asking the entrepreneurs to value the impact of a whole range of regional location factors on the firms' employment level and - if the firm is relocated - the location decision. Afterwards the impact of transport infrastructure in general on the performance of firms is discussed, and finally the impact on the firm performance of the transport corridor and the situation if this corridor was not constructed are dealt with.

Chapter 9 focuses on the impact of highway construction on urban development. The case study deals with the Amsterdam orbital motorway A10 (the A1 international transport corridor begins at this orbital motorway). Also in this case study a combination of quantitative and qualitative approaches is used. After a brief introduction regarding the need of the Amsterdam orbital motorway and the effects of the opening of this motorway on traffic volumes, traffic behaviour and congestion, the impact on economic development is discussed in two sections. First the impact of the orbital motorway on the office market prices are analyzed by distance of the office location to the nearest entry and/or exit point of the orbital motorway and by a regression analysis on basis of individual transactions on the office market. Finally entrepreneurial perceptions are studied based on survey data. Chapter 9 will round up by confronting the results achieved here with those of Chapters 7 and 8. This confrontation leads to another plea for the need to apply quantitative and qualitative approaches at various spatial levels of aggregation if one wants to reach a thorough understanding of the complex relation between transport infrastructure and spatial economic development.

Chapters 10 and 11 are devoted to location studies. In Chapter 10 the inter-urban location preferences of entrepreneurs are compared with the accessibility of those cities in the Dutch road network. In Chapter 11 the intra-urban preferences of entrepreneurs for seventeen different location profiles are analyzed by the present location of firms, the relocation pattern over the past, and by the preferences for new locations. Furthermore, trade-offs are calculated between seven locational factors.

In Chapter 10 Dutch cities are ranked by the valuation entrepreneurs attach to them as a location site for firms. We attempt to explain this valuation by a number of locational factors of those cities, among which is the location of the city in the Dutch road network. Special attention is given to the implications of various definitions of the mass of a city in the measurement of the accessibility index. Furthermore, the impact of congestion in specific parts of the road network and the proximity of foreign cities on the accessibility of Dutch cities is measured.

In Chapter 11 the attractiveness of locations within a city is investigated, by a classification of seventeen location types. Not only is the actual dispersion of firms over these seventeen location profiles analyzed, but also the changes caused by relocations over the period 1980-1994, and the preferences for new locations are inventoried. A problem with many location studies is that the results cannot be interpreted in terms of trade-offs between location factors. Therefore we finally conduct a pairwise comparison approach which does enable one to investigate trade-offs. The relative importance of seven locational factors is investigated in a pairwise way using stated preference techniques.

7 Road Infrastructure and Corridor Development: Regional Models

7.1 Introduction

Basically three approaches can be distinguished in the literature on the economic impact of transport infrastructure. The first approach is the well-known social cost benefit analysis (CBA) with *consumer surplus* as the main object (see section 3.6). This approach is firmly based on efficiency objectives in welfare economics and has been widely used in various countries.

More recently a second approach has emerged where consumer surplus is no longer the basic concept, but *productivity* is (cf. Aschauer, 1989, see section 3.2). Implicit in this approach seems to be that CBA is sometimes regarded as too broad, because its outcome may strongly depend on the travel time savings of households which do not have implications for GDP. On the other hand, CBA outcomes are sometimes regarded as too narrow, because in the calculations of the benefit the strategic long-term macro economic benefits in terms of GDP are only partially taken into account.

The third approach moves even further away from CBA by focusing on *employment* (see section 3.2 and 3.3). This happens especially in countries with high levels of structural unemployment where job creation is considered of prime interest. In this approach the spatial dimension is often relevant: transport infrastructure investments have spatially differentiated impacts on employment.

Of course the three approaches are not entirely independent; they are partially overlapping, partially conflicting, and partially complementary. In the present Chapter we will not explore these interrelationships, but mainly contribute to the third approach by focusing on employment aspects of transport infrastructure; some attention will also be paid to the second approach.

The impacts of network infrastructure are usually studied for spatial units such as cities or regions. *Cities* can be conceived as nodes in a network, the development of which is influenced by the quality of links (roads, railways, canals) and of transport nodes (airports, seaports). Thus, one can study the effect of changes in transport links (e.g., high-speed rail) on the relative position of cities, as is shown in Chapter 5 and 6.

Impacts of transport infrastructure change on *regions* can be analyzed when a transport network is connected to a spatial system in terms of zones or regions. By connecting the centroid of each region to a point in the network, changes in the network can be translated into changes in accessibility of regions. Usually, regions are defined in administrative terms. An example of such a study is given in Chapter 12.

An opportunity which has received less attention in research is to focus on *corridors*. Corridors can be defined as areas near to links in networks. The difference between corridors and regions is that regions are defined a priori (usually in an administrative way), and corridors can be defined in terms of the shape of a network. During the last 10-15 years, the availability of data with a very high spatial detail (individual, postal code) has increased substantially. As a consequence, opportunities to analyze spatial processes by means of spatial units which can be modified according to the features of the topic studied have increased considerably. This has also led to greater possibilities for studying the effects of network development on corridors. This is a welcome development, since by using more appropriate spatial units of analysis, research can be carried out in a more satisfactory way. A focus on *corridors* is also important in the eyes of policy makers. Corridors seem to be relevant elements of mental maps of countries and have an intuitive appeal. In physical planning and regional economic policies, corridors regularly appear as relevant units (RPD, 1991).

A closer look at the concept of a corridor reveals that corridors usually relate to road infrastructure. The reason is that road infrastructure has a relatively high number of entry and exit points, so that it really can be expected that land use is affected in the entire area surrounding a road or a highway. The lower the number of entry or exit points, the less relevant the notion of a corridor becomes. For most railway lines for example, distances between subsequent stops are so far that the corridor concept no longer applies. An exception may occur in the case of rail systems within metropolitan areas where stops are near to each other.

This and the next Chapter presents results of a broader study on the impact of highway development on corridors in the Netherlands (see Bruinsma et al., 1997). Two spatial approaches have been used with respect to corridors. One is to use given statistical regions (Corop-regions[1]). These regions are broader than the corridors, so that they give rise to the disadvantages mentioned above. The advantage is that they allow a rather long time series from 1970 onwards. The other approach is to use entrepreneurial surveys and their postal code areas and combine these so that they form corridors. The advantage is that this leads to an analysis of an appropriately defined spatial unit. A disadvantage is that data are available for a much shorter period.

In this Chapter we present results mainly on the basis of Corop-regions. Results of the more refined spatial data are reported in the next Chapter, where entrepreneurial surveys are used to analyse the impact of road infrastructure by concentric zones around highway entry or exit points.

In this Chapter a descriptive analysis of the economic development of the Netherland is given. In this analysis the Netherlands is divided into a core zone (the Randstad), an intermediary zone and a peripheral zone. By using a correlation analysis between employment growth and improvements in accessibility over the national road network, an introduction of the study on the relation between road infrastructure and economic development is given on the Corop-level. At this spatial level of aggregation, the economic impact of the A1 transport corridor is analyzed by a reference region approach and a regional labour market approach.

It will be shown that it is difficult to detect a positive economic impact of the construction of main road infrastructure on overall employment at the Corop-level. Therefore, in Chapter 8 an analysis at the disaggregate level of individual firms is given.

---

[1] Corop-regions are statistical areas. There are 40 of them in the Netherlands. The average population of a Corop-region is about 400.000 inhabitants (see Figure 7.1).

7.2 The Development of Accessibility in the Netherlands

7.2.1 The Development of Regional Employment in the Netherlands

Many European countries have witnessed an 'urban-rural manufacturing shift' during the past decades (Keeble et al., 1983). To analyze the urban-rural manufacturing shift in the Netherlands, it is useful to divide the country into three zones: the Randstad, the intermediary zone and the peripheral zone (see Figure 7.1).

Figure 7.1 The research area

The Randstad is a highly urbanized area, which was characterized by the highest economic and employment growth before 1970. The four largest Dutch cities are located in this region (Amsterdam, Rotterdam, The Hague and Utrecht). After 1970 a shift took place from the Randstad to the intermediary zone. This shift not only related to a migration of population out of the Randstad, but also to a low rate of employment growth in this area; employment growth in the Randstad was below the national average. The intermediary zone benefited from this trend; this area showed growth rates of employment which are above the national average (see Table 7.1).

Table 7.1 Development of employment in the Netherlands (index: 1970=100)

	1970	1975	1980	1985	1990
Netherlands	100	101	105	100	115
Randstad	100	98	102	97	109
Intermediary zone	100	104	109	104	123
Peripheral zone	100	99	105	100	112

Source: Own calculations based on: CBS, 1970-1992

The peripheral zone was characterized by a relatively strong emphasis on agriculture in the past. Economic development in this zone lagged behind the other zones before the 1970s. In the period 1970-1990, employment growth was only slightly lower than that of the national average; in most periods its growth was even higher than that of the Randstad.

A shift-share analysis reveals the extent to which these observations are caused by differences in the sectoral structure between the regions. The results of such an analysis based on 12 sectors are shown in Table 7.2.

The shifts in the Randstad are negative in all five-year periods. This means that the negative development of the Randstad is not caused by its sectoral composition, but by other factors. The shifts are clearly positive in the intermediary zone; here the locational profile appears to be clearly favourable compared with the average.

Table 7.2 Relative shifts in the different zones in the Netherlands

	'70-'75	'75-'80	'80-'85	'85-'90
Randstad	-0.04	-0.02	-0.02	-0.04
Intermediary zone	0.06	0.01	0.02	0.04
Peripheral zone	0.00	0.01	0.00	-0.02

Source: Own calculations from: CBS, 1970-1992

7.2.2 The Development of the Accessibility in the Different Zones

Highway infrastructure in the Netherlands was mainly constructed in the period 1960-1975. After 1975, the investments in road infrastructure decreased, which was partly caused by the growing deficits of the government, but also because of the growing attention to the negative external effects of road transport (see Table 7.3).

Table 7.3 The development of roads with a dual carriage way[2] outside residential areas (index: 1966 = 100)

	1966	1970	1975	1980	1985	1990
Netherlands	100	168	250	296	315	372
Randstad	100	150	189	196	196	241
Intermediary zone	100	161	252	301	313	346
Peripheral zone	100	233	474	654	791	960

Source: Own calculations based on: CBS, 1966-1992

It is interesting to analyze in which zones infrastructure construction took place most intensively. It appears that the growth rate of highway kilometres

[2] The highway system consists exclusively of dual carriage ways.

was the highest in the peripheral zone, while the network in the Randstad was not extended by much. A reason for this may be that the construction of highways started earlier in the densely populated Randstad. The construction in the intermediary zone was about the national average. By comparing Table 7.3 with the preceding Tables, it may be concluded that the strong emphasis on construction activities in the peripheral zone has not led to a clear improvement of the economic position of this area.

7.2.3 Using an Accessibility Index

The above data relate to infrastructure as an input. However, the importance of infrastructure is that it provides services to users. One way to analyze this is by using an accessibility index. Such an index is defined in Chapter 2. In this case study we used employment instead of population size as the mass indicator:

$$A_i = \Sigma_j M_j / T_{ij}^{\,c} \qquad (1)$$

where:

A_i = accessibility index for region i
M_j = mass of zone j (mass = employment)
T_{ij} = travel time from zone i to j
c = constant

The travel time is calculated by means of detailed data on speeds on many links (the Basnet system). Travel times between Corop-regions have been computed by means of a shortest route algorithm. From the definition of the accessibility measure, it is clear that there are two major reasons for changes in accessibility: changes in mass M and changes in the network itself. In the present context, we are only interested in the changes in the network. Therefore, we computed changes in accessibility while holding the masses constant.

From Table 7.4 and the Figures 7.2-7.5 it appears that accessibility in the peripheral zone grew most strongly, while that of the Randstad grew less. The differences are smaller than those of the road densities as shown in Table 7.3, because the construction of roads within a zone not only leads to an increase in its internal accessibility, but also in its external accessibility, so that other

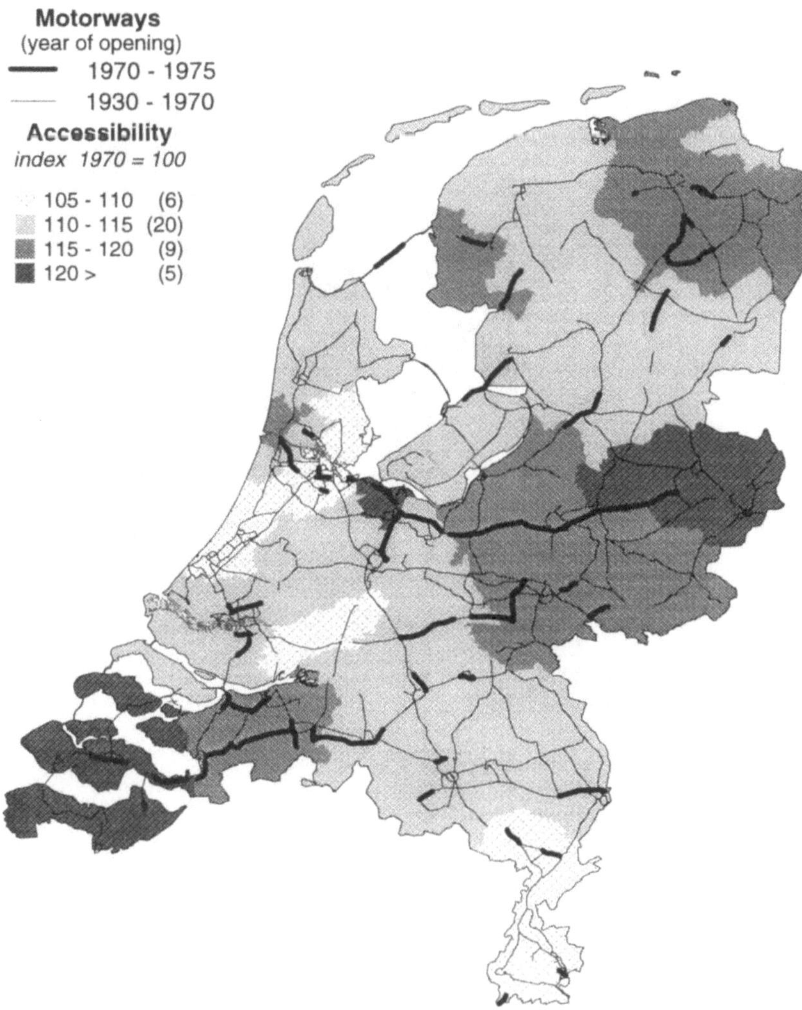

Figure 7.2 Increase in accessibility by the development of the highway network in the period 1970-1975

regions will also benefit. In addition, it should be noted that the accessibility figures given here are based partly on the underlying road network, the development of which has been less dynamic than the highway system. As a result, the increase in accessibility is less than the increase in the length of the highways. Again, we observe that even though accessibility in the peripheral zone grew most rapidly, its relative economic position did not improve. It should be noted that in this formulation of the accessibility index no attention

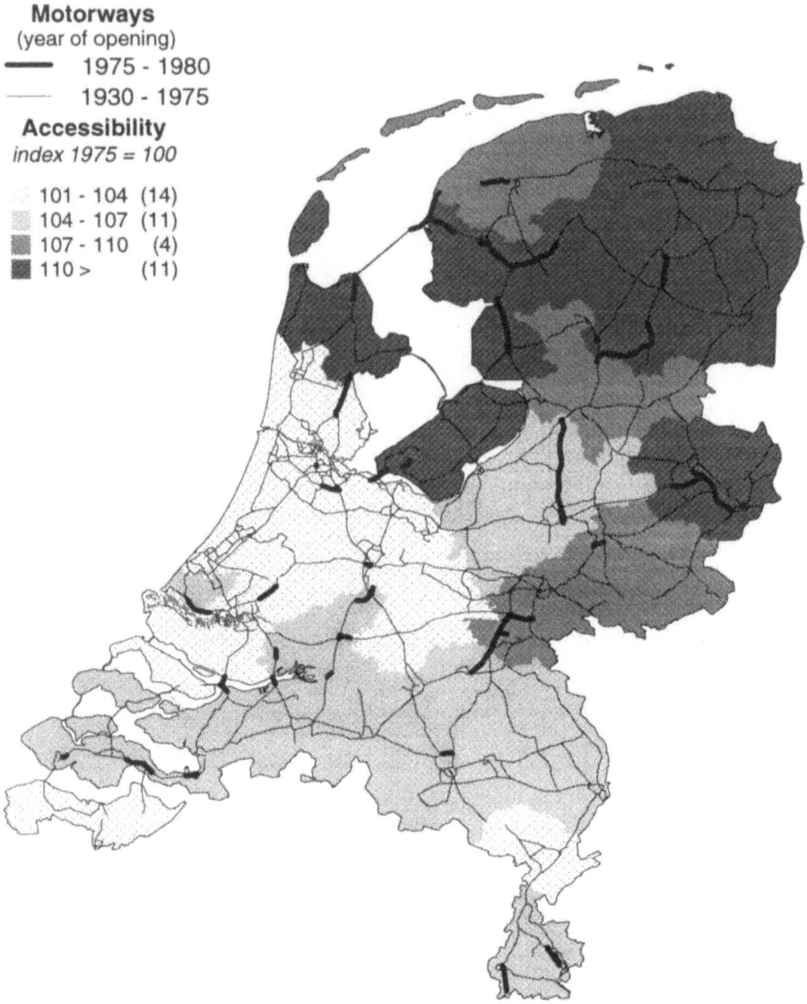

Figure 7.3 Increase in accessibility by the development of the highway network in the period 1975-1980

is given to the role of congestion. If we consider the fact that congestion has increased during the period between 1985 and 1990 in the Randstad area, it is clear that Table 7.4 may give an overly favourable picture of the development of accessibility in this part of the country. For a detailed analysis of the impact of congestion on accessibility we refer to Chapter 10.

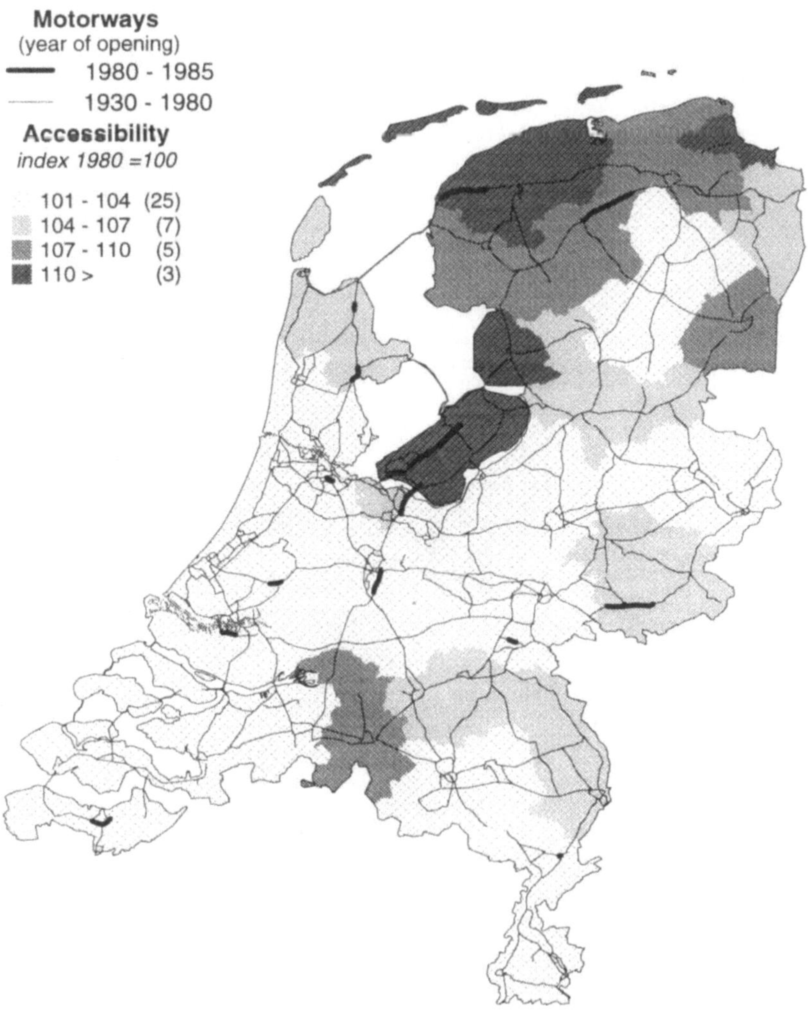

Figure 7.4 Increase in accessibility by the development of the highway network in the period 1980-1985

7.2.4 Correlations in Accessibility Growth

We consider the question whether the growth in accessibility in the different regions is correlated during various periods. The correlation coefficients between the growth in the above mentioned accessibility index in the different periods are given in Table 7.5.

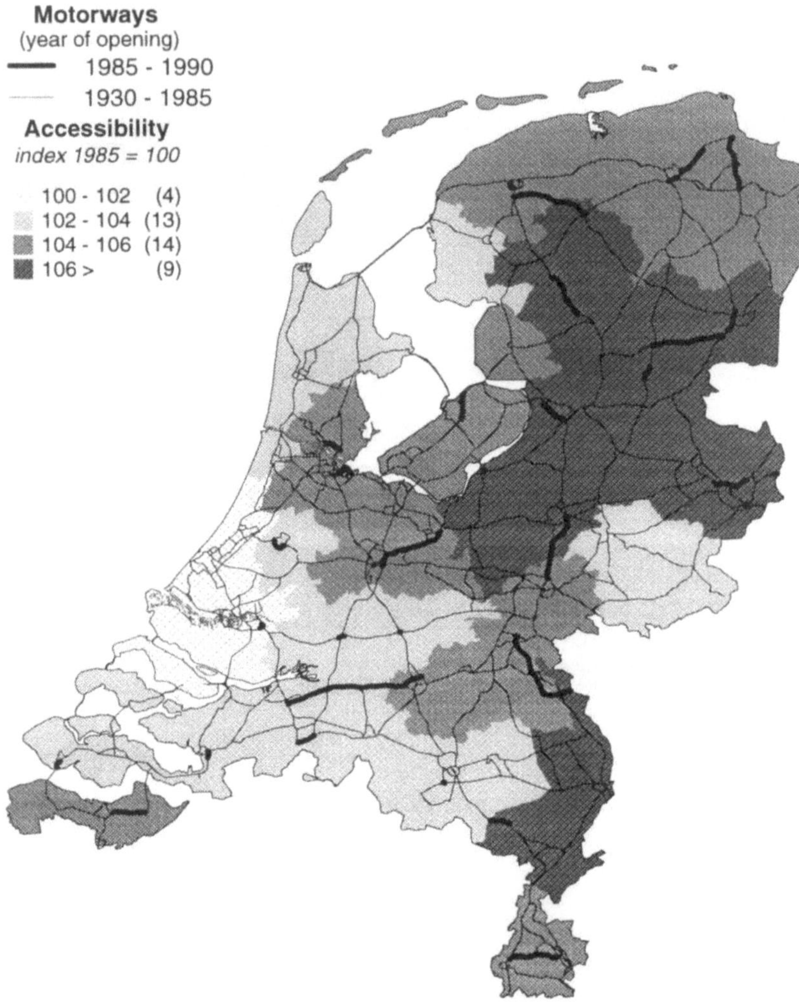

Figure 7.5 Increase in accessibility by the development of the highway network in the period 1985-1990

A positive correlation coefficient for subsequent periods means that accessibility increases do not occur in a random way, but rather there is a pattern whereby some regions experience above average accessibility growth and others have below average accessibility growth during successive periods.

It appears that between the periods 1975-1980 and 1980-1985 there is a positive and significant correlation between the growth in the accessibility indices. In the other periods the correlations are not significant and both posi-

Table 7.4 Growth of the accessibility index in different zones (1970 = 100)

	1970	1975	1980	1985	1990
Netherlands	100	113	120	125	130
Randstad	100	112	115	118	122
Intermediary zone	100	115	122	126	132
Peripheral zone	100	115	125	133	140

Table 7.5 Correlations in the growth of the accessibility index

	'70-'75	'75-'80	'80-'85	'85-'90
'70-'75	1.00	-0.08	-0.24	0.22
'75-'80	-0.08	1.00	0.58*	0.21
'80-'85	-0.24	0.58*	1.00	0.13
'85-'90	0.22	0.21	0.13	1.00

Note: * significant correlation

tive and negative signs are found. Thus, at the level of Corop-regions, the indications are not strong that infrastructure supply leads to a systematic bias for certain regions.

7.3 A Reference Region Approach

7.3.1 Introduction to the A1 Highway and the A1-Regions

The A1 is the highway which runs from Amsterdam, via Amersfoort - Apeldoorn - Almelo to the German border. In Germany this highway (E8) proceeds to Berlin and Poland. Furthermore, the A1 improved the accessibility to Northern Germany and the Scandinavian countries. In the central parts of the Netherlands it was mainly completed in 1972 and 1973; in the eastern part (Twente) several parts were opened between 1978 and 1992. The road runs through several Corop-regions: Greater-Amsterdam, Utrecht, the Veluwe, Southwest-Overijssel and Twente (Figure 7.1).

In order to study the impact of the A1-construction on the regional economy the reference region approach will be used in this section. In this approach we try to determine the impact of A1-construction by comparing the employment growth in the A1-regions with that of reference regions. These reference regions are comparable with the A1-regions in terms of economic structure and location, but in these regions little or no construction of infrastructure took place in the period(s) considered.

The study is carried out for three A1-regions situated in the intermediary zone: the Veluwe, Southwest-Overijssel and Twente. In the observed period, the A1-regions all had different economic developments. In Table 7.6 the index shows, for example, the growth of employment.

Table 7.6 Employment growth in the Netherlands and the A1-regions (1970 = 100)

	1970	1975	1980	1985	1990
Netherlands	100	101	105	100	115
Veluwe	100	111	118	117	136
Sw-Overijssel	100	100	106	98	120
Twente	100	97	99	90	104

Source: Own calculations based on: CBS, 1970-1992

The *Veluwe* is one of the biggest natural areas in the Netherlands and is a popular location region for people as well as companies; the regional economy is characterized by a large share of defence employment and by a relatively high growth of the service sectors, thus leading to a positive shift in most periods (see Table 7.7). The A1 in this region was constructed in 1972 and is one of the main transport corridors in this region.

Southwest-Overijssel is a small, mainly rural region, traditionally focused on the industry sector. The growth rate of the employment lagged behind the national average until 1985; the shift was positive, however, between 1970-1980. After 1985 the employment growth as well as the relative shift was relatively high. The A1 was constructed in 1972 and 1973 and is the only highway in this region.

Table 7.7 Relative shifts in the intermediary zone and A1-regions

	'70-'75	'75-'80	'80-'85	'85-'90
Intermediary zone	0.05	0.01	0.02	0.04
Veluwe	**0.10**	-0.00	0.03	0.05
Sw-Overijssel	**0.01**	0.02	-0.02	0.08
Twente	**0.03**	**0.02**	**-0.02**	**0.02**

Note: The period in which the A1 is opened is printed bold.
Source: Own calculations based on: CBS, 1970-1992

Finally, *Twente* was traditionally characterized by a strong textile-industry sector, which declined to a large extent in the period under analysis. Until 1985 this region experienced a negative employment growth; after 1985 the employment grew rapidly, however. When the growth is corrected for the sectoral structure, the region had a positive shift in three of the four five-year periods. The A1, which is the only highway in this region, was mainly opened between 1978 and 1992.

Next, we will carry out the reference region approach in order to investigate employment impacts of the A1-construction.

7.3.2 The A1-Regions Compared with the Remainder of the Intermediary Zone

In Table 7.7 a comparison is made between the relative shift in the three A1-regions and the remainder of the intermediary zone.

From Table 7.7, it appears that in the *Veluwe* the shift in the first period is much higher than in the intermediary zone. At first sight this seems to be the result of the A1-construction. When the sectoral structure is investigated however, it appears that the high relative shift is mainly caused by growth of the defence sector, in the form of a growth of defence-employment. It is not likely that this employment is attracted by the construction of the A1.

In the other periods during which construction took place, the relative shift in *Twente* is somewhat higher during the period 1975-1980. In the other relevant periods the shift is lower than in the intermediary zone, however. Also an investigation of lagged effects does not lead to clear results on an impact of highway construction on regional employment growth. We conclude that a positive influence of A1-construction on the shift cannot be proven by using the remainder of the intermediary zone as the reference region.

Although it is certainly relevant to compare the development of the A1-regions with that in the intermediary zone, it is clear that this zone is not an entirely satisfactory reference region. The reason is that other highway construction projects also took place in the intermediary zone. Therefore, we conducted a more detailed analysis with reference regions which did not experience extensive highway construction projects in the periods concerned.

7.3.3 The A1-Regions Compared to Reference Regions

In Table 7.8 the relative shifts of the A1-regions and those of several reference regions have been presented (see also Figure 7.1 for their location).

The Achterhoek is, like the *Veluwe*, a rural region regarded as an attractive living and location region, in which little main road infrastructure has been built in the observed period. When we compare the relative shift of both regions in the period of opening of the A1 (1970-1975), we see a higher shift in the Veluwe; as already mentioned however, this is mainly caused by a growth of defence employment. In the Achterhoek the relative shift was also high during this period, but here it was caused by market sectors, so a clear influence of A1-construction on the employment in the Veluwe is not proven during the 1970-1975 period. The same holds for a lagged effect.

Table 7.8 Relative shifts in the A1- and reference regions (%)

	'70-75	'75-'80	'80-'85	'85-'90
Veluwe	**0.10**	-0.00	0.02	0.05
ref. region: Achterhoek	*0.05*	*0.05*	*-0.04*	*0.07*
Southwest-Overijssel	**0.01**	0.02	-0.02	0.08
ref. region: NE-north-Brabant	*0.10*	*0.03*	*0.05*	*0.10*
Middle-Limburg	*0.11*	*0.09*	*0.02*	*0.06*
Twente	**0.04**	**0.02**	**-0.02**	**0.02**
ref. region: Aggl. Arnhem/Nijmegen	*-0.00*	*-0.02*	*-0.03*	*0.01*
SE-North-Brabant	*0.03*	*-0.00*	*0.02*	*0.06*

Note: The period in which the A1 is opened is printed bold.
Source: Own calculations based on: CBS, 1970-1992

Middle-Limburg and Northeast-North-Brabant are chosen as reference regions for *Southwest-Overijssel* because these are situated in the intermediary zone and also have an economic structure with an emphasis on industry. When we compare the relative shifts, it appears that the shift in both reference regions is much higher in the period of opening of the A1, so it follows clearly that no effect can be proven of the opening. The same holds for a lagged influence.

Arnhem/Nijmegen and Southeast-North-Brabant are chosen as reference regions for *Twente*, because these are characterized by medium sized cities surrounded by a rural area. In addition, Arnhem/Nijmegen was characterized by a restructuring of the industry, while Southeast-North-Brabant has a large emphasis on the manufacturing sector. Both regions are well supplied with main road infrastructure; in Twente somewhat more infrastructure was constructed. When the relative shifts of the regions are compared, we see that the relative shifts in Twente are in every period higher than in Arnhem/Nijmegen, while it is in most periods lower than in Southeast-North-Brabant. Therefore, no clear conclusion can be drawn.

We conclude that no clear impact of the construction of the A1 is found by carrying out the reference region approach. Only in some cases are the relative shifts higher in the A1 regions, but for most regions this is not the case.

7.3.4 Remarks on the Reference Region Approach

It is apparent that the results of the reference region approach are dependent on the choice of the reference region, but with other choices the conclusion seems to remain the same. Another problem is that this approach is rather crude, because no other specific features of the regions are taken into account. Clearly, in a relatively small country with only 40 statistical regions, one cannot select reference regions in such a way that they satisfy all requirements for a quasi experimental method (Isserman et al., 1989). To meet with this objection we will estimate a model where several region-specific factors are taken into account. Finally, it should be mentioned that since the spatial scale level of Corop-regions is rather high, it is very likely that an influence can be proven on a lower spatial level. This would mean that highways induce a relocation of employment *within* regions towards points of access leading to corridor effects, but that regional totals as such are not influenced.

7.4 A Regional Labour Market Approach

7.4.1 The Regional Labour Market: an Introduction

Given the limitations of the reference region approach, it is useful to estimate a model in which the various regional features are included in an explicit way. Let us take the following production function as a starting point:

$$Q_r = f(L_r, K_r, A_r, B_r) \qquad (2)$$

where:

Q_r volume of production in region r

K_r capital in region r

L_r labour in region r

A_r accessibility of region r

B_r other factors in region r, for example, other types of infrastructure and urbanization (dis-)economies.

Transport infrastructure is represented by means of the factor B, thus it enters the production function via the services it provides as measured by means of accessibility (Forslund & Johansson, 1995). This formulation of the role of transport infrastructure in the production function allows one to take into account the spatial spill-over effects associated with the supply of infrastructure.

If we may assume that resources are allocated in such a way that regional profits are maximized given the price of the output at the world market, we arrive at a labour demand function with the following arguments:

$$L_r = h(s_r, w_r, p, A_r, B_r) \qquad (3)$$

where:

s_r price of capital in region r

w_r price of labour in region r

p given price of the output

In our application we operationalize these arguments as follows:

Price of capital (s_r) The price of capital is equal for all regions with one exception: firms may receive subsidies on investments. Therefore, we take a regional policy variable to take these subsidies into account. The variable assumes the value 1 in regions receiving subsidies and the value 0 in the other regions. The sign of impact of investment subsidies on employment is uncertain. If substitution effects dominate, it will be negative, but if output effects dominate, it will be positive.

Price of labour (w_r) Because of the strong national component in the annual wage negotiations and the uniform national minimum wage policy, interregional wage differentials are very small in the Netherlands. This is one of the causes of frictions within regional labour markets. Therefore, we use the percentage of the labour force with an unemployment duration less than one year as a proxy for the difficulty to recruit labour. Long-term unemployed are not taken into account in this measure because their probability of recruitment is often low. A second labour market variable included in the equation is the level of education of the regional labour force.

Price of output (p) The impact of the price cannot be estimated with this model, because p is assumed to be the same for all regions.

Accessibility (A_r) This concept has already been defined in Section 7.2.3. We will use two accessibility indicators: one for domestic destinations and one for international destinations in neighbouring countries to test whether the process of European economic integration has an impact on regional development. In the given formulation, we may expect a positive impact of accessibility on employment. However, we note that substitution effects between labour, private capital and infrastructure may lead to a negative impact on employment. In addition, from an interregional trade perspective, an increase in accessibility may make regions more vulnerable to competition by producers from elsewhere. In order to test the latter consideration, one may investigate whether importing sectors in a region are influenced in a different way by a change in accessibility compared with exporting sectors. Whether a sector is importing or exporting can be operationalized by means of the locational quotient.

Other regional factors (B_r) The degree of urbanization is used as an explanatory variable to account for urbanization (dis-)economies.

For a detailed definition of the variables, we refer to Bruinsma et al. (1995a). We estimated the model for total regional employment as well as for the separate sectors.

7.4.2 The Results of the Model for Total Regional Employment

An empirical application of the theoretical model allows one to use a formulation in terms of changes and in terms of absolute values. We have chosen the first specification. The reason is that with a specification in terms of absolute values one may have serious causality problems. The supply behaviour of the public sector is usually such that densely populated areas have high infrastructure densities. Therefore, a positive relationship between infrastructure supply and employment does not say much when one wants to investigate the contribution of infrastructure improvement to regional growth.

In our estimation (see Table 7.9), the relative shift in employment of a region during a five-year period is explained by:

acc	the relative change in domestic accessibility
int	the relative change in international accessibility
unemp	the relative change in short-term unemployment
educ	the relative change in the level of education
urb	the level of urbanization
pol	the regional policy variable (a dummy)

Table 7.9 Results of the model for total regional employment (relative shift)

	1970-1975		1975-1980		1980-1985		1985-1990	
	Coeff	T	Coeff	T	Coeff	T	Coeff	T
acc	-0.94	-2.07*	0.74	0.81	0.22	0.54	2.12	2.11*
int	0.08	0.51	0.20	0.28	1.47	1.26	-0.67	-1.08
unemp	-0.01	-1.10	0.22	4.86*	0.05	2.21*	0.15	1.18
educ	-0.54	-3.36*	-0.09	-0.66	0.03	0.30	-0.07	-0.73
urb	-0.65	-5.04*	-0.15	-1.15	-0.19	-2.96*	-0.28	-2.14*
pol	-0.10	-1.92*	-0.04	-0.91	-0.06	-2.55*	-0.07	-1.75*
Cst	0.51	5.36*	-0.02	-0.23	0.02	0.33	0.00	0.05
R^2	0.52		0.59		0.50		0.26	

Note: * significant at the 5% level

It appears that the change in domestic accessibility had a significant negative impact on employment in the period 1970-1975, while in the period 1985-1990, this impact was positive. In the other periods no significant impact is proven. International accessibility does not have a significant impact on regional employment. So a clear impact of a change in accessibility on the shift in regional employment cannot be proven.

Of the other variables the change in the unemployment rate has the expected significant sign in two five-year periods. The level of education does not have the expected impact. The level of urbanization does have a significant negative sign in three periods, so the urban-rural manufacturing shift (see section 2.1) seems to have more influence than agglomeration effects do. The regional policy has a negative significant sign in three periods, which may mean that the substitution effect is larger than the output effect. Another possible interpretation is that the regions receiving government support are regions with structural weaknesses not captured by the other variables. Government policies are not strong enough to overcome these structural weaknesses and as a consequence, a negative sign is found for the policy variable.

7.4.3 Results for the Transport and Communication Sector

One sector where a significant impact of a change in accessibility on employment growth may be expected is the transport and communication sector. We estimated the model with the afore mentioned variables. The results are given in Table 7.10.

Table 7.10 Results of the model for the transport and communication sector (relative employment growth)

	1970-1975		1975-1980		1980-1985		1985-1990	
	Coeff	T	Coeff	T	Coeff	T	Coeff	T
acc	-0.44	-0.54	2.45	0.89	3.45	2.12*	5.54	2.37*
int	0.54	1.95*	-1.76	-0.80	10.29	2.26*	-1.81	-1.25
educ	0.08	0.29	-0.22	-0.54	0.26	0.78	-0.18	-0.74
unemp	-0.00	-0.36	0.31	2.23*	0.19	1.99*	0.74	2.51*
pol	-0.11	-1.27	0.06	0.41	0.21	-2.13*	-0.24	-2.48*
urb	-0.04	-0.17	-0.21	-0.54	-0.34	-1.39	-0.90	-2.97*
Cst	0.04	0.27	0.16	0.74	-0.04	-0.15	0.37	1.72*
R^2	0.11		0.26		0.55		0.39	

Domestic accessibility has, in the three periods between 1975 and 1990, a significant positive influence on employment in this sector. The same holds true for international accessibility in the periods 1970-1975 and 1985-1990.

We conclude that the change in accessibility has a positive impact on the employment in this sector.

The other variables give about the same results as the model for total employment: the change in the unemployment rate has a positive influence on employment growth in this sector; the regional policy has twice a significant negative sign and the level of urbanization has once a significant negative sign. The education level does not have a significant influence on employment of this sector.

The estimates for the other sectors are given in Bruinsma et al. (1995a). In general, it can be said that with individual sectors, the explained part of total variance (R^2) is lower than for total regional employment. Domestic as well as international accessibility only incidentally have a significant influence on employment growth in the other sectors.

In an alternative specification, we tested whether the accessibility variable has a different impact for importing and exporting sectors, but this did not yield better results for the individual sector estimates. The conclusion is that the transport and communication sector is the only one where an influence can be proven on the Corop-level.

7.5 Conclusions

In this contribution we have investigated the link between highway construction and the development of regional employment in the Netherlands during the past decades. In terms of employment growth, the intermediate zone has been most successful during the period considered. However, the increase in accessibility has been largest in the peripheral zone; this indicates that there is no simple mono-causal relationship between the development of accessibility and employment.

In the reference region approach, we compared A1-corridor regions with reference regions which had similar locational features and sectoral structure, but had less highway construction during the construction of the A1. By a comparison of the relevant shift in employment, no clear impact of highway construction on regional employment could be proven, and only incidentally the shift in the A1-regions was higher than in the control region.

Finally, a regional labour market model was estimated including regional features along with domestic and international accessibility as variables. For total employment no unambiguous impact could be proven; the impact of the domestic accessibility was significantly negative in one period and significantly positive in another period. Only the transport and communication sector showed a clear positive impact.

From this empirical research, we conclude that there is no evidence that the construction of main road infrastructure and the resulting change in accessibility has a clear impact on overall employment in regions. This conclusion holds for the spatial level of Corop-regions. This is an important result for regional politicians who often propose investments in road infrastructure because they believe that this will boost regional employment growth. It is very likely, however, that a positive impact on the regional employment may be found in a corridor analysis at a lower spatial level. This will be the subject of Chapter 8.

It is good to realize that our research scope has been rather narrow in this Chapter. By focusing on employment effects we have ignored the productivity and transport cost effects which receive ample attention in the cost-benefit analyses usually carried out for large infrastructure projects. The result that we did not observe substantial employment effects does of course not imply that the other effects will be small as well. Note also that a cost-benefit analysis would entail the inclusion of environmental effects which have been ignored in our study.

A point that deserves particular attention in subsequent research is the impact of traffic congestion on accessibility. Since travel time is an important component of generalized costs, accessibility becomes dependent on the time of the day (peak versus off-peak). In addition, the notion of generalized costs also includes the costs related to uncertainty of travel time. Transport infrastructure investments in congested areas do not only imply lower travel times but also lower uncertainty costs. An attempt to include congestion in accessibility measures is presented in Chapter 10.

8 Road Infrastructure and Corridor Development: Entrepreneurial Surveys

8.1 Introduction

After the - quantitative - modelling approaches at the regional aggregation level discussed in the previous Chapter, the focus of this Chapter will shift to more qualitative approaches at a disaggregate spatial level: entrepreneurial surveys. The empirical results of two studies on economic impacts of the construction of the transnational A1 highway on Dutch regions are presented in this Chapter. In both cases an entrepreneurial questionnaire was used. The first was held in 1989, the second in 1994. The questions of the surveys overlapped each other to a large extent, so changes over time in the perception of the impact of the A1 transport corridor on the firms' well-being can be measured.

A few difficulties emerge from using regional entrepreneurial surveys to measure the impact of a transport corridor on spatial economic growth:

First it is difficult, if not impossible, to make a distinction between distributive and generative effects. Are the observed effects the result of the growing internal strength, brought about by infrastructural investments? Or are these effects the consequence of a shift in employment from the surrounding areas as a result of better competiveness? Most employers tend to be unaware of such aspects.

Second, it is difficult for employers to distinguish the influence of a change in the infrastructure from the upgrading of the total regional economic production structure and environment.

Third, entrepreneurs may have the tendency to overrate the potential benefits of infrastructure projects in the hope that the research results will convince policy makers that these projects should indeed be carried out.

Finally, an obvious limitation of entrepreneurial surveys is that certain household effects are not taken into account. For example, households may experience benefits from better transport infrastructure that do not have clear impacts on economic activity, such as less time needed for social visits.

Is it possible to deal with these problems in an appropriate way? We will argue that indeed some of these problems can be solved by an appropriate questionnaire design and sampling strategy, however some of the problems remain (see also Leitham, 1996).

The first problem (generative versus distributive effects) can in principle be overcome by a broad sampling strategy so that entrepreneurs are interviewed who are located not only in the areas where positive effects occur, but also in areas where negative effects take place. The problem is, however, that the negative effects are most probably spread in a wide area. And they are small in the sense that the individual entrepreneur does not clearly notice it, even though at a collective level the negative effects may be non-negligible. Thus, a tendency may be expected that negative effects are underestimated with entrepreneurial surveys.

The second problem is a difficult one. Respondents must understand that they should not respond in a 'before versus after' fashion ('is your business more successful after the accessibility improvement than before'). The reason is that other conditions may have changed as well. As long as other relevant conditions have changed in a random way for the respondents, there is no serious problem of bias. However, other conditions will most probably not be random (general economic development in a region or country), and therefore a bias may occur. The proper way of asking the questions is in terms of ('with versus without'). A way to stimulate proper responses is a two-step approach: 1. respondents are asked which factors played a role in the development of their business, and 2. they are asked to indicate how the development would have been if the other factors had not been present. This remains a difficult question, but sources of bias have been reduced in this way. Thus we finally arrive at three relevant levels (see Figure 8.1): the 'before' level, the 'after' level, and the 'adjusted after' level. It is the difference between the former and the latter that we are looking for. In Figure 8.1 we show two cases with an equal positive infrastructure effect, but where the actual developments ('before versus after') are quite different.

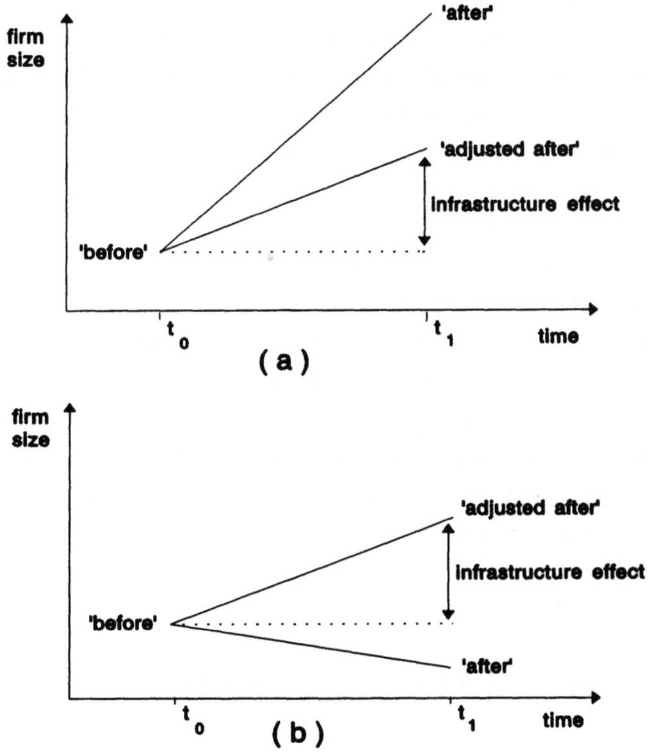

Figure 8.1 Two cases of 'before versus after' comparisons in impact analysis

The third problem is sometimes difficult to avoid. The questionnaire and the accompanying explanation should not hint at a possible importance of the outcome of the survey for a particular policy decision. In addition, the questionnaire must be balanced in that attention should not only be given to infrastructure aspects, but also the entire range of other relevant factors for the firms should be mentioned. The pertaining problem is most serious in ex-ante studies; in ex-post studies (as in the present Chapter) the incentives to give misleading responses are clearly smaller.

The last limitation is obvious: entrepreneurial surveys cannot provide the full picture of relevant effects. If one wants to offer a more complete picture households should also be interviewed to give information about the effects of infrastructure on for example, travel behaviour, and locational choices in the housing and labour market.

A final point deserving our attention is that infrastructure not only has objective effects (e.g. via transport costs); subjective effects (e.g. status considerations) are also important. Such status effects are usually difficult to touch in modelling approaches, but in the context of entrepreneurial surveys they can be accommodated.

We are aware of the limitations of the method used; the results presented here are put forward after careful interpretation of the data in the context of the specific region.

It is of utmost importance that the entire context of regional development and the role that infrastructural improvements may play within this context, is clearly indicated before employers are able to answer questions specifically dealing with those infrastructural improvements. Therefore, in the introductory part of both questionnaires the infrastructure was placed and described within the total regional context.

8.2 Entrepreneurial Surveys: an Introduction

8.2.1 The Surveys of 1989 and 1994

The survey of 1989 considered an infrastructure impact study in the Netherlands, with particular emphasis upon the assessment of its economic importance in terms of investment volumes and employment effects (direct, indirect and spin-offs). This study was undertaken to provide a more solid empirical basis for determining whether public infrastructure investments significantly contribute to the improvement of the employment situation in the Netherlands (see Bruinsma, 1990, and Bruinsma et al., 1992). The general setting of the research project was therefore broader than the impact of a transport corridor on employment growth. In three Corop-regions in different areas of the Netherlands (the western, the southern and the eastern parts), a postal questionnaire focussing on the influence of new infrastructure on the number of employees, was sent to basic sector firms with at least 50 employees. One of those regions was Twente, in which the A1 transport corridor is constructed. The results presented in this Chapter consider only Twente.

In the last week of May 1989, 255 questionnaires were sent to firms in the Twente Corop. The net response was 93 questionnaires (36.5%), which are representative for the target population according to location, sectoral composition, and size (for a more detailed analysis we refer to Bruinsma, 1990).

The second survey, which was held in 1994, is specifically focused on the theme of this Chapter: the impact of the A1 transport corridor on spatial economic development. It was included in the same research project for which the reference region approach and the regional labour market approach, as discussed in Chapter 7, were developed.

In the last week of April 1994, 1.845 questionnaires were sent to firms in the provinces of Overijssel and Gelderland having at least 10 employees. Only firms in sectors with a predominantly non-local orientation of demand have been included. The net response was 510 questionnaires (27.6%), which are representative for the target population according to location, sectoral composition, and size (for a more detailed analysis we refer to Bruinsma et al., 1995b).

8.2.2 General Results of the 1989 Survey

Before going in detail into the results of both studies considering the impact of the A1 transport corridor on employment growth, it is interesting to present the general results we found in the 1989 survey examining the impact of infrastructure on firms in the Netherlands.

In general, it is reasonable to assume that infrastructural improvements should result in a rather large improvement of the regional infrastructure complex (the synergy of all individual infrastructure components), before we may observe a spin-off effect that is sufficiently significant to be measured. The survey therefore focussed on such infrastructural projects such as motorways, railways, regional airports, and the replacement of old copper telecommunications networks with new optical fibre ones.

The employment spin-off effects of the railroad and regional airports appeared to be marginal due to the fact that those infrastructure elements are not in common use for any of the economic sectors.

The perceived employment effects of highways appear to be substantial (Table 8.1).

Table 8.1 Perceived impact of the construction of highway infrastructure on employment growth

	Leyden	Brabant	Twente
positive	27	11	22
neutral	64	74	65
negative	0	0	2
no opinion	9	15	11

In Leyden (located in the centre of the Randstad area in the western part of the country), for instance, 27 percent of the entrepreneurs indicated a positive employment development as a consequence of the improvement of the highways. In Brabant (in the southern part of the country), 11 percent and in Twente, 22 percent of the enterprises gained employment for this reason.

In another question, the employers were asked their opinion about how the firm would have acted if the new infrastructure were not realized. The results, which are presented in Table 8.2, are quite consistent with the figures shown above. In this respect, Leyden appears with 17 percent to score a little lower than would be expected, and Twente somewhat higher with 26 percent.

In both Leyden and Twente a relatively high percentage of firms indicated that their firm size would have been smaller without the new infrastructure. The effect on the investment level of firms appears to be substantial. Nearly 40 percent of the firms in Leyden, 30 percent of the firms in Twente, and 20 percent of the firms in Brabant expected lower company investments if the infrastructure had not been realized. About 15 percent of the relocated firms would not have been relocated.

The overall view from these figures is that the impact of new infrastructure in these regions on company behaviour is rather large, especially in the cases of Leyden and Twente.

Table 8.2 Effects if the infrastructure had not been realized

	Leyden	Brabant	Twente
company size smaller	10	2	17
investments smaller	37	21	30
company not relocated	17	21	22
less employment	17	8	26

What general inferences can be drawn from this case study on the Netherlands? The following factors were found to have a positive influence on the strength of the employment effects.

First, the region must possess a clear economic potential. On the labour market there should not only be a reserve quantity of labour, but it is also necessary that it contains a good quality of labour. There must also be a good entrepreneurial spirit, while the political climate has to be in favour of economic development.

Second, the new realized infrastructure has to serve the needs of all economic sectors. As a consequence, spin-off effects can be expected by an expansion of the following infrastructure elements: road infrastructure, telecommunications networks, energy, and water supply infrastructures. Energy and water supply infrastructures are basic infrastructure elements and normally already available without any capacity constraint. It is not reasonable to assume that increasing investments in those networks will lead to substantial spin-off effects, unless these networks were absent.

The construction of highways on the other hand may result in substantial structural employment effects. Through those expansions, the whole infrastructure complex of the region may be upgraded.

Numerous employment effects were found in two situations. Employment effects appeared to emerge when an essential but missing link in the infrastructure network was constructed, or when the new investments led to a capacity increase in a clearly congested network.

8.3 Infrastructure and Firm Development

The results presented in this Chapter mainly concern the 1994 survey. However, where possible, a comparison with the results of the 1989 survey is made. It will appear in section 8.4 that the perception of the impact of the construction of the transport corridor on the performance of firms located in the surrounding of this corridor has decreased over this five year period.

8.3.1 Development of Employment

The employment growth of the firms which responded was 52 % in the period 1980-1994, which is a very high growth rate. This can be explained by the fact that the companies selected have survived the recession period, or are starters during this period; so no questionnaires have been sent to companies which became bankrupt and to companies of which the number of employees fell under 10 employees. The industry sector firms, which account for most employment (48 %), grew relatively slowly (27 %) during the period considered.

Most employment in the study area (65 %) appears to be located within five kilometres of an access of a highway. Also the main population centres and largest firms are found in this area, so this is not a striking finding.

Market perspectives and internal company considerations appear to be the main factors influencing employment growth (Table 8.3). In a previous research project in 1989 (Bruinsma et al., 1992), similar results were found for the Twente region. Other factors often mentioned as 'very' important are: availability of employees, accessibility by A1, accessibility by road, parking possibilities, expansion possibilities of buildings, representativeness of the location, and the price of the location. A closer look at these factors reveals that infrastructure related factors (accessibility by road, accessibility by A1) are only considered as 'most' or 'second most' important in a small number of cases (this also holds true for the price and the representativeness of the location). Thus, although infrastructure is often very important in the eyes of the entrepreneurs, it is not often identified as 'most' or 'second most' important. Other factors are apparently more decisive when one wants to explain employment growth of individual firms.

Table 8.3 Perceived impact of (location) factors on employment growth (in %)

	very important	most important	second most important
internal company considerations	24.0	19.6	22.8
market perspectives	53.3	65.0	18.8
education employees	6.6	1.8	4.8
availability employees	12.7	2.4	13.2
telecommunication	8.8	0.2	1.1
accessibility by A1	12.9	0.7	2.2
accessibility by road	10.9	1.3	5.9
public traffic	3.5	0	0
parking possibilities	10.1	0	0.3
government	6.3	0.2	1.4
image region	8.1	0.2	2.2
expansion possibilities of buildings	25.6	6.2	13.5
representativeness location	16.6	0.7	2.2
price location	14.2	0.2	1.7
subsidies	7.6	0.4	2.0
residential environment	7.5	0.2	2.2
private factors	8.3	0.9	5.6

8.3.2 New and Relocated Firms

As also shown in the 1989 study, market perspectives and internal company considerations are not the main factors for a relocation decision; these are only found as next important factors, together with the price of buildings and location, the accessibility by road infrastructure in general and - to a lesser extent - the A1 (Table 8.4). The most important factors appear to be the expansion possibilities of buildings and the representativeness of the location.

When a distinction is made between push and pull factors, it appears that the most important push factors are poor quality and expansion possibilities of buildings and to a lesser extent, the bad accessibility by road, the residential environment, and a shortage of parking lots. On the other hand, the most important pull factors are expansion capacity, the representativeness, and to a lesser extent, the price and accessibility of the new location. An interesting conclusion that can be drawn from Table 8.4 is that road accessibility - and especially accessibility of the A1 highway - is more important as a pull factor than as a push factor in the location behaviour of firms. Only after entrepreneurs have decided to relocate do they begin to attach a role of some importance to road accessibility.

Table 8.4 Perceived impact of (location) factors on locational decisions (in %)

		push factors		pull factors	
	very	most	second	most	second
internal company considerations	24.8	9.2	7.4	7.4	2.5
market perspectives	27.0	2.0	9.6	3.4	3.3
education employees	0.9	0	0	0	0
availability employees	4.7	0	1.1	0	0.8
telecommunication	5.7	0	2.1	0	0.8
accessibility by A1	17.8	2.0	1.1	8.1	3.3
accessibility by road	21.6	7.8	12.8	13.4	9.9
public traffic	5.6	0	0	0	0.8
parking possibilities	16.2	0.7	9.6	1.3	1.7
government	4.6	0.7	2.1	0.7	0
image region	17.9	0	0	0.7	4.1
expansion possibilities of buildings	54.3	61.4	16.0	45.0	19.0
representativeness location	46.6	4.6	25.5	8.7	31.4
price location	26.0	1.3	3.2	5.4	13.6
subsidies	15.5	0	0	2.7	5.0
residential environment	14.4	7.8	6.4	2.7	0.8
private factors	10.1	2.6	3.2	0	2.5

Of the 510 firms surveyed, 100 started between 1980-1994, while 182 relocated during that period. It is noteworthy that of the 100 newly-started firms, no less than 46 have relocated at least once in this period; obviously young (successful) firms are more mobile than other firms.

The service sector appears to be the most dynamic; the percentage of starters as well as relocations is the highest for this sector. The industry sector on the other hand, is least dynamic.

When the distance to a highway is analyzed, it appears that the rate of birth of new firms is higher in the zone until 7.5 kilometres of an access compared with zones farther away (see Table 8.5). In addition, a high rate of relocation is found for firms within the 7.5 kilometres zone (36.8 %) compared with those in the longer distance zone (19.9 %). This difference cannot simply be explained by the larger presence of new firms in the 7.5 kilometres zone (new firms relocate more frequently). As shown in Table 8.5, the composition of relocating firms (in terms of already existing before 1980 and started up between 1980 and 1994) is not that different for both zones (80.5 and 76.7 % versus 29.5 and 23.3 %). Apparently, firms already existing before 1980 are also relocating more frequently in the 7.5 kilometres zone. A possible explanation is that in this - rather urbanized - zone, opportunities for expansion at the same site are limited, so that expanding firms are forced to relocate.

Table 8.5 Distance to a highway ramp and the rate of birth and relocation of firms

distance highway access	no. of firms 1994	existing before 1980	birth between 1980-1994		relocation of firms: number of relocated firms		of which birth between 1980-1994	
			abs.	%1)	abs.	%2)	abs.	%3)
< 7.5 km.	359	277	82	29.6	93	36.8	39	29.5
> 7.5 km.	151	133	18	13.5	23	19.9	7	23.3

1) the percentage concerns the firms existing before 1980
2) the percentage concerns the firms in 1994
3) the percentage concerns the share in relocating firms

When the relocated firms are analyzed in greater detail, it appears that 42 % stay in the same four digit postal code area, while 75 % stay in the same city or village. When the change in distance to a highway is measured, 41 % of the firms appear to relocate towards a highway, while 16 % of the firms relocate to a site at a larger distance. The average relative change in distance to an access due to a relocation is a decrease of 28 %. Thus, relocations of firms lead on average to shorter distances to highways. Another obvious reason why average distance to highways decreases is that new highways are constructed over the course of time (see section 11.3 for further discussions on the spatial aspects of firm relocations).

In the 1989 survey we did not find such a decrease in distance to a highway entry or exit after a relocation of the firm. This can be explained by addressing the regional government policy. In the Netherlands each square inch of space has a planned destination. In the eighties the regional policy was to expand near existing industrial sites at the outskirts of cities. The surrounding of the A1 highway should remain a green zone. By the turn of the decennium this policy changed. Large industrial sites were developed near highway ramps. As stated afore, the relocation decision of most firms is caused by a lack of expansion possibilities at the old location. Those firms can only relocate to locations where there is enough supply of space. So from 1988 onwards, firms were forced to locate near a highway ramp, simply because hardly any other location was offered. And so remains the chicken-egg problem of dynamic firms locating near highway ramps, which in turn are located near dense population centres: cities are economically more dynamic than rural areas.

8.4 Infrastructure Components

In this section the importance of various infrastructure components will be discussed. First, an analysis is presented of the infrastructure elements which are considered as most important by entrepreneurs. Next, entrepreneurial perceptions on infrastructure bottlenecks are discussed. Finally, the values given by entrepreneurs to the impacts of the A1 construction on specific elements of the firms' performance are discussed.

8.4.1 The Impact of Infrastructure in General

It seems that especially road infrastructure is valued often as 'most important' (Table 8.6). Telecommunications are found to be most important next, while public utilities are valued a little lower. It may be clear that these elements are intensively used by all firms and airports and railways are used much less; these are valued much lower, accordingly. As shown in the last column of Table 8.6, the changes in importance of infrastructure components during the past period are rather similar to the present importance, although certain shifts can be observed. Roads remain dominant, but telecommunications and airports are often mentioned as infrastructure components of increasing importance.

Table 8.6 The importance of infrastructure (in %)

	very	most	second	increased 1990-1994
roads	76.2	79.4	9.1	66.4
railways	2.3	0.7	11.6	7.3
waterways	2.6	1.3	4.0	3.8
airports	5.2	0.4	4.5	17.0
telecommunications	39.3	3.6	29.5	62.1
public utilities	27.3	4.7	18.2	24.2
terminals	4.3	0.4	4.8	8.9
no opinion		9.4	18.2	
total		100	100	

We also investigated to what extent the present location of firms has an impact on the valuation of the importance of infrastructure (Bruinsma et al., 1995b). It appears that there are no significant differences between the valuations according to distance to highways. This also holds true for the valuation of highways: entrepreneurs with firms located farther from the A1 did not value road infrastructure lower than entrepreneurs with firms nearby.

8.4.2 Bottlenecks in the Infrastructure Networks

Among the entrepreneurs, 44 % mention that they suffer from bottlenecks in infrastructure networks, especially in road infrastructure and to a lesser extent in telecommunications. Concerning road infrastructure, bottlenecks at all spatial levels are mentioned. At the *national* level especially congestion - primarily in the western part of the Netherlands (the Randstad) - is experienced as a bottleneck. Thus congestion in the most highly urbanized part of the country not only disturbs firms located there, but also firms in the more peripheral parts of the country. Striking is that bottlenecks in the *international* road network are only mentioned in 1 % of the cases; in particular the connection with the German Ruhr area is mentioned. This may be considered - especially for a border region - as very low. It is an indication that there is in this case no serious mismatch between supply and demand for infrastructure in border regions. A similar result was found for a much broader set of European border regions in Chapter 5.

The valuation of the importance of infrastructure components depends on the intensity of use as well as on bottlenecks experienced. This is the reason that road infrastructure and telecommunications figure prominently in Table 8.4. The low level of bottlenecks in services of public utilities is an explanation why these receive lower scores in this Table.

8.4.3 Impact of the A1 on the Performance of Firms

In the questionnaire the entrepreneurs were asked to consider the hypothetical case of the A1 highway *not* having been constructed. The impacts of the construction of the A1 highway on firms are described in Table 8.7.

The impact is especially found in increased accessibility and shorter travel times and to a lesser extent in the punctuality of goods supply. These results are not very striking since the research area is not suffering from serious congestion. Striking however, is that the least positive impact is found for the size of the international market area. For a transnational transport corridor like the A1, these findings are rather disappointing. The impact of the construction appears to be smaller when the company is located farther away from the A1. This holds true for the impact on sales as well as on accessibility. The impact on travel time is especially small when the site is located more than 7.5 kilometres away from the highway.

When a sectoral distinction is made, we find that for the transport and communication sector the most positive impacts are found; this is consistent with the findings in the regional labour market approach (Chapter 7). For the service - and to a lesser extent the industry sector - relatively small impacts are obtained.

Table 8.7 Impacts on firms of the A1 construction

	strong improvement	improvement	no impact	no opinion
accessibility	38.7	31.7	20.7	9.3
travel time	31.9	38.9	20.1	9.2
punctuality deliveries	13.2	33.0	38.8	15.0
sales	7.9	19.5	55.2	17.4
national market	9.8	19.6	54.4	16.2
export market	5.6	12.3	63.0	19.2
costs per unit	6.2	23.6	51.1	19.0

On the question of how the situation would be if the highways would not have been constructed, some 10-20 % of the entrepreneurs indicates that this would have had a negative impact on employment and the level of investments (see Table 8.8). These answers are mainly given by entrepreneurs located at less than 7.5 kilometres from a highway access.

Table 8.8 Expected impacts on firms if the A1 was not constructed

	agree	disagree	no opinion
company closed down	0.8	78.9	20.4
company size enlarged	0.5	74.4	24.7
company size smaller	6.9	66.0	27.2
investments enlarged	17.0	56.8	26.3
investments smaller	2.1	69.8	28.1
expansion employment	14.5	58.7	26.8
less employment	2.1	69.8	28.1

In comparison to the 1989 study, a substantially smaller share of the entrepreneurs indicate negative impacts of the non-construction. It appears that the positive impacts of the construction which are reported by entrepreneurs just after the construction are less evident to entrepreneurs when they are interviewed some years later.

8.5 Conclusions

In the relation between transport infrastructure and economic development the perception of entrepreneurs about the value of transport infrastructure is of major significance.

A high consensus exists among entrepreneurs about the importance of road infrastructure and, to a lesser extent, telecommunications and public utilities. This common feeling about the importance of road infrastructure is partly reflected by actual location behaviour (see also Chapter 11). The average distance to a highway access is decreased by 28 % after the relocation of a firm. However, it is important to note that a decrease in the distance to a highway access is not a main reason to relocate a firm. The main reason to relocate is the unsuitability of the old location. Moreover, the choice of the new location depends on the supply of industrial sites. In the period 1988-1994, large industrial sites were developed nearby highway ramps. This might partly explain the decrease in the average distance to a highway access of the relocated firms.

Another indication for an impact of highway construction on economic development is found in the high percentage of entrepreneurs who stated that their companies' accessibility has improved and construction of the A1 has led to a decrease in travel times. That fewer entrepreneurs experienced an increase in the reliability of delivery times seems convincing since there was hardly any congestion in the area before the A1 was constructed. The reliability of delivery times was and still is relatively high.

Other relationships between transport infrastructure and the vitality of firms are less convincing. Although entrepreneurs indicate that infrastructure in general and the construction of the A1 highway in particular are of a certain importance for the development of employment, they value market perspectives and internal company considerations clearly higher. In addition, there is no evident indication of a tendency of decreasing impacts with increasing distances to an access of a highway.

The questionnaire shows that the valuation of transport infrastructure by entrepreneurs is not only based on solid rational and objective reasons such as transport costs for instance. Subjective reasons such as image effects are also involved. The impact of these subjective reasons might be considerable. If an improvement of the perceived attractiveness of an area by the construction of a highway leads to an increase in private investments, then public investments could be used to provoke private investments when applied in a proper manner. In 1994 the entrepreneurs indicate the impacts of the A1 construction to be considerably lower than in 1989. This means that over time there is a reduction in the perceived importance of the A1.

9 Infrastructure and Urban Development: the Amsterdam Orbital Motorway

9.1 Introduction

In the 1930s the first orbital motorways were planned around cities such as Berlin, Munich and London. It took a long time to complete these orbital motorways and some of them were never completed. As Hall (1990) indicates, the major reason for the planning of the European orbital motorways was not just the desire to remove traffic congestion. At that time roads were not yet very congested in Europe. Other incentives to build orbital motorways were the desire to reveal and reinforce the organic spatial structure of cities and make monumental artifacts which could serve nationalistic purposes.

After the Second World War, road traffic began to grow at a very rapid rate in many European countries and this has induced the creation of orbital motorways in many countries. Orbital motorways have various types of impacts on metropolitan areas including direct impacts on traffic and impacts on the spatial structure of activities.

We know much more about the direct impacts of the construction of orbital motorways on traffic than about impacts on the spatial structure of activities (cf. Bruinsma, 1994). It is for this reason that we have decided to focus on the second type of effects in this Chapter.

This Chapter is structured as follows. In section 2 we give an overview of results of empirical studies on orbital motorways. In section 3 we give information on the completion of the Amsterdam orbital motorway in 1990 and the observed effects on traffic flows since then. Effects of the orbital motorway on office rents are discussed in section 4 (see Bruinsma et al., 1996). Responses of firms, based on a stated preference approach are discussed in section 5. In section 6 the conclusions are drawn. Finally, in section 7 the findings of Chapters 7, 8 and 9 are compared to analyze the differences in results of the presented infrastructure impact studies achieved by model and survey approaches.

9.2 Analyzing Impacts of Orbital Motorways

An important feature of orbital motorways is that they make route choice much more flexible. When traffic is blocked on one part of an orbital motorway one can still use other parts of this road to reach any destination desired. In the case of orbital motorways three uses can be identified:

1. use for intra-urban trips,
2. use for trips from outside the urban area to inside, or vice versa,
3. use for trips with both origin and destination outside the urban area.

With type 1 effects, attention is focused on the role of orbital motorways within the urban network. These effects depend strongly on the existing intra-urban network. In the case of type 2 effects, accessibility of the urban area from the surrounding regions is the major concern. In the case of type 3 effects, the role of an orbital motorway in a larger inter-regional network is taken into account.

The relative importance of the three types of use will vary greatly among cities. In some cases an orbital motorway may be very important for the accessibility of regions in a large inter-regional network. In other cases its dominant importance may be as a relief for congestion within cities. The balance between the two effects depends strongly on the radius of the orbital motorway: when it is small the type 1 use will be most important; when it is very large the type 3 use will become important. Scant attention paid to the radius of orbital motorways may be one of the reasons why research on the effects of the construction of these motorways does not lead to unambiguous outcomes.

Another important factor which has to be remembered is that it is not always well defined how one measures the impact of an orbital highway.

Does one take the impact of the entire motorway, or only of the last part which was completed many years after important parts of the motorway were already put into use. In some countries the time between the use of the first part and the final completion of an orbital motorway may be as long as 25 years. Still another factor which makes it difficult to compare the impacts of orbital motorways at different places is that the zero-situation may vary strongly. In one case there may have already been a reasonable road connection before the orbital motorway was constructed. In other cases the construction of an orbital motorway may have a much more revolutionary impact, for example, when it involves the replacement of ferry services by a tunnel or bridge.

Choice of transport mode is an interesting theme in the case of the construction of orbital motorways. Rail transport connections usually have a star-shaped structure. As a consequence many rail transport users must travel via the centre thus giving rise to substantial detour factors and time losses when one has to change trains. The construction of orbital motorways leads to an adjustment of the road network away from a star-shaped structure. This means that travel time gains may be large, especially when the destination of the trip is not in the centre of the city. Thus, the construction of orbital motorways leads to a deterioration of the competitive position of rail transport, especially in criss-cross trips.

A study on the accessibility effects of the M25 London Orbital Motorway was carried out by Linneker and Spence (1992). The M25 was built between 1975 and 1986. It is located in the green belt around London and, with a radius of some 26 kilometres, it clearly has the character of an outer ringroad. In terms of accessibility based on travel times, the construction of the M25 has led to relatively large improvements in those areas of South East England through which the road passes (an increase of more than 10%). For inner London there is almost no change in accessibility. For regions farther away a moderate but certainly not negligible improvement can be observed (1-5%). Changes in route choice due to the construction of the M25 usually lead to longer though quicker routes, so that when generalized transport costs are taken into account the improvements in accessibility are less pronounced and may even become negative. These results depend strongly on the coefficients for time and vehicle operating cost used in the generalized transport cost calculations.

Given the result that accessibility changes due to orbital motorways may be substantial, one may wonder which relocation effects will occur. Such effects are not studied by Linneker and Spence (1992), but in other studies attempts have been made to estimate them. Guiliano (1986) reports on a broad survey of effects of beltways in USA metropolitan areas. The effects are on average smaller than one might expect: the strongest conclusion possible is that beltways can have small though significant effects on regional development patterns and the economies of central cities.

Rather indecisive results are found in a comparative study of integrated transport-land-use models (Webster et al., 1988). Simulations with models developed for cities in a number of countries (Germany, Great Britain, Japan, Spain) indicate that the overall decentralizing effects of orbital motorways on the location of residences and workplaces is small compared with the effects of autonomous changes in land use patterns. The uncertainty in the outcomes may be due partly to the differences in the radii of the ringroads in the cities.

Table 9.1 Effects of ringroad investments (percentage change)

	new investment in	
	outer ringroad	inner ringroad
mode share:		
car	0.3	0.2
public transport	-0.2	0.0
walk	-0.8	-0.3
mean trip:		
distance	0.9	0.0
time	-1.2	-0.7
road speed	0.2	-0.1
distribution of:		
population	?	?
employment	?	weak centralization

Source: Dasgupta and Webster (1992)

A more precise analysis is carried out in Dasgupta and Webster (1992), where a distinction is made between an outer ringroad with a large radius and an inner ringroad with a small radius. Based on a comparative study in three cities (Leeds, Bilbao and Dortmund), the authors find no clear results on a centralizing or decentralizing effect of an outer ringroad on the distribution of employment or population (see Table 9.1). A possible explanation is that the results of the model simulations depend on where the new road is located in relation to the boundary of the study area. This underlines the importance of having a sufficiently large study area in analyses of this type. An inner ringroad appears to have a weak centralizing tendency. As Dasgupta and Webster note, these rather small effects may be because the three cities considered already have relatively well-developed orbital links. One should be aware that the effects reported in the Table relate to the average outcome for the whole area. So although ringroads only have small effects on the average degree of decentralization in an urban area, their effects on particular zones may of course be much larger.

The result on trip distances and trip times indicates that outer ringroads lead to longer trips in terms of distances travelled, but to shorter travel times, so that average road speeds increase.

Much larger impacts of a ringroad project on urban development are found by Morisugi et al. (1993) for the city of Gifu, Japan. In this case, the modelling simulations indicate that quite substantial changes will occur in urban land prices, leading to a high benefit incidence for absentee land owners at the expense of households and private firms. In spatial terms, the construction of the ringroad, of which the radius is about four kilometres, leads to a sizable change in structure. The model predicts a suburbanization of population and a growth of employment in the central part of the area.

Lathrop and Cook (1990) indicate that the spatial distribution of economic activities in the USA has been influenced by major factors such as the desire to live in spacious single-family houses and the growth in car ownership. This has induced suburbanization tendencies, which would have occurred even if no beltways would have been built, but beltways certainly have acted as catalysts for such a development.

9.3 The Amsterdam Orbital Motorway

9.3.1 Introduction

Like most capital cities in European countries, Amsterdam suffers from traffic problems. Not only is the city confronted with large flows of traffic, but especially during peak hours the roads near Amsterdam are also heavily congested. During the past decades the inner city of Amsterdam with its rings of canals has proved to be insufficiently equipped to handle the growing car traffic. Partly as a result of these traffic problems companies moved out of the inner cities to the suburbs, where large office buildings were constructed. Especially the south-eastern, southern and south-western part of the agglomeration were favourably regarded as locations for those companies. Governmental policies to guide the suburbanization of population proceeded into another direction, however. By means of its own housing programs, as well as restricted allowances for building elsewhere, the government planned entire residential cities at some distance from large cities in the Netherlands for over two decades. Especially locations north of Amsterdam were selected for the construction of large numbers of dwellings. However, the growth of employment in these new cities straggled far behind the expectations of the government. This led to an increasing imbalance in local labour markets and an increase in commuting distances. As a consequence, one of the main traffic problems is the crossing of the river IJ, which is splitting Amsterdam just north of the inner city into two parts. This crossing is important since a large flow of commuters travels every day from the residential areas north of the river IJ to the employment centre in the southern part of the agglomeration. But severe traffic problems have also arisen on other roads giving access to Amsterdam.

So both the housing program and the shift of employment from the centre of the city to the southern part of the agglomeration caused the need for an orbital motorway in the Amsterdam region to make the necessary cross agglomeration trips possible. The orbital motorway was constructed in several phases and partly used existing motorways. Major sections were already completed in the 1970s and 1980s. In September 1990, the last part of the Amsterdam orbital motorway - the Zeeburgertunnel under the river IJ - was completed (see Figure 9.1). With a radius of about 5 kilometres, the Amsterdam orbital motorway has the character of an inner ringroad: important parts of the agglomeration are located outside the orbital motorway circle.

The orbital motorway has three functions:

- improvement of the accessibility of Amsterdam and the province of North Holland to the region north of the river IJ,
- relieving the secondary road network in the Amsterdam region,
- creating improved conditions for the economic development of the Amsterdam region.

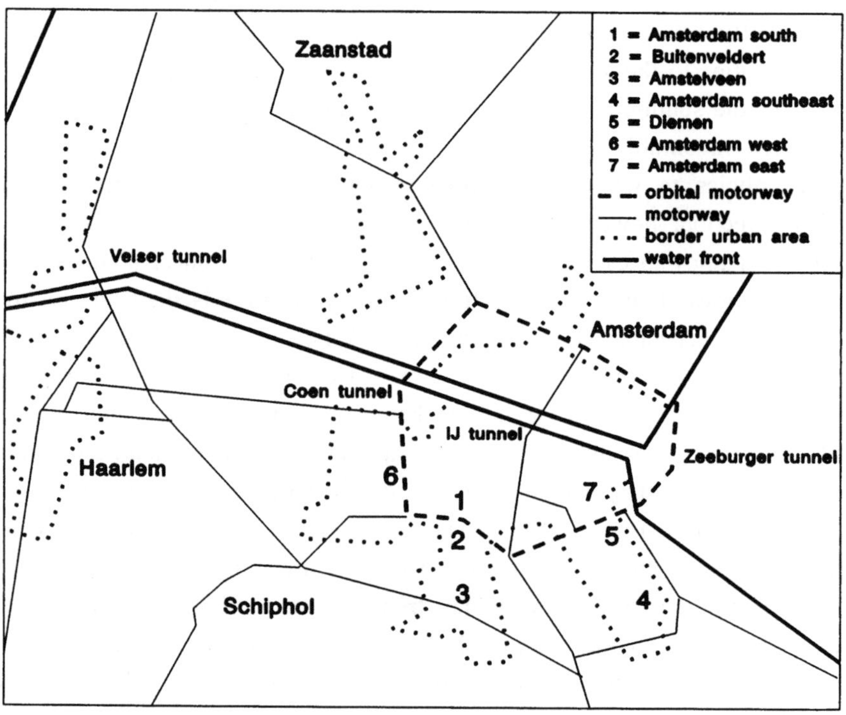

Figure 9.1 Map of the Amsterdam agglomeration

In this section we will concentrate on the first two topics[1]. They concern effects which can measured in the short-run. Research on these topics has

[1] In this section intensive use has been made of: Rijkswaterstaat (1992).

been done by several institutes and traffic consultants under the supervision of the Ministry of Transport and Public Works. The latter sections focus on the last topic and contain our own research.

In this section the following effects of the completion of the orbital motorway on traffic flows will be discussed:

- effects on the traffic volumes
- effects on travel behaviour
- effects on congestion

These results are based on large scale measurements in the road network in Amsterdam before and after the opening. In addition, several surveys have been conducted among residents in the region.

9.3.2 Effects of the Opening of the Orbital Motorway on the Traffic Volumes

The opening of the Zeeburger tunnel as the final part of the orbital motorway leads to a huge growth in capacity of the regional road network, especially concerning the crossing of the river IJ. The new connections make it possible to pass Amsterdam on the eastern side. For a number of relations this means a decrease in travel distance and/or travel time.

The total number of kilometres driven in the Amsterdam region increases slightly faster compared with the induced traffic volume for the whole of the Netherlands (0.5 to 1.0 % in the Amsterdam region compared with 0.1 to 0.2 % in the Netherlands). This gain is the result of an rise in the number of kilometres driven on the main road network. The secondary road network of the Amsterdam region has been relieved. For instance, the number of kilometres covered by traffic crossing the river IJ on the highways increased by 13 %, while the number of kilometres made by river crossing traffic on the other roads decreased by 33 %.

In the Amsterdam region one can observe a clear shift of traffic flows in the easterly and northerly directions. The traffic density on the southern and western highway decreases within a range of 9 till 29 %. On all existing river IJ waterway crossings, traffic density dropped as a result of the opening. Per day, 58.000 motor vehicles pass through the new Zeeburger tunnel.

9.3.3 Effects of the Opening of the Orbital Motorway on the Traffic Behaviour

The possibility of passing Amsterdam on the eastern side has led to a major shift in route choice. Of the car drivers who cross the river IJ, 25 % have changed their choice of waterway crossing point. The Zeeburger tunnel now has a share of 19 % of all motor vehicles that cross the river IJ. Because of the changing routes towards the use of the orbital motorway, Amsterdam's city road network has been relieved.

Of the car drivers who cross the river IJ, 31 % changed their time of departure (either earlier or later). Before the opening they were forced to travel at times less suitable for them mainly due to traffic congestion. Since the opening of the orbital motorway, they can select their time of departure more according to their individual preferences. This means that individually, drivers are benefitting. The changes in departure times have led to a 16 % increase in the number of crossings of the river IJ by drivers in the morning rush hour between 7.00 and 9.00, the so-called 'return-to-the-peak' effect.

The opening of the orbital motorway has caused minor changes in the frequency of journeys, the destination of journeys, and the transport mode choice for journeys crossing the river IJ. For instance, the opening has led to an increase of 1 % for commuter traffic and 5 % for traffic with other purposes (shopping, leisure, social visits).

The opening of the orbital motorway has made visible a latent demand for crossings of the river IJ in the rush hour. Here it largely concerns changes within existing traffic patterns. Travellers mainly adjust their route and time of departure, often in combination. The number of daily journeys on the waterway crossings increased by 4.5 % due to the opening. Of this, 1.5 % are journeys which existed before the opening; 3.0 % were caused by journeys generated by the opening.

Only small changes were observed in the use of the various transport modes (car driver, car passenger, user of public transport). This limited change in the transport mode corresponds to daily changes in behaviour that would have occurred without the opening. The public traffic flows across the river IJ have hardly been influenced, or have not at all been affected by the opening of the orbital motorway.

9.3.4 Effects of the Opening of the Orbital Motorway on Congestion

Considerable journey time gains have been recorded for passing journeys using the main road network in the Amsterdam region. The largest gains refer to journeys between the Province of North Holland north of the river IJ and the central part of the Netherlands. Travel via the 'old' routes also takes less time as a result of a congestion level that has decreased almost everywhere.

The reduction in the total time lost due to congestion on the entire main road network in the region is 20 %. This is the result of extra capacity and of the changes of route due to the opening. Because of the changes in the choice of departure time ('return-to-the-peak' effect), this reduction is less than what might have been expected on the basis of the increase in capacity. In the existing tunnels - Coen, Velser and IJ - the total loss of time due to congestion decreased by 39, 58 and 100 %, respectively.

One must bear in mind however, that the total journey time losses due to congestion are relatively small in the Amsterdam region; we guess that they are not larger than 5 % of total travel time (Bruinsma et al., 1993). Thus, the gain in total journey time due to the reduction of congestion is about 1 %. One interpretation of this outcome is that it is rather low. On the other hand, the reduction in time losses due to congestion of 20 % may be considered as rather high. From interviews with road users in the area, it appears that many of them perceive the change as big (see also section 5).

9.4 Effects of the Orbital Motorway on the Office Market

An indication for the potential of suburban areas to attract new firms and to develop in a broader economic sense can be found in their popularity as location sites for offices. The development of prices of absorbed office space is a proper indication for this popularity, because it reflects to a certain extent the willingness of entrepreneurs to pay for specific sites. The advantage of analyzing prices of absorbed office space relative to office construction decisions is the fact that the first prices reflect latest changes in market preferences, while the latter are in this respect delayed. The opening of the orbital motorway might increase the willingness to pay in some specific areas along it, because of cost reductions from improved logistic organization, more punctuality in deliveries of goods, an extension of the geographical labour market, promotional advantages in the case of sites that are visible from the motorway (sight locations), and so on.

For a period of three years - two before the completion (1987 and 1989) and one after the completion of the orbital motorway (1991) - data on transaction prices of offices larger than 500 m^2 on the Amsterdam office market were collected. These prices were analyzed in relation to the distances of the buildings to the nearest entry of the orbital motorway. It would have been more accurate if the accessibility had been measured, including both travel costs and travel time and combining different modes, but unfortunately, the data required for such a measure were not available.

Two techniques for measuring the impact on office prices were used: an impact analysis and a regression analysis.

In order to assess the effects of the orbital motorway on office prices, it is important to make a distinction between suburban districts in which (a part of) the orbital motorway was already existent, and suburban districts that were made accessible by the new sections. The first group of districts includes Amsterdam-West, -South, Buitenveldert, and Amstelveen; the second group includes Amsterdam-North, -East, -South-East and Diemen (see Figure 9.1).

Areas in which distances to the orbital motorway did not change average prices of office space - classified according to the distance by road from the respective sites to the nearest ramps of the orbital motorway - as shown in Table 9.2. A distinction has been made between locations at less than 2 kilometres and locations at more than 2 kilometres distance from the orbital motorway.

In the areas where distances did not change, cost of office space between 1987 and 1991 was significantly higher at sites within a limited distance from the orbital motorway compared with sites farther away. Office prices increased considerably in these areas between 1989 and 1991, which was when the orbital motorway was opened.

The development of office prices in zones where the distance to the orbital motorway had changed due to its completion, is given in Table 9.3. This Table shows the average office prices for various zones indentified according to the decrease in distance to the orbital motorway.

Table 9.2 Number of transactions and average prices per m² (in Dutch guilders) of new offices at locations in which the distance by road to the orbital motorway did not change by the time of its opening, per city district

	1987		1989		1991	
distance orbital motorway	no.	price	no.	price	no.	price
< 2 km.	22	250	35	256	28	294
> 2 km.	36	212	61	220	53	264
total	58	230	96	239	81	273

Table 9.3 Number of transactions and average rent per m² (in Dutch guilders) of new offices at locations in which the distance to the orbital motorway decreased by the time of its completion

decrease in distance	1987		1989		1991	
1991 compared with 1987	no.	price	no.	price	no.	price
< 1 km	7	125	9	204	4	147
1 - 2 km	12	237	27	238	16	238
> 2 km	4	197	7	211	7	217
total	23	208	43	222	27	224

What is most noteworthy in Table 9.3 is the fact that office prices in zones that profited most by the new orbital motorway practically did not rise between 1989 and 1991, when it was completed. They increased at a lower growth rate than the prices in areas where accessibility was unaffected by the new segment of the orbital motorway. Apparently, an improvement in the accessibility of certain areas because of the orbital motorway construction has, until 1991, not led to a stronger competitive position of these areas in relation to the western and southern parts of the agglomeration. However, as mentioned earlier, the timing of this research may have been too early after the completion to detect these effects.

One might argue that an anticipation of office prices could have already occurred during the construction phase of the orbital motorway. Indeed, the possibility of such an anticipation is quite relevant for multi-year contracts on renting office space. However, if one compares the totals of Tables 9.2 and 9.3, no sign of such an anticipation is found for the three years before the opening. The increase in office prices of locations of which the distance to the orbital motorway decreased is of about the same order as for locations where the distance to the orbital motorway did not change. It is difficult to ascertain whether an anticipation took place before 1987.

The conclusion is that in so far as the completion of the orbital motorway has influenced office prices, this influence is not observable in areas where the new segments were constructed.

The impact of infrastructure on office prices can also be studied by means of a regression analysis on the basis of individual transactions within the office market. Together with the influence of the orbital motorway, the influences of the existing rail infrastructure and the metro network have also been analyzed, including the extension of the metro network in 1990. In the regression model, the assumption has been made that office prices are dependent on the following location factors:

- the distance by road from the office location to the nearest orbital motorway ramp
- the walking distance from the office location to the nearest railway station
- the walking distance from the office location to the nearest metro station.

One has to take into account the influence of status aspects. The quality of office buildings is also an important factor determining office prices. In this context the following must be mentioned:

- quality of the construction
- possible alternative use of the building
- the ratio between gross and net office space
- the representativeness of the building
- the level to which the building meets specific user requirements. For example, telematics infrastructure adds a new dimension to the user value of the building.

Especially the latter two have become more and more important in recent years. Together with governmental policies such as subsidies and land taxes, these factors may lead to high variations in office prices in the Amsterdam area.

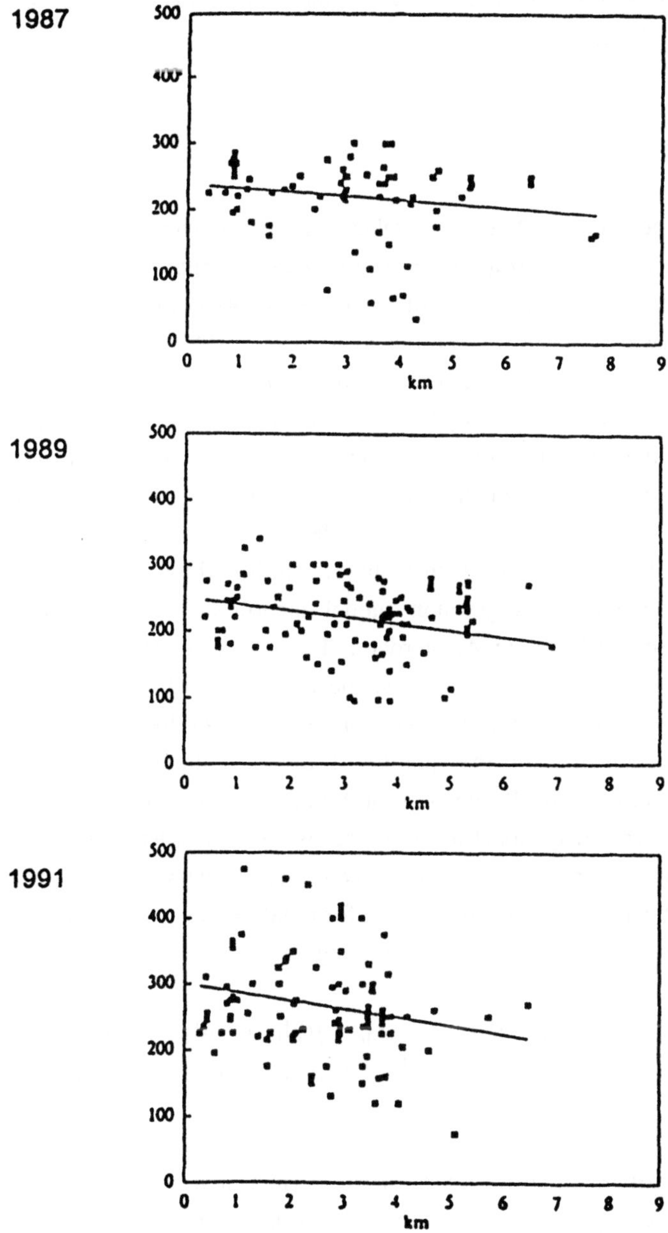

Figure 9.2 Variations in office prices at locations in the Amsterdam agglomeration, by road distance to the nearest ramp of the orbital motorway (1987, 1989, 1991)

A simple linear regression model, having only the distance by road to the nearest orbital motorway ramp as an explanatory factor for office prices, has the following form:

$$y_n = \alpha_0 + \alpha_1 x_n \tag{1}$$

whereby

y_n = price per m^2 of office space taken into use on site n

x_n = distance by road from site n to the nearest orbital motorway ramp.

In Figure 9.2 the variation in office prices relative to their position to the orbital motorway are shown for the years 1987, 1989 and 1991. Except for the fact that the variation becomes wider during the investigated period, it appears that in all three years office prices are negatively correlated with the distances to the orbital motorway. The angle of the regression line slightly increases in this period, which could stress a growing influence of the orbital motorway. In 1991, when larger price differences occur than during 1989, top level prices clearly increase when distances to the orbital motorway decrease.

A more sophisticated regression model incorporates the influences of the rail and metro networks in Amsterdam. To correct for the influence of non-infrastructure factors on office prices, the quality of office buildings and area status influences are also accounted for. The age of the buildings has been taken as an indicator for the quality of office buildings. A distinction has been made between (old) existing buildings that get new users and (new) buildings, either newly constructed or renovated. In order to deal with status aspects, a distinction has been made according to the various districts. These factors can be added to the model by formulating them as dummy variables. The resulting model then has the following form:

$$\begin{aligned} y_n = \quad & \alpha_0 + \alpha_1 x_{1n} + \alpha_2 x_{2n} + \alpha_3 x_{3n} + \alpha_4 d_{1n} + \alpha_5 d_{2n} + \alpha_6 d_{3n} \\ & + \alpha_{7d} d_{4n} + \alpha_8 d_{5n} + \alpha_9 d_{6n} + \alpha_{10} d_{7n} + \alpha_{11} d_{8n} \end{aligned} \tag{2}$$

whereby:

y_n = price per m^2 of office space at site n (in Dutch guilders)

x_{1n} = distance by road from site n to the nearest ramp of the orbital motorway (in kilometres)

x_{2n} = walking distance from site n to the nearest railway station (in metres)

x_{3n} = walking distance from site n to the nearest metro station (in metres)

$d_1..d_7$ = dummy variables for the district in which the office is located
d_8 = dummy determining whether the concerned buildings are old existing or newly constructed/renovated buildings.

Table 9.4 Regression results on office prices

	1987		1989		1991	
	coef.	stand. error	coef.	stand. error	coef.	stand. error
constant	245.27*	51.16	231.35*	41.14	299.16*	58.80
orbital motorway	-17.27*	10.72	-10.49	6.53	-23.21*	9.15
railway station	0.43	11.87	3.23	5.32	-0.19	10.18
metro stops	1.81	10.60	-4.09	4.34	8.62	6.96
dummy variables						
South	7.64	31.37	57.87*	15.55	63.26*	23.25
Buitenveldert	7.87	38.00	43.07*	22.74	-1.81	26.25
Amstelveen	46.23	46.96	54.99*	23.95	68.37*	32.02
South-east	53.80*	29.94	25.51	18.39	6.52	33.29
Diemen	16.66	29.54	-12.63	19.41	-58.52*	33.81
West	-46.08	56.43	-18.86	22.91	-92.76*	32.04
East	-21.01	30.73	-5.85	20.21	-45.23	29.76
New buildings	24.12*	12.77	31.55*	8.49	36.86*	12.77
R^2	0.27		0.46		0.44	

* = significant at 10 % level

The more complete model is superior to the simple model formulated above (see Table 9.4), although not all variables appear to be significant and the R^2 values are rather low. The orbital motorway appears to be a robust explanatory factor. The coefficient of -23 in 1991 for example, means that - other factors kept constant - the price of office space near to a ramp of the orbital motorway is 23 guilders per m^2 higher than on locations 1 kilometre away. The role of public transport networks is not clear. In 1991, railway stations have a slightly negative coefficient, while metro stations have a positive coefficient. Standard errors of both coefficients are high. The coefficients of the dummy variables for the various districts confirm the higher status of Amsterdam-South, Buitenveldert and Amstelveen in relation to other office sites in the Amsterdam area. For example, the average office price per m^2 in Amsterdam-South in 1991 is 63 guilders higher than in the Centre, other factors keeping constant. The influence of the quality of office buildings is also clearly confirmed. In 1991, the average price of new (or renovated) office buildings is 37 guilders higher than the price of existing offices.

The overall conclusion to be drawn from this quantitative analysis of prices on the Amsterdam office market is that although the effects are not directly visible in the areas where the new segments were constructed, in general, the orbital motorway is an important location factor for office firms.

9.5 Effects of the Orbital Motorway upon Entrepreneurs

9.5.1 Introduction

In Spring 1992 a questionnaire was sent to 516 entrepreneurs - with a labour force of at least 10 persons - divided over three types of zones. The first zone consists of areas which are located near segments of the orbital motorway which already existed. The second zone consists of areas which became accessible after the opening of new segments of the orbital motorway. The third zone consists of areas which are located rather at a distance from the orbital motorway (either the inner city or the remote suburbs). The first zone is entitled the *old accessible zone*, the second zone the *new accessible zone* and the third zone the *remote zone*.

The questionnaire was sent to entrepreneurs in four economic sectors: manufacturing, distribution, services, and the office sector. The first two sectors have a strong orientation on the transport of goods, the latter on business and commuting traffic.

The net response of 25 % was evenly spread across zones and sectors. In such an over-surveyed area as the Amsterdam region, the net response rate of 25 % is relatively high. Nevertheless, this level of response rate does not give a full guarantee of the representativeness of the questionnaire for the economic sectors involved in the survey.

In the old accessible and remote zone the office sector is the dominant sector with shares of 74 and 59 %, respectively. In the new accessible zone the manufacturing sector (41 %) and distribution sector (30 %) are the main economic sectors. These shares are close to the overall sectoral composition in those zones. Results will be presented for the whole agglomeration as well as divided at the zonal or sectoral level.

9.5.2 The Use of the Amsterdam Orbital Motorway

The first part of the questionnaire deals with the actual use of the orbital motorway. It seems that all 44 companies in the old accessible, 44 out of 45 companies in the new accessible, and 30 out of 41 companies in the remote zone use the orbital motorway. The least use of the orbital motorway is made by the service sector in the remote zone. An explanation might be that those companies are oriented at the local market, and thus make no use of the orbital motorway which is at some distance. A second argument concerns the activities of the service sector in general. The service sector generates more traffic in the form of customers rather than its own traffic.

Table 9.5 shows that the use of the orbital motorways for all traffic activities distinguished is high, especially for commuting and contacts with customers.

At the zonal level (9.5A), the scores of the new accessible zone are above the average use for all activities except commuting. This can be explained by the strong specialization of this zone in the distribution and manufacturing sector. As shown below, commuting is relatively unimportant for those sectors. The scores of the remote zone are below average for all traffic activities. The scores of the old accessible zone are more diverse: above average for commuting and contacts with customers and below average for the delivery of goods. In the old accessible zone - and the remote zone - the office sector is dominant. This explains the importance of commuting for this zone. The use of the motorway for visits by customers, business travel and commuting in the old accessible zone is much higher compared with the remote zone. These differences explain the lower use of the orbital motorway in the remote zone compared with the old accessible zone.

At the sectoral level (9.5B), some interesting results are found. The relatively low use of the orbital motorway for the delivery of goods is explained by

Table 9.5 The importance of the orbital motorway for several transport activities (in %)

A. Zonal	All respondents			New accessible		
	Imp.	Neutral	Unimp.	Imp.	Neutral	Unimp.
inflow of goods	60	9	31	81	5	14
outflow of goods	58	8	34	80	3	18
visits by customers	72	16	12	74	19	7
visits to customers	71	13	16	74	14	12
business trips	66	18	16	67	19	14
commuting	78	12	10	72	14	14

	Old accessible			Remote		
	Imp.	Neutral	Unimp.	Imp.	Neutral	Unimp.
inflow of goods	47	10	43	38	14	48
outflow of goods	41	14	45	38	10	52
visits by customers	83	5	12	50	31	19
visits to customers	73	11	16	63	15	22
business trips	67	19	14	59	15	26
commuting	88	12	0	73	10	17

B. Sectoral	Manufacturing			Distribution		
	Imp.	Neutral	Unimp.	Imp.	Neutral	Unimp.
inflow of goods	68	16	16	89	5	5
outflow of goods	75	13	13	84	11	5
visits by customers	76	20	4	78	11	11
visits to customers	88	13	0	75	10	15
business trips	88	8	4	60	35	5
commuting	72	12	16	76	10	14

	Offices			Services		
	Imp.	Neutral	Unimp.	Imp.	Neutral	Unimp.
inflow of goods	26	6	69	92	8	0
outflow of goods	26	6	69	70	0	30
visits by customers	68	16	16	67	20	13
visits to customers	71	13	17	33	17	50
business trips	66	16	18	36	21	43
commuting	85	12	4	67	20	13

the low scores of the office sector for these activities, which is the dominant sector in our sample. All other sectors score above average for the delivery of goods. The scores of the sectors for the use of the orbital motorway for the activities distinguished are fairly robust. For instance, the scores for the delivery of goods are low for the office sector and the service sector scores low for visits to customers and business travel.

9.5.3 Hindrance Before the Opening

When asked about the hindrance that companies have experienced, over two third of the companies had complaints about the use of the infrastructure network before the opening of the motorway in 1990. The complaints were highest in the new accessible zone (which was not accessible by motorway before the opening) and by the distribution sector.

In Table 9.6, the scores for different kinds of hindrance are given. At the zonal level (9.6A), the fact that the new accessible zone scores below the average, except for the lack in punctuality of deliveries, is interesting. It seems that the companies incorporated the expected delays in their behaviour. This is something which cannot be said for the companies in the old accessible zone. They give very high scores for time delay, late arrival of personnel and staff, and the lack of punctuality to appointments. The remote zone also gives a high score for delays in travel time.

At the sectoral level (9.6B), it seems strange that the office sector scores higher for time delay than the distribution sector. However, the transport activities of the office sector mainly concern commuting within the peak hours. The activities of the distribution sector are more evenly spread over the day. In addition, the distribution sector makes more detours to avoid congested areas. The concentration of activities of the office sector in the rush hours also explains the relatively high scores for late arrivals of staff and personnel. The lack of punctuality of deliveries has serious consequences for the productivity of the distribution and industry sector.

Only 40 % of the companies took measures to relieve the hindrance. Most measures were taken in the new accessible zone and by the manufacturing sector. Least measures were taken in the old accessible zone. Over two third of the measures concern adjustments in route planning, working hours or a combination of both.

Table 9.6 Hindrance before the opening of the orbital motorway (in %)

A. Zonal	Total	New acces.	Old acces.	Remote
delay in travel time	90	79	100	94
detour	58	59	56	59
lack of punctuality to appointments	50	44	56	53
personnel too late to work	46	35	67	35
lack of punctuality of deliveries	27	35	22	18
staff too late to work	14	6	26	12

B. Sectoral	Manufacturing	Distribution	Offices	Services
delay in travel time	88	76	97	90
detour	59	65	56	40
lack of punctuality to appointments	29	65	56	30
personnel too late to work	29	41	59	40
lack of punctuality of deliveries	24	47	9	60
staff too late to work	0	18	25	0

9.5.4 The Effects of the Opening for the Companies

Most companies stated that the opening of the orbital motorway relieved the traffic problems for their company (Table 9.7). A clear majority noticed a reduction in travel time, a better accessibility and/or a decrease in annoyance participating the traffic. Although the effect reported by the majority is neutral, a significant percentage of companies still noticed a rise in turnover, a decrease in the costs per unit product, and/or improved punctuality in the delivery of goods.

At the zonal level (9.7A), only a few differences from this general pattern occur. As before, the major contrasts are between the new and the old accessible zone. For instance, in the new accessible zone, 64 % of the companies noticed a rise in the punctuality in the delivery of goods, whereas in the old accessible zone, 74 % noticed no difference at all.

Table 9.7 Effects of the orbital motorway (in %)

A. Zonal	All respondents			New accessible		
	Imp.	Neutral	Worsen.	Imp.	Neutral	Worsen.
accessibility	80	19	2	91	10	0
travel time	76	14	10	76	12	12
annoyance in traffic	75	20	5	74	19	7
inflow/outflow of goods	43	50	7	64	28	8
costs per unit product	19	78	3	23	71	6
turnover	12	87	1	15	85	0

	Old accessible			Remote		
	Imp.	Neutral	Worsen.	Imp.	Neutral	Worsen.
accessibility	74	24	2	72	24	3
travel time	71	21	7	83	7	10
annoyance in traffic	78	20	3	72	21	7
inflow/outflow of goods	24	74	3	35	52	13
costs per unit product	9	89	3	28	72	0
turnover	16	82	3	4	96	0

B. Sectoral	Manufacturing			Distribution		
	Imp.	Neutral	Worsen.	Imp.	Neutral	Worsen.
accessibility	80	20	0	84	16	0
travel time	80	12	8	80	10	10
annoyance in traffic	73	23	4	74	11	16
inflow/outflow of goods	50	38	13	65	18	18
costs per unit product	32	68	0	31	63	6
turnover	10	86	5	18	82	0

	Offices			Services		
	Imp.	Neutral	Worsen.	Imp.	Neutral	Worsen.
accessibility	75	22	4	87	13	0
travel time	76	18	6	64	14	21
annoyance in traffic	79	21	0	75	17	8
inflow/outflow of goods	24	74	2	60	40	0
costs per unit product	9	86	5	9	91	0
turnover	9	91	0	27	73	0

At the sectoral level (9.7B), it is important to note that a relatively high percentage of companies in the manufacturing and distribution sector experienced a decrease in the costs per unit product. Those sectors are - more than the other sectors - used to working in terms of costs per unit product and actually could notice changes in the costs structure of their products. Relatively many companies in those sectors also noticed an improvement in the punctuality of the delivery of goods. On the other hand, a relatively high number of companies experienced a decrease in punctuality. The fact that the office sector is not accustomed to working with physical products leads to very neutral scores for changes in costs per unit product and in the delivery of goods.

The developments in the service sector are interesting. The relatively high score for the rise in turnover could indicate a better accessibility that could have resulted in a rise in the number of customers. Second, the strong reduction in travel time has led to an expansion of the market area. So not only has the accessibility improved, but so has the reach. And third, the rise in punctuality in the delivery of goods has led to more efficient operations.

9.5.5 Expected Consequences if the Orbital Motorway had not been Completed

In Table 9.8 the results are given to questions about the effects which would have occurred in case the orbital motorway had not been completed.

As shown, this would especially have had effects on the investments of the companies. About 10 % stated that they would have invested less. A minor number of companies expected to have fewer personnel or stated that they would have moved to another location. These negative effects would have been felt hardest in the new accessible zone and in the office and distribution sector.

9.6 Conclusions Amsterdam Orbital Motorway

The completion of the Amsterdam orbital motorway has had substantial impacts on the choice of routing and timing of trips. The impact on modal

Table 9.8 Expected effects if the orbital motorway had not been completed[1]

A. Zonal	Total	New acces.	Old acces.	Remote
firm closed	1	-	1	-
firm smaller	4	1	1	2
firm relocated	7	2	3	2
less invested	11	8	2	1
less personnel	6	4	0	2
Total	130	45	44	41

B. Sectoral	Total	Manufac-turing	Distri-bution	Offices	Services
firm closed	1	-	-	1	-
firm smaller	4	-	1	3	-
firm relocated	7	-	1	6	-
less invested	11	2	4	4	1
less personnel	6	-	2	3	1
Total	126	26	22	60	18

1 * since no entrepreneur mentioned an enlargement of the firm, or expected that their investment or employment figures would have been higher had the orbital motorway not been constructed, these factors are not included in the Table.
* four observations have been deleted in part B because the sector is unknown.
* In some cases a firm may have indicated more than one option (for example: firm smaller, less invested).

choice and trip frequencies have been small. Time losses due to congestion in the Amsterdam area have been reduced by about 20 %.

There are no indications that office prices in zones directly benefitting from the completion of the orbital motorway increased more strongly than in other zones. On the contrary, office prices at locations near already existing parts of the orbital motorway displayed the largest increase after the completion. Thus, the orbital motorway seems to have reinforced the position of zones which already had a strong competitive position in the region.

A statistical analysis of office rents reveals that distance to the nearest orbital motorway ramp has a significant negative impact on office prices. For the railway and metro system no such effects are found.

In a survey among entrepreneurs, approximately 60 - 70 % indicated that the orbital motorway is important for transport activities such as inflow and outflow of goods, visits of customers and commuting. Before the motorway was completed, firms experienced hindrance in the form of delays (90 %), lack of punctuality to appointments (50 %), and personnel arriving too late (40 %). According to about 80 % of the respondents, the completion of the orbital motorway led to an improvement of accessibility. About 10 - 20 % of the firms revealed that they experienced an increase in turnover and/or productivity as a consequence of this. Almost 10 % of the firms indicate that they would have invested less had the orbital motorway not been completed.

9.7 Models Versus Surveys?

In Chapters 7, 8 and 9 we presented the results of multi-approach research efforts at the regional and urban level of spatial aggregation to achieve greater insight into the impacts of the construction of transport infrastructure - in particular road infrastructure - on spatial economic development.

At both the regional and urban level, quantitative (models) and qualitative (entrepreneurial surveys) approaches are applied to achieve understanding of the complex relation between the construction of transport infrastructure and spatial economic development. In this section we will analyze the suitability of quantitative and qualitative approaches in this field of research.

9.7.1 Results Achieved by Model Approaches

The model approaches applied to measure the impact of highway construction on spatial economic development have been investigated with data at different spatial levels. At the highest level of spatial aggregation - the Corop level - modelling approaches are used to measure the impact of the A1 highway on regional employment development. When the reference region method is applied, it appears that no convincing indications are found that the A1 construction had a positive impact on total regional employment. The regional labour market approach showed that only the transport and communication sector benefits from the increased national and international accessibility caused by the construction of the A1 highway. For this sector a priori the biggest impacts were to be expected.

It may then be concluded that the impact of constructing highways on regional employment growth is not significant (except for the transport and communication sector). An explanation for this result may be that the spatial level of analysis used is too high for finding clear spatial economic impacts. It is possible that spatial effects do exist, but that these relate to distribution effects *within* regions. Highway construction may, for example, lead to relocation of firms in a region to sites near an access point. In this case one does not observe an impact at the level of the region as a whole, but if data were available at a lower spatial level one might find relocation effects.

At this low spatial level, impact analysis and regression analysis are used to measure the impact of the opening of the Amsterdam orbital motorway on office rents. It appears that the office rents of locations near the already existing parts of the orbital motorway are higher; they increased at a faster rate than the rents of offices near the newly constructed parts of the orbital motorway. Thus, the office areas near the orbital motorway which had already strong positions in the region benefitted more than the office areas which had a weaker starting position, although their accessibility clearly improved due to the orbital motorway. An important lesson is that the major impacts of the construction of a link in a network will not necessarily occur near that link.

9.7.2 Results Achieved by Entrepreneurial Surveys

The survey approach makes it possible to carry out an analysis at the level of individual firms and investigate spatial effects within regions. In the A1 transport corridor study it appears that transport infrastructure might attract firms. There is a tendency for firms that are relocating to move nearer to the highway. A positive relation between the growth of companies and the distance to the A1 is also found. Zones at a distance of up to 7.5 kilometres from the A1 have on average higher employment growth than zones at longer distances. A difficulty is that all major urban areas in the region are located within a distance of 7.5 kilometres from the highway, so that distance to highway and degree of urbanization are strongly correlated. The zone with a distance of up to 7.5 kilometres from a highway is in all respects most dynamic: in total employment growth, having new companies and relocating companies.

The rather uniform questionnaire used in the three surveys discussed in Chapters 8 and 9 allows us to compare some results of these studies (see Table 9.9). The results for Twente are given for the years 1989 and 1994. The results of 1994 are selected from the data of the A1 transport corridor study, which was held for the eastern part of the country.

Table 9.9 Impact of highway construction on firm performance

	Corop-regions				orbital motor-way	A1 transport corridor
	Twente		Brabant	Leyden		
	1989	1994	1989	1989	1992	1994
Hindrance use highway	28	37	36	56	68	40
Improvement from construction						
- accessibility	--	--	--	--	80	74
- travel time	--	--	--	--	76	78
- inflow/outflow of goods	--	--	--	--	43	54
- costs per unit product	--	--	--	--	19	37
- turnover	--	--	--	--	12	33
Effect of non-construction						
- firm smaller	17	9	2	10	4	8
- less invested	30	22	21	37	10	19
- less personnel	26	20	8	17	6	17

A first finding is that a relatively high percentage of entrepreneurs experience hindrance in the business use of the highway network. The percentage of firms stated that the experience of hindrance decreases with the degree of peripherality. A relatively high percentage of companies in the Amsterdam agglomeration experience hindrance compared with the peripherally located firms of Twente. This result is of course according to the actual congestion levels on the highway network in the Netherlands. It is remarkable that regardless of the opening of the A1 transport corridor, an increasing percentage of firms in Twente experience hindrance in the use of the highway network in 1994 compared to 1989. The explanation is that entrepreneurs in Twente show a stronger orientation to the Randstad area after the opening of the A1 transport corridor, and that in the Randstad itself, congestion has increased. In Twente there is not much congestion on the highway network.

There is a strong similarity in the results of the orbital motorway study and the transport corridor study considering the impact of the construction of the highway and the improvement of accessibility and travel times. In Twente a higher percentage of firms than in Amsterdam experience an improvement in the delivery time of goods, a decrease in costs per unit product and an increase in turnover. These differences can partly be explained by the sectoral structure of both regions. In the Amsterdam region the service and office sector are dominant, whereas the industry and construction sector are dominant in Twente. The industry and construction sector have a much stronger orientation to goods than the office and service sector have.

A final comparison considers the expectations of the firms if the highway had not been constructed. The entrepreneurs stated that in the case of non-construction they expected a strong negative impact on the investment level and the number of employees of their firm. Depending on the region - and the highways constructed in those regions - about 10 to almost 40 percent of the firms expected to have invested less had the infrastructure not been constructed. In 1994 a considerably lower percentage of firms stated an expectation of negative impacts due to non-construction in Twente compared to 1989. In 1989 the highway was new and the entrepreneurs had high expectations of the opportunities this new highway could offer their firms. In 1994 the firms are accustomed to the availability of the highway and their expectations are levelled to more realistic proportions. Nevertheless, it is very important to note that entrepreneurs perceive the construction of a highway as a new opportunity to expand their activities (in investment level and number of employees). This is an indication that government public investments can be effective in provoking private investments. Moreover, it is an indication of the fact that entrepreneurial perceptions of the opportunities of the new infrastructure contribute to the impact of infrastructure on spatial economic development. It is problematic to cover such perceptions in model exercises, so qualitative approaches can enrich the information necessary to reach to a balanced analysis. Moreover, where models often consider a regional level of spatial aggregation, a qualitative approach can be applied at a low level of spatial aggregation and produce useful information on distributive and generative effects.

We plea for an approach in which quantitative and qualitative applications are combined to gain a deeper understanding of the impact of transport infrastructure on spatial economic development.

10 Accessibility and the Valuation of Cities

10.1 Introduction

The scientific interest in the location of firms has a long history. The initial impetus for the development of the classical location theory for the agricultural sector is given by Von Thünen (1821). Later on, Launhardt (1882), Weber (1909) and Predohl (1925) elaborated the theory for the industrial sector and Christaller (1933) and Lösch (1940) for the services and trade sector. In the 1950s those sector-specific location theories were replaced by more general theories; examples are integrated theories of interregional trade and regional factor movements (Isard, 1956), the cumulative causation approach (Myrdal, 1957), the growth pole theory (Perroux, 1958), and the regional development potential theory (Biehl et al., 1975). Many elements of these theories are brought together in the work of Krugman (1991), who pays much attention to the role of economies of scale in regional development patterns.

One of the main assumptions in traditional location theory and modelling is that decisions of actors are based on perfect information. This information serves as an input to the actor's objective function to be maximized. However, in reality information on locational alternatives will be far from perfect; decisions are based on partial information only (see also section 3.3.2). Moreover, actors might show satisfying instead of maximizing behaviour. Such behaviour has impacts on the actual location decisions of firms, since entrepreneurs are not informed about the pros and cons of all location sites which are interesting as locations for his/her firm. Empirical research on the spatial knowledge c.q. perception of entrepreneurs is limited (Pellenbarg,

1982). The impact of transport infrastructure on the attractiveness of a location as an industrial site is important both from a scientific and a policy viewpoint. When it is shown that transport infrastructure is one of the critical success factors for the image/reputation of a region, it becomes appealing to improve the image of certain regions by the construction of transport infrastructure. Of course, the impact of the construction of transport infrastructure on the image of the region will depend on the already existing infrastructure networks and the type of infrastructure to be constructed. A related issue is to which extent the impact of the construction of transport infrastructure on the spatial pattern of economic activities is influenced by the prevailing image of actors concerning the region in which the infrastructure is constructed.

The attractiveness of cities as a location of firms is closely related to productivity. Economic theory suggests a number of reasons for productivity differences among cities, such as differences in the quality of the labour force, the sectoral composition, economies of agglomeration, and the quality of local and non-local public infrastructure.

In the present Chapter we will focus on the final factor mentioned above, but closer inspection will reveal that the attractiveness of cities is related to the agglomeration factor. During the last decade in economic research much attention has been upon the contribution of public infrastructure to productivity at various spatial levels. This has led to a wide range of production function based studies leading to rather mixed results (see Chapters 3 and 4). A tendency can be observed that the original high contributions of public infrastructure to investments at the national level such as found by Aschauer (1989) are now supplemented with lower estimates based on dynamic analysis at lower spatial levels (Kelejian & Robinson, 1997). See also Gramlich (1994) and Sturm and de Haan (1995).

An important feature of most studies is that infrastructure is dealt with in an aggregate way (various types of transport modes plus telecommunications, plus public utilities). Spatial spill-overs are usually ignored. The quality of public infrastructure is typically measured as the size of the public capital stock. A way to overcome some of these limitations would be to introduce accessibility as an indicator of the services offered by non-local infrastructure. Accessibility can be measured in many ways (see Chapter 2 for a review); a common element is that the accessibility of a city is a measure of the potential interactions with other cities. Factors playing a role here are the size of other cities and the transport costs of getting there. For particular firms more specific definitions are needed to define accessibility with respect to potential customers or suppliers of inputs.

In order to deal with accessibility, network features of infrastructure have to be taken into account. This obviously leads to a demand for detailed network data, but with current GIS software, this is no longer a major bottleneck.

An important issue in the measurement of accessibility of cities is the delimitation of the relevant set of potential destinations. Especially given the relatively high rate of growth of international trade between EU members, it is advisable not to ignore destinations abroad. The importance of foreign destinations for city systems depends of course on the particular purpose of the study. In our analysis for the Netherlands, we will compare two approaches, one where the city system is confined to Dutch cities only, and one where larger cities in neighbouring countries are also included.

A particular feature of the accessibility notion is that when measuring a city's accessibility, the accessibility of the city with respect to itself has also to be added. As will be explained and illustrated in the next sections, neglect of the internal accessibility term would have counter-intuitive results.

The inclusion of the internal component of accessibility has an important implication for the study of another productivity-enhancing factor mentioned above: agglomeration economies. Agglomeration economies consist of localization economies which accrue to firms of the same sector because of being in close proximity, and urbanization economies which accrue to firms when they are located in urban areas. In a recent article, Ciccone and Hall (1996) find that agglomeration economies as represented by density of employment are significant. Their result is that in the United States, a doubling of employment density would imply an increase of average labour productivity by around six percent. A background of this agglomeration effect is that in high density areas, average distances are low so that transport costs are lower as well (note that when cities would be seriously plagued by congestion the agglomeration advantage could become a disadvantage). In addition, levels of specialization in high-density areas are higher, leading to higher quality of intermediate inputs and services. Essentially both explanations of the agglomeration effect are based on costs of transport and communication. Hence, the internal component of the accessibility measure is closely related to agglomeration economies. There is a considerable overlap between the two, which has not yet been noted in the literature as far as we know. In the empirical part of this Chapter this overlap will receive further attention.

In this Chapter the relation between the attractiveness of cities as a location for firms and the accessibility via the road network is analyzed in an empirical study for the Netherlands (see Bruinsma & Rietveld, 1998b). In section 2 the valuation of the attractiveness of 67 cities by entrepreneurs is given. In section 3 the accessibility of those cities via the Dutch road network is measured. In section 4 correlations between these measures are studied. Regression analysis is used to explain the valuation of cities by a number of factors of which the accessibility via the road network is one of the explanatory variables.

10.2 The Valuation of Dutch Cities as Industrial Sites

Meester and Pellenbarg (1986) investigated the subjective valuation of cities as location sites by entrepreneurs in the Netherlands. Their research was focused on the spatial cognition of entrepreneurs. Cognition is defined as the knowledge of spatial structure, entities and relations. The knowledge need not be complete or correct to be regarded as cognition (cf. Hart & More, 1973). Spatial cognition is often confused with perception. However, perception is limited to the subjective interpretation of objective reality as directly experienced by the individual, whereas cognition also involves processes such as thinking, representation, arguing, judging, and remembering.

This 'total knowledge' is called cognition in psychology. Especially the visualization of cognition in mental maps has attracted the attention of geographers. A mental map is defined as the representation of the spatial structure of a city or region in the human brain (Dietvorst et al., 1984). More complex is the term 'image', in which non-spatial characteristics (social, political, economic) are considered as well as spatial characteristics of geographical entities.

In 1983 Pellenbarg and Meester distributed 1800 questionnaires having a national orientation among firms with at least 10 employees. In the questionnaire a map of the Netherlands was enclosed upon which 67 cities were marked. The question was '*Suppose you have to relocate your firm - whatever the reason might be - how do you value the cities on the map as new locations for your firm?*'. The entrepreneurs had to give their values on a five-point scale (++, +, 0, -, --). The chosen cities are well-known and as evenly spread as possible over the country. From the 642 returned questionnaires, 388 were retained after a correction for completeness of the answers, number of employees, and spatial orientation of the firm (only firms with more than 10 employees and having a national orientation are included). An index is constructed in which the city receiving the highest mean score (Utrecht, in the centre of the Netherlands) received the value 100, and the city having the lowest mean score (Winschoten, in the province of Groningen) got the value 0 (Table 10.1, column 1)[1]. See Figure 10.1 for the location of the Dutch provinces.

The general picture is that cities in the province of Utrecht score best, followed by the cities of the Randstad and the corridor Breda - Den Bosch - Eindhoven in Noord-Brabant near the Belgian border. The Randstad area is defined as the area between the cities Utrecht - Amsterdam - Haarlem - Den Haag - Rotterdam - Gouda - Utrecht. However, more interesting is the sharp decline in scores in the provinces of Zeeland and Limburg in the south, the upper part of Noord-Holland, and the four northern provinces of Groningen, Friesland, Drenthe, and Overijssel. Those sharp declines show a rather sudden decrease in the valuation of cities as attractive location sites (Meester & Pellenbarg, 1986). In their analysis of the mental maps of the entrepreneurs of each province, Meester and Pellenbarg arrive at two intriguing conclusions. First, there appears to be a kind of neighbourhood effect, meaning a preference for cities within each entrepreneurs own province. Second, entrepreneurs located in rather peripheral cities give higher valuations to other peripherally-located cities than their collegues in centrally-located cities do.

The scores of the cities express the mean valuation of the cities. The variance in the valuation of cities by entrepreneurs may of course be substantial. To analyse the basic dimensions underlying the valuations by the entrepreneurs, Meester and Pellenbarg (1986) used a principal component analysis with varimax rotation. Factor analysis replaces a large number of - linked - variables by a limited number of constructed variables, which are by definition independent of each other. By varimax, the factors are rotated in

---

[1] A more accurate analysis could have been conducted by means of an ordered logit model based on individual observations. However, since the original data are no longer available at the micro level, the analysis has been carried out at an aggregate level, i.e., the average score of cities across all respondents on a scale from 0 to 100.

Table 10.1 Valuation indices of cities as a location site for firms

Pro.	City	1	2	3	4	Pro.	City	1	2	3	4
Gr	Groningen	14	.716	15	25	NH	Alkmaar	28	.491	28	32
Gr	Delfzijl	3	.833	2	16	NH	Den Helder	12	.720	9	18
Gr	Winschoten	0	.780	1	17	NH	Hoorn	25	.514	22	29
Gr	Veendam	1	.788	3	18	NH	Hilversum	79	-.165	78	72
Fr	Leeuwarden	9	.769	8	19	ZH	Den Haag	61	.181	55	68
Fr	Harlingen	6	.813	0	16	ZH	Leiden	57	.119	52	59
Fr	Sneek	6	.777	1	14	ZH	Gouda	73	-.045	71	69
Fr	Heerenveen	9	.747	4	16	ZH	Gorinchem	69	.043	62	69
Fr	Drachten	8	.771	5	17	ZH	Dordrecht	65	.194	57	63
Dr	Assen	16	.681	12	24	ZH	Rotterdam	82	.171	73	82
Dr	Emmen	14	.691	13	27	Ze	Middelburg	8	.751	6	22
Dr	Hoogeveen	20	.620	20	30	Ze	Vlissingen	11	.748	7	24
Dr	Meppel	22	.554	19	31	Ze	Terneuzen	5	.764	5	22
Ov	Zwolle	47	.364	48	46	Ze	Goes	8	.753	5	24
Ov	Ommen	27	.423	30	34	Ze	Zierikzee	8	.752	5	22
Ov	Almelo	28	.487	30	43	NB	Den Bosch	64	.099	69	74
Ov	Hengelo	30	.503	32	47	NB	Oss	52	.091	61	60
Ov	Enschede	28	.512	30	45	NB	Helmond	43	.294	51	59
Ov	Deventer	49	.248	54	54	NB	Eindhoven	54	.294	59	69
Ge	Arnhem	68	.003	69	72	NB	Tilburg	51	.255	52	63
Ge	Ede	71	-.148	73	80	NB	Breda	53	.283	54	67
Ge	Harderwijk	57	.092	60	62	NB	Roosendaal	35	.462	34	44
Ge	Apeldoorn	66	.013	75	74	NB	Bergen op Z.	26	.535	5	38
Ge	Zutphen	43	.245	51	54	Li	Maastricht	20	.585	25	39
Ge	Winterswijk	23	.500	27	40	Li	Heerlen	18	.580	25	41
Ge	Doetinchem	37	.260	42	48	Li	Sittard	18	.584	23	40
Ge	Nijmegen	60	.119	66	60	Li	Roermond	27	.514	37	46
Ge	Tiel	65	-.105	67	68	Li	Weert	28	.485	35	48
Ut	Utrecht	100	-.244	100	98	Li	Venray	35	.342	43	49
Ut	Amersfoort	90	-.246	94	89	Li	Venlo	34	.436	42	53
NH	Amsterdam	68	.148	61	69	Fl	Almere	72	.023	67	62
NH	Haarlem	57	.223	50	60	Fl	Lelystad	54	.266	48	46
NH	IJmuiden	38	.463	34	42	Fl	Emmeloord	22	.518	18	33
NH	Zaandam	47	.318	43	42						

Pro. = Dutch province (see Figure 10.1)

1 = 1983 (n=388) 2 = loads on first factor 3 = 1983 (n=152) 4 = 1989

Figure 10.1 The Dutch provinces

such a manner that only heavy loads remain. So each factor is linked to a limited number of variables. This factor analysis results in three components. The first - explaining 25 % of the variance - is interpreted by Meester and Pellenbarg as an accessibility component in the notion of access to markets (see Table 10.1, column 2). The second and third components - explaining 22 and 13 % of the variance, respectively - are interpreted as quality of life and agglomeration advantage. The interpretation of the first - accessibility - component is significant in this context. If this interpretation is correct, the loads of this factor should correlate stronger with the scores of the accessibility via the road network than with the scores of the valuation of cities as location site. This will be tested in section 4.

The exercise is repeated in 1989. In total, 152 entrepreneurs from the 388 entrepreneurs of the first project participated in this second exercise (see

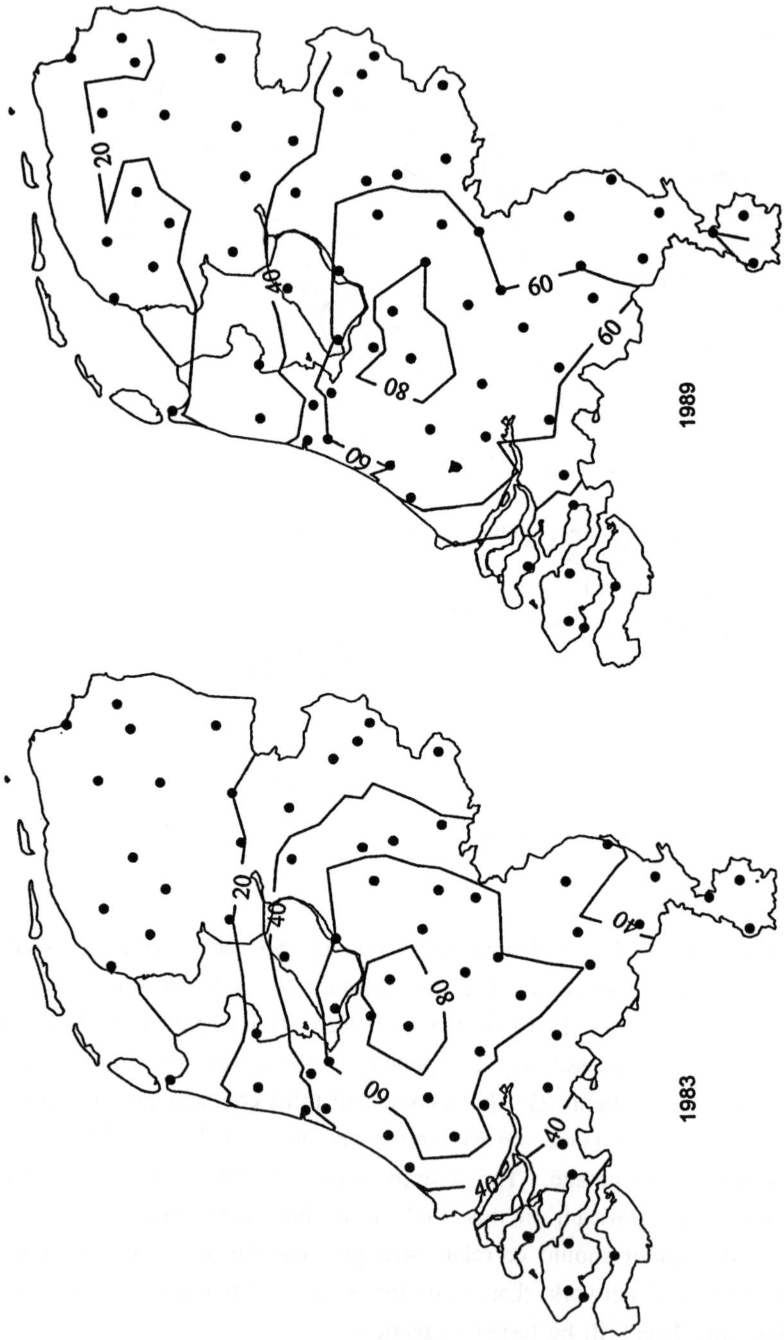

Figure 10.2 Valuation of cities as location sites of firms (1983 lower picture, 1989 upper picture)

Pellenbarg et al., 1993). To compare the first and second exercise, the results of the first project were re-measured only for the subset of entrepreneurs also involved in the second exercise. This has led to minor changes in the score of some cities (see Table 10.1, column 3). Next, the scores for 1989 are re-measured to the scale of 1983 (Table 10.1, column 4). By using the same scale, a direct comparison between the results of 1983 and 1989 is allowed.

The results are shown in Figure 10.2. Compared to 1983, the peak in 1989 is lower and might be explained by the increasing congestion in the most preferred area. Not only is the peak lower but the isolines also tend to move outward. This is an indication of a general rise in the valuation of cities as location sites by entrepreneurs. In the period of 1983-1989, the average score per city increased by nearly nine percent.

The studies described above grant the opportunity for analyzing the relation between transport infrastructure and the subjective location preferences of entrepreneurs in the Netherlands for the period 1983-1989. In the next section the accessibility via the road network is operationalized and accessibility indices are constructed.

10.3 Accessibility of the Cities via the Road Network

10.3.1 Operationalization Accessibility

Accessibility concerns the geographical location of a city in relation to all other cities. If the mean distance to all other cities is relatively short, then the location of the city within the network is relatively favourable. A reasonably well-located city within the city network has a better potential to attract national or internationally-oriented firms compared with an unfavourably located city within the network. For the conceptualization and operationalization of the concept and a review of accessibility concepts, we refer to Chapter 2. In the present Chapter we give a short summary of the approach we apply in this case study. Our conceptualization of accessibility considers the travel times or distances between cities via the road network. The following simple gravity approach is used:

$$A_i = \Sigma_j \; 1/T_{ij}^c \qquad (1)$$

The accessibility (A_i) of a city is measured as the summation (in the sense of a harmonic mean) of all travel times to all relevant destinations (T_{ij}). The gravity parameter c in general receives the value 1. Since the interaction between cities increases when the population size of the cities increases, it is reasonable to introduce weighting by the population size of the cities. This leads to the following equation in which P_j represents the population size of the destination j;

$$A_i = \Sigma_j \; P_j/T_{ij}^c \qquad (2)$$

The fact that the interaction within the city i is ignored would imply that relatively large cities score relatively low. This can be corrected by including the internal interaction potential into the gravity model. The internal interaction potential is depending on the size of one's own city (P_i) and the mean internal travel time (t_i). The gravity model can now be described in the following form:

$$A_i = P_i/t_i^c + \Sigma_{j\neq i} \; P_j/T_{ij}^c \qquad (3)$$

The internal travel time depends on the surface of the city (O) and the average travel speed (s). Conforming to Rich (1980), the average internal distance (d) can be measured as follows (see also section 5.2):

$$d = \sqrt{(O/\pi)}/2 \qquad (4)$$

After measuring the internal distance, the mean internal travel time (t_i) can be calculated on the basis of the average internal travel speed. As we will show in our empirical application, the outcomes for the accessibility index depend considerably on the way the internal accessibility is modelled. This is a topic that has not received much attention in the literature. Therefore, we deal with this issue in our application.

10.3.2 The Accessibility Measures

The accessibility of the 67 Dutch cities via the road network is measured for the years 1983 and 1989. Conforming to the approach of Meester and Pellenbarg (1986), the results were scaled from 100 for the most accessible city to 0 for the least accessible city.

Centrality indices. In a first series of calculations only the distance via the road was included (equation 1). The results are shown in Table 10.2 (column 1) and Figure 10.3. Since the index is completely dependent on the distances via the road network, we will call this index the centrality index. If we compare this index with the index in which the entrepreneurs value the cities as location sites, a few differences appear. Amsterdam is the best scoring city in terms of centrality instead of Utrecht. Moreover, the peak of the centrality index is flatter; the 80 and 60 isolines cover a much larger area. An objection to this centrality index might be that international relations are excluded. This might be a disadvantage for cities in border regions. To analyze the impact we added all foreign cities within 100 kilometres via the road network from the border and with a population size at least comparable to the Dutch cities. The cities added were Antwerp, Brussels, Liege and Gent in Belgium and Aken, Köln, Bonn, Duisburg, Düsseldorf, Dortmund, Wuppertal, Osnabrück, Munster and Essen in Germany. Conforming to Chapter 5, the interactions with foreign countries are reduced by .333 to indicate the barrier of an international border. The impact of the foreign cities on the scores of the Dutch cities is rather small and limited to cities in Limburg and to a lesser extent Noord-Brabant (see Table 10.2, column 2). No impacts are measured for the northern cities near the border. This is caused by the fact that there are no large German cities within 100 kilometres of the border in the northern area of the Netherlands.

Over the period 1983-1989, the extensions in the Dutch main road network were limited to quality improvements or upgrading of existing roads. The centrality index for this period remained the same.

As mentioned in section 3, Pellenbarg et al. (1993) concluded that the flattening of the peak in the central part of the Netherlands between 1983-1989 might be explained by increasing congestion in this area. We added congestion to the model by increasing travel time according to the share of each link between the cities that was located inside the Randstad. When the

Table 10.2 Centrality indices

Pro.	City	1	2	3	4	Pro.	City	1	2	3	4
Gr	Groningen	28	27	30	29	NH	Alkmaar	47	47	39	38
Gr	Delfzijl	9	8	11	10	NH	Den Helder	11	10	10	9
Gr	Winschoten	15	14	17	16	NH	Hoorn	49	49	41	40
Gr	Veendam	27	26	29	28	NH	Hilversum	91	92	91	90
Fr	Leeuwarden	34	33	35	35	ZH	Den Haag	50	50	42	42
Fr	Harlingen	21	21	23	22	ZH	Leiden	56	56	46	45
Fr	Sneek	42	42	44	43	ZH	Gouda	69	69	60	60
Fr	Heerenveen	48	48	50	49	ZH	Gorinchem	80	81	75	75
Fr	Drachten	38	38	40	39	ZH	Dordrecht	67	68	64	64
Dr	Assen	38	38	40	39	ZH	Rotterdam	70	70	65	64
Dr	Emmen	30	30	32	31	Ze	Middelburg	39	39	39	39
Dr	Hoogeveen	58	58	60	59	Ze	Vlissingen	38	37	38	37
Dr	Meppel	68	68	69	69	Ze	Terneuzen	0	0	0	0
Ov	Zwolle	79	79	80	79	Ze	Goes	28	28	27	27
Ov	Ommen	71	71	72	72	Ze	Zierikzee	24	25	24	24
Ov	Almelo	54	55	56	56	NB	Den Bosch	80	82	81	82
Ov	Hengelo	66	67	68	68	NB	Oss	78	80	80	81
Ov	Enschede	47	48	48	49	NB	Helmond	69	72	70	73
Ov	Deventer	95	95	96	96	NB	Eindhoven	68	71	69	72
Ge	Arnhem	92	94	92	93	NB	Tilburg	60	63	61	62
Ge	Ede	88	90	88	89	NB	Breda	62	64	57	59
Ge	Harderwijk	84	84	84	84	NB	Roosendaal	54	55	50	51
Ge	Apeldoorn	99	100	100	100	NB	Bergen op Z.	43	46	40	42
Ge	Zutphen	89	90	90	90	Li	Maastricht	6	14	8	15
Ge	Winterswijk	30	31	31	32	Li	Heerlen	8	16	10	18
Ge	Doetinchem	53	54	54	54	Li	Sittard	20	25	22	26
Ge	Nijmegen	76	78	76	77	Li	Roermond	32	37	34	38
Ge	Tiel	70	71	69	69	Li	Weert	39	42	40	43
Ut	Utrecht	98	99	92	92	Li	Venray	46	51	48	52
Ut	Amersfoort	97	97	93	93	Li	Venlo	37	43	38	45
NH	Amsterdam	100	100	82	81	Fl	Almere	76	76	70	70
NH	Haarlem	82	82	64	63	Fl	Lelystad	63	63	61	60
NH	IJmuiden	64	63	50	49	Fl	Emmeloord	58	58	59	59
NH	Zaandam	83	68	67	66						

Pro. = Dutch province (see Figure 10.1)

1 = 1983/1989 2 = foreign cities added 3 = congestion added 5 = both added

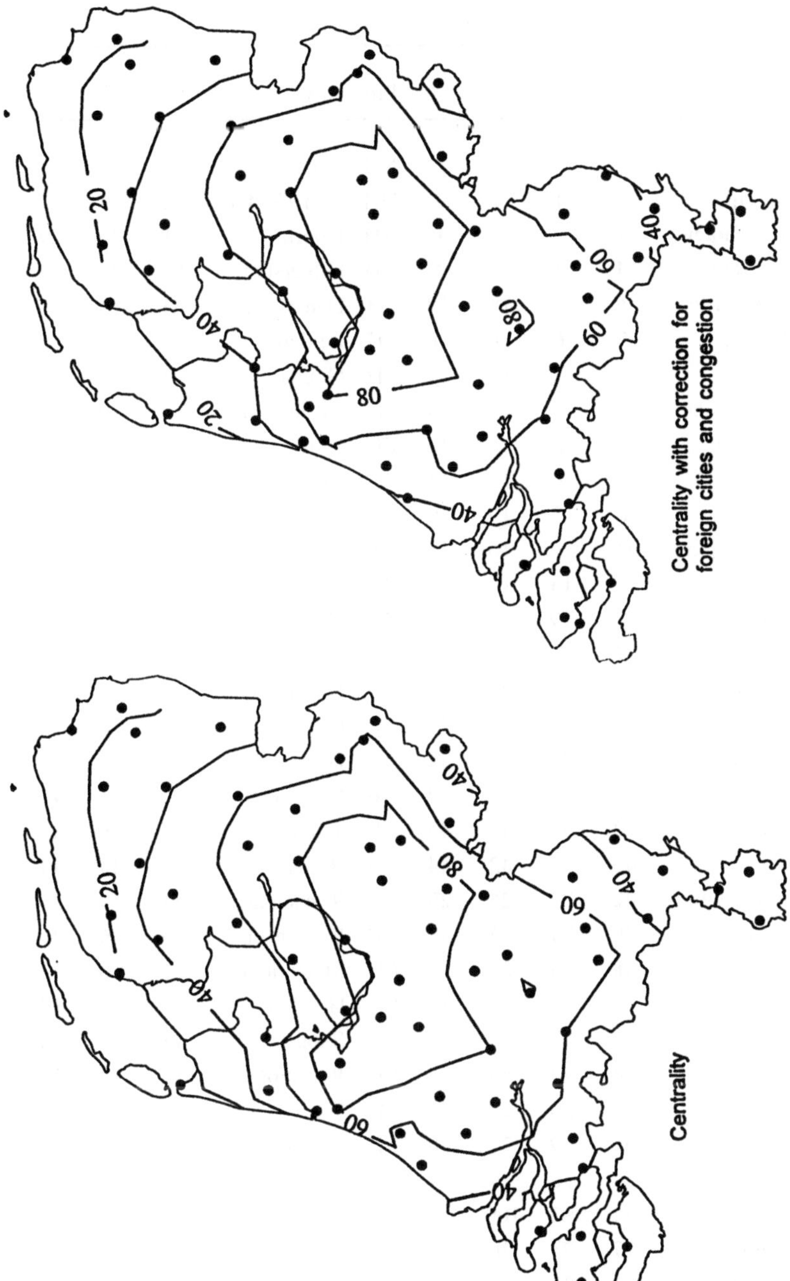

Figure 10.3 Centrality of cities via the Dutch road network (centrality 1989 lower picture, centrality with correction for foreign cities and congestion 1989 upper picture)

Table 10.3 The impact of the mass indicator

Pro.	City	1	2	3	4	Pro.	City	1	2	3	4
Gr	Groningen	1	24	27	27	NH	Alkmaar	38	33	36	38
Gr	Delfzijl	0	1	1	1	NH	Den Helder	11	14	17	17
Gr	Winschoten	1	2	0	0	NH	Hoorn	38	26	29	32
Gr	Veendam	4	1	5	5	NH	Hilversum	65	42	53	53
Fr	Leeuwarden	6	14	18	19	ZH	Den Haag	41	100	78	77
Fr	Harlingen	9	6	3	4	ZH	Leiden	64	56	55	56
Fr	Sneek	13	11	10	10	ZH	Gouda	71	47	51	52
Fr	Heerenveen	13	5	12	13	ZH	Gorinchem	60	30	37	38
Fr	Drachten	11	2	9	10	ZH	Dordrecht	61	39	54	54
Dr	Assen	10	8	13	14	ZH	Rotterdam	41	77	87	86
Dr	Emmen	6	5	19	20	Ze	Middelburg	10	9	11	11
Dr	Hoogeveen	16	9	16	16	Ze	Vlissingen	8	10	12	11
Dr	Meppel	22	13	13	14	Ze	Terneuzen	2	0	5	5
Ov	Zwolle	26	19	30	31	Ze	Goes	11	4	9	10
Ov	Ommen	26	7	13	14	Ze	Zierikzee	17	4	5	5
Ov	Almelo	21	19	23	23	NB	Den Bosch	45	33	42	42
Ov	Hengelo	30	22	31	31	NB	Oss	42	26	32	33
Ov	Enschede	15	20	33	33	NB	Helmond	40	24	33	34
Ov	Deventer	39	27	33	34	NB	Eindhoven	29	41	47	47
Ge	Arnhem	43	33	47	47	NB	Tilburg	35	36	46	46
Ge	Ede	47	21	42	44	NB	Breda	41	33	44	45
Ge	Harderwijk	42	20	28	29	NB	Roosendaal	31	19	27	28
Ge	Apeldoorn	40	23	47	48	NB	Bergen op Z.	25	16	21	22
Ge	Zutphen	38	23	26	26	Li	Maastricht	4	20	22	22
Ge	Winterswijk	14	4	10	10	Li	Heerlen	6	18	19	20
Ge	Doetinchem	23	12	18	19	Li	Sittard	11	16	20	20
Ge	Nijmegen	34	44	44	44	Li	Roermond	15	12	14	14
Ge	Tiel	42	20	27	28	Li	Weert	22	9	18	18
Ut	Utrecht	61	69	70	70	Li	Venray	23	8	17	18
Ut	Amersfoort	58	36	49	51	Li	Venlo	15	15	19	20
NH	Amsterdam	52	92	100	100	Fl	Almere	58	22	37	45
NH	Haarlem	77	71	70	70	Fl	Lelystad	31	12	27	29
NH	IJmuiden	61	33	45	45	Fl	Emmeloord	22	6	18	18
NH	Zaandam	100	58	80	80						

Pro. = Dutch province (see Figure 10.1)

1 = no mass 2 = surface 3 = population density '83 4 = population density '89

percentage of the link between two cities inside the Randstad is 10 to 24 %, we increased the travel time by 5 %. If the percentage is 25-49, 50-74 or over 75 %, we then increased the travel time by 10, 15 and 20 %, respectively. There are still two roads (near Breda and Vianen) outside the Randstad which were in the top 20 of the most congested roads in 1989. Links using these roads also received a penalty of 5 %. The results of this adding of congestion is twofold. The scores of cities inside or close to the Randstad area decrease; the relative accessibility of cities in Limburg slightly increases (see Table 10.2, column 3 versus column 1).

A final accessibility index is calculated in which foreign cities and congestion are added (see Figure 10.3 and Table 10.2, column 4). Compared with the index without those additions, the change in scores of the Randstad cities is heavily depends on the congestion, whereas the scores of the cities in Limburg and to a lesser extent Noord-Brabant, are increasing by both corrections. An exception has to be made for Breda in Noord-Brabant. The congested road near that city has a negative impact on its accessibility which is larger than the positive impact of being located near some Belgian cities.

Accessibility indices. Until now, the accessibility of the cities is measured only by distances via the road network. A first extension is given in equation (2), where weighing by the population size takes place (Table 10.3, column 1). In this equation the weighing is restricted to the population size of the destinations. The internal interaction within the city itself is neglected. There are two objections against this procedure. The first objection is that it leads to an underestimation of the importance of large cities. When a city has a relatively large share in the total population, the population of all other cities to interact with is relatively small, so the final score will be rather small. This effect is shown in Chapter 5 for the accessibility of Paris and London in the European infrastructure networks. However, in this case study there is no city with an unevenly large share in total population, so the impact will be limited. The second objection has a larger impact. This objection regards to the fact that the accessibility of relatively small cities located near a large urban area is overestimated. The impact is shown for Zaandam. This city is located near Amsterdam (11 kilometres, nearly 700,000 inhabitants) and is one of the best accessible cities. The share of Amsterdam in the total score of Zaandam is 46 %. Therefore, the accessibility of Zaandam heavily depends on its nearness to Amsterdam. However, in calculating the accessibility of

Amsterdam we neglect its size.

Because of these problems, equation 3 is used to calculate the relative accessibility of cities within the Dutch road network when the internal interaction potential of cities is also taken into account. The mean internal distance has first been quantified by using the surface of the municipalities in equation 4 (Table 10.3, column 2). This appeared to be an unlucky choice, since there are large differences in the share of the built-up area within municipalities. For example, the built-up area of Den Haag covers nearly all the surface of the municipality. Using the surface of the municipality to calculate the mean travel distance leads to a share of the internal interaction potential in the total score of 73 % in the case of Den Haag. This makes Den Haag the best accessible city. On the other hand, the share of the internal interaction potential in the total score of the municipality of Apeldoorn is only 27 % and it is ranked in 26th place. The built-up area of this municipality is relatively small compared to its surface.

It is important to note that two factors influence the size of the share of the internal interaction potential: the size of the city and the relative position of the city within the road network. The population size of Den Haag is larger than that of Apeldoorn, what suggests a justification for the larger share of Den Haag. However, Den Haag is also more closely located to a number of other large cities within the Randstad. Because relatively short distances lead to strong interactions in spatial interaction models the external accessibility might also be expected to be higher than that of Apeldoorn. This conflicts with the relatively high share of the internal interaction potential for Den Haag. We conclude that the use of the surface of municipalities does not lead to satisfactory results. Corresponding with Chapter 5, the size of the urban area is calculated by assuming an urban population density of 2,000 inhabitants per km^2 (see Table 10.3, column 3). Den Haag is now ranked fourth with a share of the internal interaction potential of 59 %, and Apeldoorn is ranked fourteenth and the share of the internal interaction potential has increased from 27 to 45 %.

The same approach is used to measure the accessibility of cities for 1989 (Figure 10.4 and Table 10.3, column 4). The differences in the indices are not explained by improvements in the road network (see the centrality index), but by different growth rates of cities during the period 1983-1989. The differences in population growth rates influence both the external accessibility and the internal interaction potential. A growth rate which is higher compared

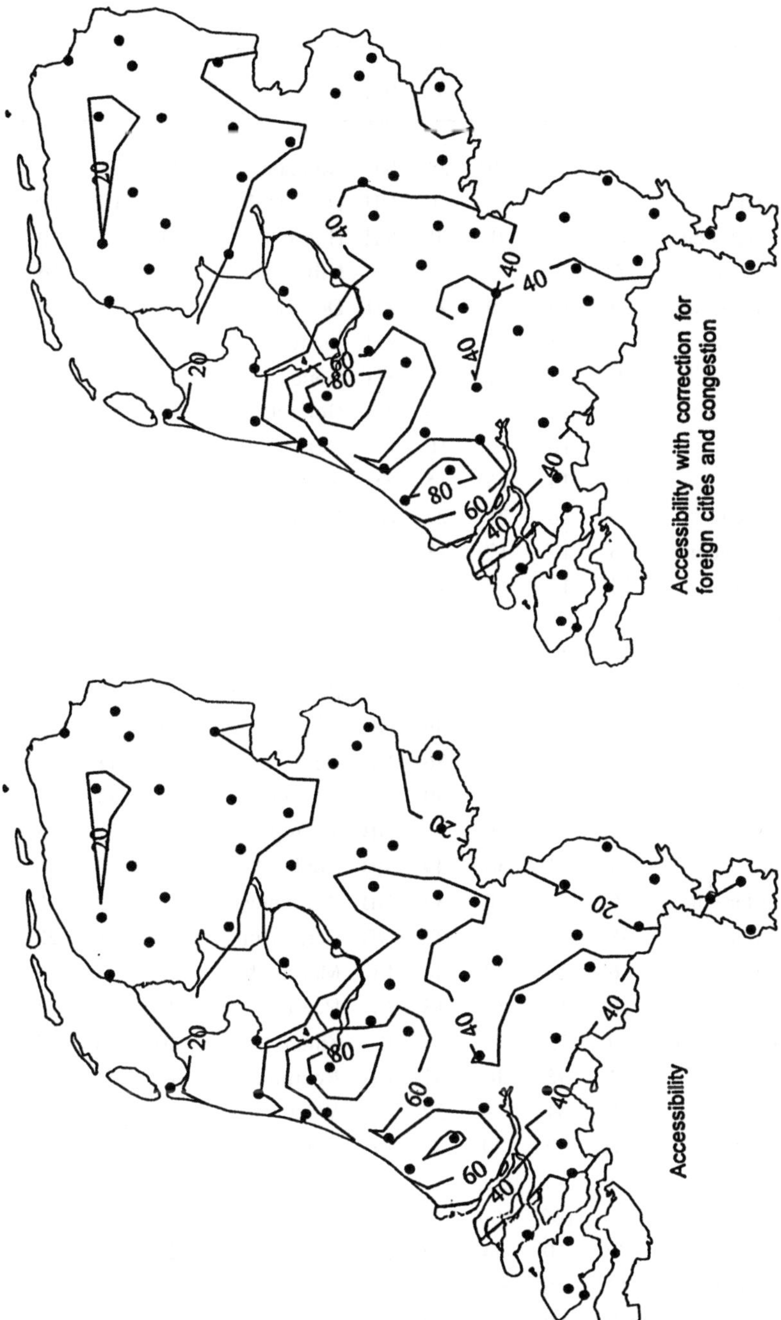

Figure 10.4 Accessibility of cities via the Dutch road network (accessibility 1989 lower picture, accessibility with correction for foreign cities and congestion 1989 upper picture)

Table 10.4 Accessibility indices

Pro.	City	1	2	3	4	Pro.	City	1	2	3	4
Gr	Groningen	27	27	29	29	NH	Alkmaar	38	38	35	35
Gr	Delfzijl	1	1	1	1	NH	Den Helder	17	17	18	17
Gr	Winschoten	0	0	0	0	NH	Hoorn	32	32	29	30
Gr	Veendam	5	5	5	6	NH	Hilversum	53	55	56	58
Fr	Leeuwarden	19	19	20	20	ZH	Den Haag	77	78	79	79
Fr	Harlingen	4	4	4	4	ZH	Leiden	56	57	55	56
Fr	Sneek	10	11	11	11	ZH	Gouda	52	54	51	53
Fr	Heerenveen	13	14	14	14	ZH	Gorinchem	38	41	37	41
Fr	Drachten	10	10	10	10	ZH	Dordrecht	54	57	56	59
Dr	Assen	14	14	15	15	ZH	Rotterdam	86	88	90	91
Dr	Emmen	20	20	21	22	Ze	Middelburg	11	14	11	14
Dr	Hoogeveen	16	17	17	18	Ze	Vlissingen	11	13	12	13
Dr	Meppel	14	16	15	16	Ze	Terneuzen	5	7	5	7
Ov	Zwolle	31	33	33	35	Ze	Goes	10	12	10	12
Ov	Ommen	14	16	15	17	Ze	Zierikzee	5	8	5	7
Ov	Almelo	23	24	24	26	NB	Den Bosch	42	47	44	49
Ov	Hengelo	31	32	32	34	NB	Oss	33	37	35	40
Ov	Enschede	33	34	35	36	NB	Helmond	34	41	36	43
Ov	Deventer	34	37	36	39	NB	Eindhoven	47	53	50	56
Ge	Arnhem	47	51	49	53	NB	Tilburg	46	51	48	54
Ge	Ede	44	47	46	49	NB	Breda	45	50	46	52
Ge	Harderwijk	29	32	30	33	NB	Roosendaal	28	32	28	33
Ge	Apeldoorn	48	50	50	153	NB	Bergen op Z.	22	28	22	29
Ge	Zutphen	26	29	27	30	Li	Maastricht	22	33	24	36
Ge	Winterswijk	10	13	11	13	Li	Heerlen	20	31	21	33
Ge	Doetinchem	19	22	20	23	Li	Sittard	20	28	22	30
Ge	Nijmegen	44	48	46	51	Li	Roermond	14	21	15	23
Ge	Tiel	28	31	28	32	Li	Weert	18	25	19	26
Ut	Utrecht	70	72	70	73	Li	Venray	18	25	19	27
Ut	Amersfoort	51	53	52	55	Li	Venlo	20	30	21	33
NH	Amsterdam	100	100	100	100	Fl	Almere	45	46	43	45
NH	Haarlem	70	70	64	65	Fl	Lelystad	29	30	29	30
NH	IJmuiden	45	46	40	41	Fl	Emmeloord	18	20	19	20
NH	Zaandam	80	80	74	74						

Pro. = Dutch province (see Figure 10.1)

1 = 1989 2 = foreign cities added 3 = congestion added 4 = both added

with the growth rate of the best accessible city (Amsterdam) results in an increasing share of the internal interaction potential. An example is Almere. The share of the internal interaction potential of this city is increased from 24 to 31 % over this period. In all other cities much smaller changes occured. Along with an improved accessibility from a relatively rapid increase in its population size, a city can improve its accessibility when it is near a fast-growing city. In this case the share of the external accessibility increases.

A comparison of the centrality index with this weighted accessibility index shows a clear shift of the most accessible centrally located area towards the major cities, especially within the Randstad. The accessibility of the largest population centres (Amsterdam, Rotterdam and Den Haag) is strongly dependent on the share of the internal interaction potential (61, 62 and 59 %, respectively). The accessibility of Groningen also heavily depends on its population size. The share of this city's internal interaction potential in total accessibility is the highest with 63 %. On average, the share of the internal interaction potential is 40 %.

Conforming to the centrality approach, the weighted accessibility index also has been calculated when foreign cities, congestion, and the combination of foreign cities and congestion are added to the gravity model (see Table 10.4).

By adding the foreign cities, the accessibility of the cities in the province of Limburg increases even more as was the case in the centrality index. The cities of Limburg are not only close to a number of German cities, but the German cities also have relatively large population sizes. The impact of congestion is less dramatic for the accessibility of the Randstad cities. The addition of the congestion factor in the Randstad area has no impact on the internal interaction potential of cities in this area. In the calculation, where both factors are added, the impact of foreign cities is most important (see Figure 10.4). The accessibility of all cities in Limburg is now above the 20 isoline. More important - contrary to the centrality index - is that in the case of Breda, the positive impact of its location near Belgian cities now exceeds the negative impact of a congested road nearby.

10.4 Road Infrastructure and the Valuation of Location Sites

In this section the relation between the accessibility of cities via road infrastructure and their valuation as a location site for firms is investigated by correlation and regression analysis.

10.4.1 Correlation Analysis

In Table 10.5 the correlations are given between the valuation of cities as location sites for firms and their relative positions within the Dutch road network. The correlation coefficient shows the relative strength of the relation between the two factors. In Table 10.5 the Pearson correlations, based on the cities' index value, and the Spearman correlations, based on the ranking of the cities, are given. The correlations found are all rather high. In general, the correlations are slightly higher when rank numbers are considered compared with the quantitative values.

The loads on factor 1 of the factor analysis of Meester and Pellenbarg (see section 2) correlates better with the centrality index than with the accessibility index. This underpins the interpretation that Meester and Pellenbarg (1986) gave: factor 1 reflects the physical accessibility. The impact of the addition of foreign cities and/or congestion into the model only leads to minor changes in the correlation coefficients. In case of the centrality indices, the correlation coefficient - after adding foreign cities - is slightly higher and - after adding congestion - slightly lower. When both are added the impact of foreign cities and congestion neutralize one another. This result holds true for the indices based on the ranking of the cities as well as for the indices based on the scores of the cities. Thus, the valuation of cities is not merely influenced by their position in the Dutch road network; international links receive a certain weight as well. Congestion does not seem to play a role in the valuation in the years our research examined.

The impacts of foreign cities and/or congestion are less straightforward in the case of the accessibility indices. So no real indication is found that internationalization and congestion changed the valuation of cities as location sites for companies in the period 1983-1989. However, the situation might

Table 10.5 Correlations between the valuation of cities as location sites and their relative position in the road network

		Valuation of cities as location sites for firms			
		1983 (n=388)	factor 1 (n=388)	1983 (n=152)	1989 (n=152)
Centrality indices					
1983 = 1989	index	.832	.829	.818	.763
	rank	.844	.843	.831	.773
1989 foreign cities included	index				.790
	rank				.794
1989 congestion included	index				.742
	rank				.751
1989 both included	index				.766
	rank				.775
Accessibility indices					
1983	index	.807	.708	.746	
	rank	.868	.802	.835	
1989	index				.802
	rank				.834
1989 foreign cities included	index				.802
	rank				.856
1989 congestion included	index				.790
	rank				.845
1989 both included	index				.827
	rank				.869

have changed rather drastically since then. The ongoing growth of congestion, especially in the Randstad area, and the creation of one common European market by the end of 1992, might have had strong impacts on the valuation of cities as location sites for firms in the period after 1989.

The correlation analysis shows that although there are some differences in the correlation coefficients, those differences are rather small. The correlations found appear rather insensitive to changes in definition. However, one has to be aware that until now, only mono-causal relations have been explored. In the next subsection the focus is on a multivariate approach to investigate the valuation of cities as location sites for firms.

10.4.2 Regression Analysis

In the regression analysis the valuation of cities as location sites for firms is explained by a number of variables. The explanatory variables are chosen as close as possible to the factors Meester and Pellenbarg (1986) deduced from open interview sessions with entrepreneurs. In those open sessions entrepreneurs had to state the most important factors in the search for a new location site for their company. The factors mentioned are given and explained in Table 10.6. Of all factors mentioned 41 % refers to 'location', 'infrastructure', and 'accessibility'. Other factors often mentioned are 'bond with the region' and 'characteristics of the building/site' (both about 10 %).

Table 10.6 Location factors according to entrepreneurs

location	strategic, central location to suppliers and customers
infrastructure	roads, rail roads, parking lots
accessibility	accessibility not related to 'location' and 'infrastructure'
bond with the region	historical bond, personnel not prepared to relocate
image building/site	size, quality, representativity
prices real estate	land prices, construction prices, rents
government	subsidies, construction regulations, licenses, civil servants
labour market	quality and quantity of personnel
agglomeration advantages	service sector
competitive considerations	competitive advantages and disadvantages
mentality population	labour mentality
environmental quality	natural parks, landscape, crime

Source: Meester and Pellenbarg, 1986

In Table 10.7 is the translation of those factors into variables of which data is available. For the first three factors deduced by Pellenbarg and Meester, the centrality and accessibility indices given in section 3 are used. It is difficult to find data at the disaggregated level of municipalities for most of the other factors. For 1983 only data on population size (as an indication of agglomeration advantages) and investment subsidies is available. Fortunately, for 1989 there is also data available on unemployment and vacancies at the municipality level, so we can calculate the unemployment/vacancy ratio. The average price of office buildings - in m^2 prices - at the municipality level is measured for 46 cities on the basis of supply-side information given by a real estate journal (VGM, 1990). The crime variable is based on county court district data (CBS, 1989).

Table 10.7 Operationalization of location factors

location factor	operationalization
location	▼
infrastructure	centrality and accessibility indices
accessibility	▲
bond with the region	-
image building/site	-
prices real estate	square metre prices of office buildings
government	investment subsidies
labour market	unemployment/vacancy ratio
agglomeration advantages	population size
competitive considerations	-
mentality population	-
environmental quality	crime in county court district

The regression analysis is applied to:

$$y_n = \alpha_0 + \alpha_1 x_{1n} + \alpha_2 x_{2n} + \alpha_3 x_{3n} + \alpha_4 x_{4n} + \alpha_5 x_{5n} + \alpha_6 x_{6n} \tag{5}$$

where:

y_n = value of city n as location site
x_{1n} = centrality or accessibility of city n
x_{2n} = population size of city n
x_{3n} = unemployment/vacancy ratio of city n
x_{4n} = m^2 price of offices in city n
x_{5n} = percentage investment subsidy in city n
x_{6n} = crime in county court district in which city n is located

For the year 1989 sixteen regression analyses have been carried out. The centrality index and the accessibility index have also been used when foreign

Table 10.8 Regression analysis of valuation of cities by various definitions of roads

Centrality index, without foreign cities and congestion

	office prices excl. (n=62)		office prices incl. (n=45)	
	coefficient	standard error	coefficient	standard error
constant	20.65	5.56	11.63	13.38
road infrastructure	0.510	0.069	0.439	0.087
population size[1]	0.039	0.014	0.018	0.017
unemployment/vacancy ratio	-0.608	0.248	-0.588	0.312
square metre office price	--	--	0.101	0.067
investment subsidy	0.081	0.135	0.176	0.161
crime in county court district[1]	0.140	0.365	0.084	0.400
R-squared	0.690		0.687	

Centrality index, foreign cities and congestion added

	office prices excl. (n=62)		office prices incl. (n=45)	
	coefficient	standard error	coefficient	standard error
constant	17.62	5.24	9.88	12.63
road infrastructure	0.540	0.064	0.483	0.084
population size[1]	0.043	0.013	0.025	0.015
unemployment/vacancy ratio	-0.565	0.231	-0.558	0.294
square metre office price	--	--	0.090	0.064
investment subsidy	-0.009	0.123	0.073	0.052
crime in county court district[1]	0.035	0.330	0.241	0.383
R-squared	0.734		0.722	

Accessibility index, without foreign cities and congestion

	office prices excl. (n=62)		office prices incl. (n=45)	
	coefficient	standard error	coefficient	standard error
constant	30.82	5.24	20.52	14.85
road infrastructure	0.968	0.155	0.721	0.208
population size[1]	-0.049	0.022	-0.041	0.024
unemployment/vacancy ratio	-0.626	0.267	-0.677	0.350
square metre office price	--	--	0.108	0.077
investment subsidy	0.154	0.148	0.220	0.185
crime in county court district[1]	0.543	0.440	-0.261	0.532
R-squared	0.642		0.603	

cities and/or congestion are added. Since office prices are available for only the largest 46 cities, all regression analyses are made with and without this variable. In the analyses in which office prices are included, very few cities in the peripheral areas are included.

The results of the regression analyses are satisfactory. The R-squared varies from .60 to .77. The results of a subset of the regression analyses are given in Table 10.8. The negative sign of the unemployment/vacancy ratio (significant at the 5 % level in most cases) means a positive impact on the valuation of a city as a location site when there are relatively many vacancies in relation to unemployment. A relatively large number of vacancies is seen as an indicator of a dynamic urban economy instead of an indication that it is difficult to recruit personnel. The Dutch labour market was very much in a situation of excess supply during 1989. Hence, even with a relatively low unemployment/vacancy ratio in a certain city, there is not much danger of recruitment bottlenecks.

In a strict sense, one would expect a negative sign for the office space price level: a rational entrepreneur would prefer a low cost location above a high cost one. However, it may well be that the office space price level functions here are a proxy for other unobserved variables. An example of such an unobserved variable might be the strong presence of dynamic firms leading to relatively high office prices. In such a case, firms might perceive high office prices as a positive signal. Note that real estate studies often interpret low office prices as a poor indicator of urban economies.

The crime variable is not significant. This means that at this spatial level of analysis, criminality does not affect the perception of locations. It is not impossible however, that at a more detailed spatial level, criminality would play a role in the perception of entrepreneurs.

For the investment subsidy, in the majority of the cases a positive coefficient is found. However, the coefficient is not significant. Thus, our analysis does not support the hypothesis that investment subsidies make entrepreneurial views on particular locations more favourable.

The empirical results show that in a multivariate context centrality is also a quite significant explanatory factor of entrepreneural valuations of locational sites. This result is no surprise given the similarity between Figures 10.3 and 10.4 and also the high simple correlation coefficients in Table 10.5.

The specification where congestion and trans-border effects are taken into account has a better fit. Hence, in the multivariate context, our results indicate that congestion and internationalization are factors that do matter in the entrepreneur's view.

An interesting property of the upper four results in Table 10.8 is the positive coefficient for the city size variables. Thus, agglomeration advantages do matter in entrepreneurial valuations. An alternative interpretation might be that larger cities are better known than small cities, which might have led to an upward bias in the responses of the entrepreneurs for large cities.

The introduction of the accessibility variable in conjunction with city size leads to an unexpected result: the sign for the city size coefficients is negative. In the upper part of Table 10.8 where centrality was used as a road infrastructure indicator, a positive urban size effect was found. The background of these diverging results is that the accessibility of a city depends strongly on its urban size. Thus, the accessibility variable combines two notions: internal agglomeration advantages and external proximity. In the context of a study where one is interested in both agglomeration advantages and external accessibility, it is better to represent these notions by means of separate indicators.

The impact of the location of the city in the road network on the valuation of cities as location sites is considerable. The variable is clearly significant; t-values vary from 3.5 till 10.6. The coefficient of .51 in the case of the centrality index (m^2 office prices excluded) implies that an increase of 1 point on the centrality index translates into a .51 point increase on the index in which cities are valued as attractive location sites. To compare, a growth of the population size of a city by 100,000 inhabitants implies an increase of 3.9 points on this index.

10.5 Conclusion

In this survey among Dutch entrepreneurs the cities in the province of Utrecht are valued as the best location site. Other cities which score relatively high are the Randstad cities and to a lesser extent the corridor of cities in Noord-Brabant near the Belgian border. Cities in the northern and southern provinces score poorly. However, in the period 1983-1989, there is a tendency that the differences between the highly and poorly accessible cities had become smaller.

To measure the *centrality* of cities, a gravity model is used based on the distances via the road network as formulated in equation (1). The model by which the *accessibility* of cities is measured has two additions: the internal interaction potential and a weighing by the population size of the destinations (see equation (3)). In the centrality index cities in the northern part of the Randstad and cities in the central part of the country score relatively high. In the accessibility index the large cities in the Randstad score extremely well. Both the centrality and the accessibility measure can be refined by introducing foreign cities and congestion. For example, the impact of introducing foreign cities into the analysis has a larger impact on the accessibility index than on the centrality index. The foreign cities which are located within 100 kilometres by road from the Dutch border have relatively large numbers of inhabitants. In both types of indices the scores of the cities in Limburg and to a lesser extent Noord-Brabant improve most by adding foreign cities.

Correlation and regression analysis are used to investigate the relation between the valuation of cities as location sites by entrepreneurs and the accessibility/centrality of those cities in the Dutch road network. The correlations found are relatively high and rather insensitive to changes in definition. In a multivariate analysis, both the centrality and the accessibility indices contributed most to the explanation of the valuation of the cities as location sites for firms.

We also included city size as an indicator of agglomeration economies. Indeed, larger cities are valued by entrepreneurs as more attractive than smaller cities. However, an alternative interpretation would be that larger cities are better known by entrepreneurs than small cities, and may therefore lead to biased responses.

The inclusion of city size as an indicator of agglomeration economies may interfere with the use of accessibility as an indicator of the quality of infrastructure. The reason is that the internal component of accessibility tends to receive a relatively high share. Thus, in empirical studies where accessibility is used, it represents not only the quality of the external network of a city, but also the internal agglomeration economies. This may give the accessibility indicator a somewhat hybrid character.

The correlation and regression analysis both indicate a strong coherence between the valuation of cities as location sites for companies and the relative position of those cities in the Dutch road network.

In this Chapter the external attractiveness of cities was under scrutiny; in the next Chapter the attractiveness of location sites within cities is the central theme.

11 Urban Location Preferences of Entrepreneurs

11.1 Introduction

In locational studies much interest is focused on empirical research on the location behaviour of firms at the regional or urban level of spatial aggregation. In general, two approaches are used: revealed preference models and stated preference models. An example of revealed preference models are regional location models. These models attempt to explain the investment level of firms or the development in employment by a number of regional locational factors such as agglomeration effects, labour costs, investment subsidies, sectoral structure, accessibility of markets etc. See, for instance Chapter 7, section 4 or Mills and Carlino (1989).

The second type of research concerns surveys in which entrepreneurs are asked to give their views on the importance of a number of location factors on the choice of location or the performance of their company. Those surveys often have a local - urban - orientation.

This Chapter focuses on the second type of research: entrepreneurial surveys at the urban level of spatial aggregation. In section 2 we give a brief discussion of some applications in this field. In sections 3 and 4, we make use of the 1994 entrepreneurial survey held in the eastern part of the Netherlands to analyze the actual relocation pattern of firms from 1980-1994 and draft an inventory of preferences of those firms for a future relocation (section 3). In section 4 we estimate trade-offs between seven location factors.

11.2 Different Applications of Entrepreneurial Surveys

In this section we will discuss a number of applications of entrepreneurial surveys in which the importance of location factors on the location and/or development (in terms of employment or investments) of firms is studied. We identify three different ways entrepreneurs can be asked to formulate their views on the importance of location factors.

1. entrepreneurs express the importance of an entire list of locational factors by either ranking them or giving them marks;
2. entrepreneurs give their views on the impact of a major change in one or more locational factors on the location and development of firms;
3. entrepreneurs give their views on desired location types.

The characteristics of the applications of entrepreneurial surveys are summarized in Table 11.1 by emphasizing the most important pros and cons of each approach. Ranking or rating a complete list of locational factors has the advantage of a clear ordering of locational factors. Although this ordering is interesting by itself, one does not know exactly what this ordering represents. The perceived impact of changes in major locational factors is important because they directly relate to the key question addressed in this book. However, it is difficult to isolate the impacts of the change in the locational factors from the regional economic context. An example of such a study has been given in Chapter 8. In section 8.3 the attempt is made to filter the effects of the construction of the A1 transport corridor on employment growth and relocation decisions from the regional context where the firms are settled. The information grounded on a comparison of location types is useful for spatial planning processes. However, when the list of potentially relevant location factors is large, this method of comparing location types may be difficult to apply. In this Chapter we concentrate on the third application by analyzing actual and future changes in location patterns of firms as well as the estimation of trade-offs between seven location factors.

In Table 11.1 we focused on differences between methods based on entrepreneurial survey data. We must be aware that these methods share a number of problems. For example, the responses may depend somewhat on the information level of the entrepreneurs. Some respondents may, in their answers, also consider the possible responses of others. In order to reduce the size of these problems, it is essential that respondents are given precise information about the features of the decision problem, including what has to be assumed about the behaviour of other respondents. A general problem is that since the outcomes are associated with stated behaviour, rather than to actual behaviour, one is not entirely sure whether the entrepreneurs will behave according to their statements. This is a rather fundamental problem which cannot entirely be overcome, although it can be reduced by a number of research designs to correct for various biases (Hoevenagel, 1994). The reason stated preference approaches are nevertheless used is because they can help overcome problems with revealed preference approaches. Among these problems are a possible lack of variability among explanatory variables, multicollinearity and the lack of observations. The ideal approach would be to combine where possible stated preference approaches with revealed preference approaches.

Table 11.1 Strengths and weaknesses of types of entrepreneurial surveys

	strength	weakness
ranking or rating of location factors	complete listing of locational factors; easy to apply	it is not clear what this rating precisely represents; no information on trade-offs between location factors is given
impact of major change in locational factor	insight into how impacts of improvements are perceived	difficult to isolate impact from total context
comparison of location types	useful for spatial planning; information on trade-offs.	less easy to apply, especially in the case of a long list of location factors

In the following sections we will carry out analyses of the third type. In section 3 the actual relocation pattern - revealed preference - over the period 1980-1994 and the future location desires - stated preferences - of firms are analyzed by a classification of seventeen different location types. In section 4 special attention will be paid to estimating trade-offs between seven location factors. The costs of - willingness to pay for - location factors will be explicitly taken into account.

11.3 Relocation of Firms

In this section the location preferences of entrepreneurs are analyzed. A sample of 510 observations is drawn from firms with more than 10 employees in the eastern Netherlands (details on sample design can be found in Chapter 8). This part of the country contains both rural areas and a good number of medium sized cities (50-150 thousand inhabitants). Seventeen location types are first distinguished. The typology is used to study the actual - 1994 - location of firms, the changes caused by relocations over the period 1980-1994, and the locations preferred in the event that the entrepreneur must relocate the firm in the near future. So the actual location and the relocation pattern are based on revealed preference, whereas the future location desiderata are based on stated preference. The combination of the two techniques allows us to compare the future location desiderata with the actual relocation tendencies. An important question is: are the preferences for future locations consistent with the actual relocation tendencies over the period 1980-1994?

11.3.1 The Location of Firms in 1994

As mentioned above, a typology of seventeen different location types are identified on the basis of the urban profile, the orientation of the site on specific economic activities, and the access of the site by transport infrastructure (see Figure 11.1). The urban profiles are: inner city, residential quarter, suburban area, and countryside. The identified sites are heavy industry, light industry and office sites. The transport infrastructure elements under scrutiny are highway, railway and waterways.

A total of 443 entrepreneurs provided us with the actual location of their firms according to the typology given (see Table 11.2). It appears that most firms are located on sites oriented to either light or heavy industry. Only a small selection of firms on those industry-oriented sites have access to a

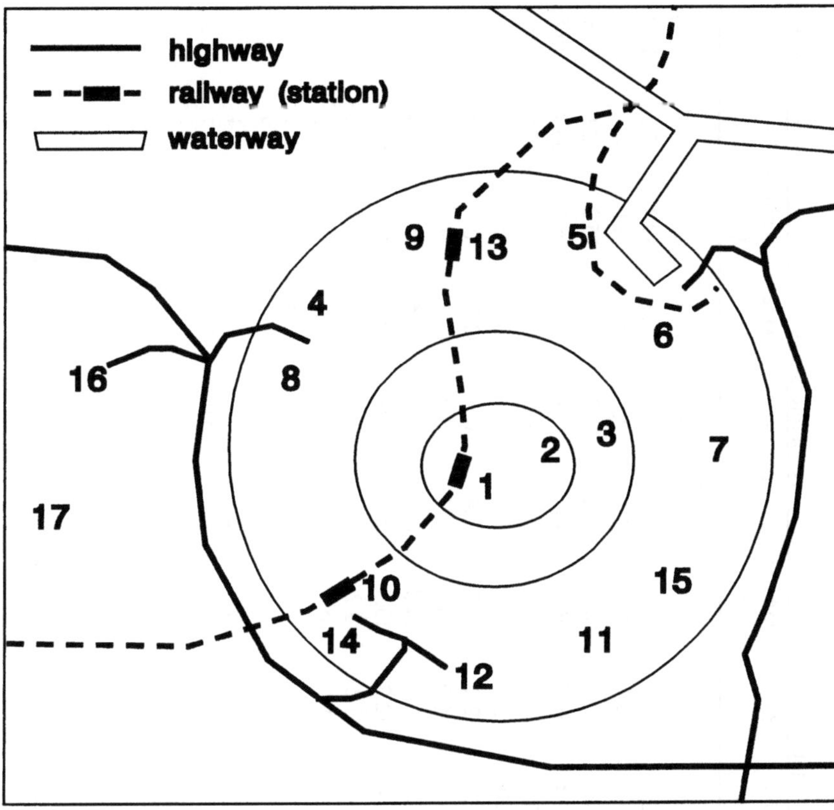

1-2 City centre	9 walking distance railway station
1 walking distance railway station	10 near highway ramp and station
2 not at walking distance station	11 no highway ramp or station
3 residential quarter	*12-15 office location in suburb*
4-7 site for heavy industry	12 near highway ramp
4 near highway ramp	13 walking distance railway station
5 rail and waterway connection	14 near highway ramp and station
6 near highway, rail and waterway	15 no highway ramp or station
7 no highway, rail or waterway	*16-17 countryside*
8-11 site for light industries	16 near highway ramp
8 near highway ramp	17 no highway ramp

Figure 11.1 Classification of location sites

Table 11.2 Location of firms by economic sector and firm size (in %, 1994)

code figure 11.1		total abs. 443	total %	economic sectors: industry 185	trs/com. 37	constr. 111	trade 75	services 33	size company: 10-49 276	> 50 167
1-3	*total urban area*	*103*	*23.3*	*14.6*	*32.4*	*28.8*	*13.4*	*60.7*	*23.5*	*22.6*
1	- inner city, near station	39	8.8	4.3	21.6	6.3	4.0	39.4	7.2	11.2
2	- inner city, no station	18	4.1	2.2	2.7	7.2	2.7	6.1	4.3	3.4
3	- residential quarter	46	10.4	8.1	8.1	15.3	6.7	15.2	12.0	7.8
4-7	*total heavy industry*	*174*	*39.3*	*50.8*	*40.5*	*27.9*	*42.7*	*6.1*	*38.4*	*40.8*
4	- near highway ramp	71	16.0	21.6	16.2	9.9	16.0	6.1	17.0	14.4
5	- near railway and waterway	10	2.3	2.7	0	0	6.7	0	2.2	2.4
6	- near high-, rail- and waterway	9	2.0	2.7	5.4	0	2.7	0	2.2	1.8
7	- no high-, rail- or waterway	84	19.0	23.8	18.9	18.0	17.3	0	17.0	22.2
8-11	*total light industry*	*109*	*24.7*	*24.3*	*13.5*	*25.4*	*34.6*	*15.2*	*25.4*	*23.4*
8	- near highway ramp	54	12.2	13.5	5.4	9.9	17.3	9.1	12.7	11.4
9	- near railway station	14	3.2	3.2	5.4	3.6	2.7	0	2.2	4.8
10	- near highway and station	6	1.4	0	0	4.5	1.3	0	1.1	1.8
11	- no highway or station	35	7.9	7.6	2.7	7.2	13.3	6.1	9.4	5.4
12-15	*total office location*	*12*	*2.7*	*0.5*	*2.7*	*1.8*	*3.9*	*15.1*	*2.9*	*2.4*
12	- near highway ramp	7	1.6	0.5	2.7	0.9	1.3	9.1	1.4	1.8
13	- near railway station	3	0.7	0	0	0.9	1.3	3.0	0.7	0.6
14	- near highway and station	1	0.2	0	0	0	0	3.0	0.4	0
15	- no highway or station	1	0.2	0	0	0	1.3	0	0.4	0
16-17	*total countryside*	*45*	*10.1*	*9.7*	*10.8*	*16.2*	*5.3*	*3.0*	*10.0*	*10.8*
16	- near highway ramp	13	2.9	2.7	2.7	3.6	4.0	0	4.0	1.2
17	- no highway	32	7.2	7.0	8.1	12.6	1.3	3.0	6.0	9.6

railway or waterway. About half of the firms are located near a highway ramp. It is noteworthy that a relatively high percentage of firms is located within residential quarters. Suburban office sites have hardly been developed in the eastern part of the Netherlands in 1994.

There are important differences if we study the sectoral diffusion over those locational sites (Table 11.2). Half of the industrial companies are located on heavy industry sites and one quarter of the remainder are on light industry sites. Of the firms in the trade sector, about 75 percent are located on industrial sites, however, the distribution between heavy and light industry sites is more balanced. Firms in the transport and communications sector are relatively well represented in the inner city - particularly within walking distance of a railway station - and in residential areas. The firms in the construction sector are most dispersed over the different location types. Remarkable for firms of this sector and of the service sector as well is the relatively high percentage of firms located within residential areas. The locational pattern of the service sector is clearly different from that of the other sectors; especially the share of locations near a railway station (39 %) is much higher than the average share of 9 %.

The differences in location sites between small firms (10 - 50 employees) and large firms (over 50 employees) are relatively small (Table 11.2). Small firms are more often located in residential areas, whereas large firms are more strongly represented in the inner city by being within walking distance of a railway station.

To summarize the situation in 1994: most companies are located on industrial sites or within the old urban area (the inner city and the residential quarters). Relatively few companies are located in the countryside and in particular at office locations in the suburban area. Between the economic sectors there are rather large differences in the types of sites upon which most firms are located.

Table 11.3 Relocation pattern of firms over the period 1980 - 1994

		no. of companies		migration balance		immigrants		stayers		emigrants	
		1980	1994	abs.	% '80	abs.	% '80	abs.	% '80	abs.	% '80
1-3	*total urban area*	*151*	*103*	*-48*	*-31.8*	*10*	*6.6*	*13*	*8.6*	*58*	*38.4*
1	- inner city, near station	58	39	-19	-32.8	3	5.2	8	13.8	22	37.9
2	- inner city, no station	26	18	-8	-30.8	3	11.5	1	3.8	11	42.3
3	- residential quarter	66	46	-20	-30.3	4	6.1	4	6.1	24	36.4
4-7	*total heavy industry*	*141*	*174*	*33*	*23.4*	*43*	*30.5*	*21*	*14.9*	*10*	*7.1*
4	- near highway ramp	47	71	24	51.1	24	51.1	10	21.3	0	0
5	- near railway and waterway	10	10	0	0	2	20.0	2	20.0	2	20.0
6	- near high-, rail- and waterway	7	9	2	28.6	2	28.6	0	0	0	0
7	- no high-, rail- or waterway	77	84	7	9.1	15	19.5	9	11.7	8	10.4
8-11	*total light industry*	*87*	*109*	*22*	*25.3*	*29*	*33.3*	*11*	*12.6*	*7*	*8.0*
8	- near highway ramp	38	54	16	42.1	18	47.4	6	15.8	2	5.3
9	- near railway station	11	14	3	27.3	3	27.3	0	0	0	0
10	- near highway and station	4	6	2	50.0	2	50.0	0	0	0	0
11	- no highway or station	34	35	1	2.9	6	17.6	5	14.7	5	14.7
12-15	*total office location*	*10*	*12*	*2*	*20.0*	*3*	*30.0*	*2*	*20.0*	*1*	*10.0*
12	- near highway ramp	6	7	1	16.7	2	33.3	2	33.3	1	16.7
13	- near railway station	3	3	0	0	0	0	0	0	0	0
14	- near highway and station	0	1	1	∞	1	∞	0	0	0	0
15	- no highway or station	1	1	0	0	0	0	0	0	0	0
16-17	*total countryside*	*55*	*45*	*-10*	*-18.2*	*0*	*0*	*1*	*1.8*	*10*	*18.2*
16	- near highway ramp	15	13	-2	-13.3	0	0	1	6.7	2	13.3
17	- no highway	40	32	-8	-20.0	0	0	0	0	8	20.0
		443	443	0		85		48		85	

11.3.2 The Relocations in the Period 1980-1994

Of the 443 entrepreneurs who gave the actual location of their company, 133 stated that their firm has relocated over the period 1980-1994. In Table 11.3 the consequences for the location pattern of 1994 are calculated by placing those relocated firms into their old - 1980 - location type. A first eye-catcher is the fact that just over one third (36 %) of the firms relocated from an identical type of location site. The remaining 64 % concerns a change of location type. Table 11.3 shows us the balance of relocation.

It is obvious that the relocation of firms over the period 1980-1994 have had rather dramatic consequences for the number of firms located in the urban area (inner city and residential quarters). The large outflow of firms from this area is by far not compensated by the inflow of firms. It is remarkable that sites at walking distance of a railway station are not more resistant to this outflow than sites without such a railway access. However, note that in this research project only firms with over 10 employees are included. It is reasonable to assume that the locations left behind by the relocating firms are filled by small - start up - firms: the so-called breeding place function of the inner city. Such developments were outside the scope of this research project[1].

The migration balance is also negative in the countryside, particularly on sites without access to a highway. So there appears to be a double shift: not only are firms from the inner city and residential areas relocating in the direction of suburban areas, but firms in the countryside are also moving toward suburban areas.

The sites for heavy industry and light industry have a positive migration

[1] Research of Van Steen & Van der Velde (1993) on the expansion space needed by companies showed that 52.5 % of the 1086 companies included in their research panel stated that the firms' employment developed by over 10 % in the period 1988-1992. Nearly 17 % of all firms expected to be unable to realize another growth in employment by 10 % at the site where the firm is presently located. Often there is no space available to expand the firm. Those figures were significantly higher with firms located in inner cities or residential quarters. In such areas, over one quarter of the firms expected to be unable to realize a 10 % employment growth at the present location. In our research project, a relatively large share of companies stated a fast growth in employment over the period 1980-1994. It is reasonable to assume that the inability to expand at locations within the urban area contributes to the explanation of the large outflow of companies from the inner city and the residential areas as found in this study.

balance. Industrial sites having an access to a highway show the largest inflow of firms. However, industrial sites without an access by highway, railway or waterway also received a large share of the relocating firms over the period 1980-1994.

The increased orientation to highway locations is also found when - at four digit postal code level - the distances to the nearest highway ramp of the old and the new location are compared. It appears that with 59 % of the relocations, the old postal code zone is abandoned (conforms to Chapter 8). For those relocations, the average distance of the firm to the nearest highway ramp decreased by 28 percent. This might indicate a strong orientation to highways in the event of a relocation of the firm. However, this statement should be noted with caution. As shown in Chapter 8, the main reason to relocate is the need to expand, and this expansion cannot be undertaken at the present location. Those firms can only relocate wherever enough supply is offered. The supply side can restrict the opportunities of the expanding firms to relocate to the most optimal location. Considering the supply of industrial sites and their proximity to a highway ramp in the eastern part of the Netherlands, it is important to realize that regional government strongly intervenes in the supply of industrial sites. Until 1988 the supply of industrial sites near highway ramps was restricted. Supply was offered nearby existing sites in the urban outskirts. The area surrounding the highway war earmarked to remain a green zone. However, around 1988 this policy radically changed. Large new industrial sites directly located in the neighbourhood of highway ramps were developed.

A final remark concerning Table 11.3 is that although office locations in the suburban areas show a positive migration balance, the number of firms on those sites remains low. It seems that until 1994, there is scarcely any demand for such locations in the area under study. However, it is also possible that the supply of office locations is too low, and there is a latent demand for those sites. This will be studied in the next subsection where locational preferences are analyzed.

11.3.3 Locational Preferences of Firms

In Table 11.4 the preferred locational sites are compared with the actual - 1994 - sites upon which the firms are located. From the total number of 443

Table 11.4 Location preferences of the firms

		no. of companies		migration balance		immigrants		stayers		emigrants	
		1994	prefer	abs.	% '94	abs.	% '94	abs.	% '94	abs.	% '94
1-3	*total urban area*	*80*	*21*	*-59*	*-73.7*	*2*	*2.5*	*19*	*23.8*	*61*	*76.3*
1	- inner city, near station	33	15	-18	-54.6	2	6.1	13	39.4	20	60.6
2	- inner city, no station	11	3	-8	-72.7	0	0	3	27.3	8	72.7
3	- residential quarter	36	3	-33	-91,7	0	0	3	8.3	33	91.7
4-7	*total heavy industry*	*117*	*140*	*23*	*19.7*	*89*	*76.1*	*51*	*43.6*	*66*	*56.4*
4	- near highway ramp	43	94	51	118.6	59	137.2	35	81.4	8	18.6
5	- near railway and waterway	6	4	-2	-33.3	4	66.7	0	0	6	100
6	- near high-, rail- and waterway	6	29	23	383.3	24	400.0	5	83.3	1	16.7
7	- no high-, rail- or waterway	62	13	-49	-79.0	2	3.2	11	17.7	51	82.3
8-11	*total light industry*	*77*	*115*	*38*	*49.4*	*76*	*98.7*	*39*	*50.6*	*38*	*49.4*
8	- near highway ramp	38	69	31	81.6	40	105.3	29	76.3	9	23.7
9	- near railway station	9	3	-6	-66.7	3	33.3	0	0	9	100
10	- near highway and station	4	34	30	750.0	30	750.0	4	100	0	0
11	- no highway or station	26	9	-17	-65.4	3	11.5	6	23.1	20	76.9
12-15	*total office location*	*8*	*20*	*12*	*50.0*	*15*	*187.5*	*5*	*62.5*	*3*	*37.5*
12	- near highway ramp	6	8	2	33.3	4	66.7	4	66.7	2	33.3
13	- near railway station	1	1	0	0	1	100	0	0	1	100
14	- near highway and station	1	11	10	1000	10	1000	1	100	0	0
15	- no highway or station	0	0	0	0	0	0	0	0	0	0
16-17	*total countryside*	*31*	*17*	*-14*	*-45.2*	*6*	*19.4*	*11*	*35.5*	*20*	*64.5*
16	- near highway ramp	11	13	2	18.2	6	54.5	7	63.6	4	36.4
17	- no highway	20	4	-16	-80.0	0	0	4	20.0	16	80.0
		313	313	0		188		125		188	

firms that had given their actual location type, 313 gave information about the location type preferred if a relocation should be necessary. Almost 40 % (125) of the entrepreneurs stated a preference for a location type identical to the actual location type. This percentage is comparable to the percentage of firms that relocated to a similar location type during the period 1980-1994. It appears that the percentage of firms that indicate their desire to relocate to a identical location type increases as the quality of the actual location site increases. Firms located on sites for heavy industry are relatively loyal to their location, which may partly be due to the inertia of those firms. However, a higher percentage of the firms located on sites for light industry prefer to relocate to a similar location type. The highest percentage of loyalty is found by firms located on office sites, however the total number of companies located on such location types is low.

The fact that 60 percent of the entrepreneurs prefer another location type indicates a large latent relocation demand. The large outflow of firms in the period 1980-1994 from the inner city and residential quarters, is found in even a larger extent in the desires of the firms remaining in these areas. Three quarters of the firms located in the old urban areas prefer a location outside these areas. In particular, firms want to leave the residential quarters. This is another indication of the failure of government policy to integrate the work and residential function within cities. One of the most important reasons that firms wish to leave residential areas is the stricter environmental legislation on noise and air pollution with which firms must comply. Only two companies wished to relocate to the inner city (at walking distance of a railway station). In the old urban area, locations at walking distance of railway stations are most resistant to this outflow of companies. However, as stated earlier in this study, only relatively large firms are involved. The old urban area may still have an important function as a breeding place for new, starting firms.

Other location less favoured by entrepreneurs are sites with a bad access to transport infrastructure (highway, railway - persons and goods -, and waterways). However, in the period 1980-1994, a relatively large number of firms relocated to such poorly accessible sites. A logical explanation is the supply and price level of location sites. The firms might have chosen these bad accessible location cites simply because no better locations were available for a reasonable price. As stated above, expansion of the firm is the main reason to relocate. A firm can only relocate to locations where adequate space is offered at an affordable price for the firm. It is possible that either the supply of good accessible locations is too low, or the prices of such locations are too high for the relocating firm.

Although the bad accessible industrial sites show an outflow of firms, the migration balance of industrial sites in general stays positive, due to a large inflow of firms that wish to relocate to the good accessible industrial location types.

Apparently, there is a relatively large desire to relocate to office locations, particularly accessible by highway and - to a lesser extent - a railway station.

Even countryside locations accessible by highway appear to attract new firms. However, countryside locations without an access to a highway will show a large loss of firms when the relocation desires become reality.

In summary, it appears that the location desires of firms to a large extent are comparable with the actual relocation pattern during the period 1980-1994. The old urban area and the countryside experience a rather dramatic decrease in the number of firms in favour of the suburban - industrial - locations. The main differences between the location preferences and the relocation pattern during the period 1980-1994, concerns the wish to locate on a good accessible site, whereas the actual relocation pattern shows a strong growth of the less accessible sites and the desire to locate on an office site in the city outskirts, which in reality are scarcely developed.

This study on locational preferences of firms is not the only study in this field in the Netherlands. In 1987, Lukkes used his firm panel to investigate the preferences for new locations (Lukkes, 1988). Some minor adjustments make the studies comparable[2].

In the study of Lukkes, 52 % of the firms stated a preference for a location similar to the actual location. In our study, the percentage was lower: 40 %. However, there are large differences between the Dutch provinces in the Lukkes study. The major part of our study area is located within the province of Overijssel. The percentage of firms in the Overijssel province that desire to relocate to a corresponding location is only 36 %: a percentage which is close to our finding.

---

[2] A difference between the studies is that our study only concerns the eastern part of the Netherlands, whereas the Lukkes firm panel covers the entire country. Further, our study only deals with firms with at least 10 employees in the basic sectors of the economy. No such division is used by Lukkes.

Table 11.5 Balance of migration of the various location site types

	The Netherlands Lukkes (1988)	Eastern part our research (1993)
Inner city		
railway station at walking distance	10.7	-54.6
station not at walking distance	-42.4	-72.7
Residential quarter	-78.0	-91.7
Urban area	*-43.6*	*-73.7*
Heavy industry sites	32.3	19.7
Light industry sites	-15.2	49.4
Office locations	130.2	50.0
Countryside	-13.3	-45.2

Source: own calculations based on Lukkes (1988)

The migration balances of the various location types are given in Table 11.5. Although there are some differences, the general development is rather similar. The old urban area and the countryside lose firms, and the suburban areas gain. In the case of two location types, opposite migration balances are found. Lukkes found that inner city locations at walking distance of a railway station have a positive migration balance, but we found a negative one. Two explanations may deepen the understanding of this contradiction in findings. Some railway locations in the major urban agglomerations of the Netherlands (Amsterdam, Rotterdam, The Hague and Utrecht) are indeed successful in attracting new firms. In the eastern part of the Netherlands there are no such large agglomerations. Second, there has been a shift in the perception of railway locations in the period between the study of Lukkes (1988) and our study (1995). In 1988, the concept 'railway station location' was 'hot'. These locations were expected to have plenty of potential; they received much political attention, therefore railway locations had a positive image. In the mean-time however, the concept has lost some of its mystification and the perceptions of entrepreneurs for these locations has become more modest.

The second difference in both studies concerns the positive migration balance we found for light industry sites versus the negative balance of the study of Lukkes. A first explanatory factor is the sectoral structure of the eastern part of the Netherlands. In this part of the country the industry sector is overrepresented and the services underrepresented. This has a strong impact on the demand for location types. Industry is located on light or heavy industry sites; services are situated on office locations. Furthermore, the negative balance in the Lukkes study is related to the strong positive balance he found for office locations. These office locations flourished especially in the western part of the Netherlands. The relocation of firms from a light industry site to an office location is seen as an upgrading of the location profile. Since in the eastern part of the Netherlands hardly any office location has been developed, the desire to relocate to such a location is not so strong in this part of the country. In this area most firms upgrade their locational profile by relocating towards a site for light industry.

In this section an analysis has been given of the relocation pattern of firms as well as their locational preferences. One of the differences that appeared between the actual relocation pattern and the locations preferred by firms concerns the access to a highway. A location near a highway access is clearly preferred by a large majority of the firms (82 %). Such a tendency can also be observed for actual relocations between 1980 and 1994, but to a smaller extent (54 % of the firms that moved actually relocated to a place near a highway ramp). This difference can be explained by factors such as insufficient supply of industrial sites in the past and perhaps price differences between locations which have good versus bad access to a highway. In order to give greater insight into the role of various locational factors in the relocation decision of entrepreneurs, we will calculate trade-offs between a number of locational factors in the next section.

11.4 A Stated Preference Analysis of Locational Profiles

In this section, a stated preference analysis will be used to investigate the trade-offs entrepreneurs make between seven locational factors (see Bruinsma & Rietveld, 1997). Each respondent was asked to answer five questions in which two location types are compared. The location types consist of a profile with seven location factors. Each pair of location types compared has equal scores for five of the location factors and different scores for the two remaining factors. Examples of the location types compared are given in Table 11.6. The entrepreneur had to select one of the two locations described in each question. Each location is represented by two location factors in which in location A, the first factor has a relatively favourable value and the second factor a relatively unfavourable one. For location B the opposite holds.

Table 11.6 One of the five sets used in the analysis

Imagine you have to relocate your company, whatever the reason may be. Select for each hypothetical combination of location factors the location A or B which you prefer most. The other location factors are assumed to be equal for A and B.		encircle preferred location
1 Location A	Access to a highway at 2 kilometres distance	
	Price of the land is dfl 150 per square meter	A
Location B	Access to a highway at 10 kilometres distance	B
	Price of the land is dfl 60 per square meter	
2 Location A	Investment subsidy of 25 %	
	Most suppliers/customers at 30 kilometres distance	A
Location B	No investment subsidy	B
	Most suppliers/customers at 10 kilometres distance	
3 Location A	Price of the land is dfl 60 per square meter	
	A one hour bus service	A
Location B	Price of the land is dfl 150 per square meter	B
	A fifteen minutes bus service	
4 Location A	Access to a highway at 2 kilometres distance	
	Large city (100.000 inhabitants) at 30 kilometres distance	A
Location B	Access to a highway at 10 kilometres distance	B
	Large city (100.000 inhabitants) at 10 kilometres distance	
5 Location A	Distance to a railway station 500 metres	
	A one hour bus service	A
Location B	Distance to a railway station 10 kilometres	B
	A fifteen minutes bus service	

Respondents were requested to compare locational profiles for a limited number of pairs. For example, 20 % of the respondents compared the five pairs of profiles given on Table 11.6. Other sets of pairs of profiles were generated to be judged by other respondents.

In section 11.4.1 a qualitative analysis is given based on a matrix in which the scores between each pair of location factors are presented. In section 11.4.2 a logit model is used for a further analysis of trade-offs between the distinct location factors.

11.4.1 Qualitative Analysis

Table 11.7 presents for each combination of the location factors the percentage of respondents which prefers the alternative with a favourable score for the first location factor, compared with the alternative with a favourable score for the second location factor, assuming that the other five location factors are constant. The values which have been used in the questionnaire are presented in Table 11.8. From Table 11.7 it can be concluded, for example, that 35 % of the respondents indicate that a change in distance to a highway access from 2 to 10 kilometres is not compensated for by lowering the land prices from 150 to 60 Dutch guilders per square meter. The remainder (65 %) thinks this reduction of land price is a sufficient compensation for the longer distance. It should be acknowledged that the percentages mentioned in Table 11.7 also depend on the values and the difference in the positive and negative values of Table 11.8[3].

Taking the values in Table 11.8 as a point of departure, it appears that the price of land is considered to be more important than the other location factors (Table 11.7). The rank order of the other factors is[4]: investment subsidies, the distance to customers and suppliers, the distance to a highway access, the distance to a railway station and finally, the number of bus services, which is thought to be the least important location factor.

---

[3] Note that with perfect knowledge of preferences of entrepreneurs, one could construct the ranges in Table 11.8 in such a way that the entrepreneurs would be indifferent between any pair of alternatives, so that all figures in Table 11.7 would equal .50.

[4] The ranking is based on a weighted average of row scores.

It appears that the most important location factors - while applying the used favourable and unfavourable reference values - are those which increase the costs of the entrepreneur in a direct way (both the price of land and an investment subsidy). The least important factors are public transport facilities (both bus and rail). The opinions on the other three location factors are more varied.

Table 11.7 Percentage of entrepreneurs that in a pairwise comparison chooses for an alternative with a favourable value for the location factor mentioned in the first column (compared with a favourable value in the first row).

	highway access	price of land	subsidy	suppl. custom.	bus service	large city	railway station
distance to a highway access	---	35 n=415	30 n=80	49 n=91	82 n=83	53 n=80	78 n=81
price of land	65	---	82 n=81	70 n=80	93 n=80	75 n=91	93 n=83
investment subsidy	70	18	---	74 n=80	88 n=91	53 n=83	81 n=91
distance to suppliers/customers	51	30	26	---	64 n=81	53 n=83	74 n=80
frequency bus services	18	7	12	36	---	34 n=80	41 n=80
distance to a large city	47	25	47	47	66	---	58 n=81
distance to a railway station	22	7	19	26	59	42	---

n = number of responses. Distance to highway access and price of land is compared by all respondents. The other pairs are only compared by about 20% of the respondents.

Table 11.8 Values of location factors as used in the questionnaire

	positive value	negative value
distance to a highway access	2 kilometres	10 kilometres
price of land	dfl 60	dfl 150
investment subsidy	25 percent	none
distance to suppliers/customers	10 kilometres	30 kilometres
frequency bus services	4 times per hour	one per hour
distance to a large city (100.000 inhabitants)	10 kilometres	30 kilometres
distance to a railway station	500 metres	10 kilometres

Two final remarks can be made on the relation between the two highest and lowest valued location factors. First, the investment subsidy - which is the second important location factor - is valued much lower than the highest location factor (the percentage is only 18.5 %). Another interesting result is that 41 % of the respondents favours a four-times-per-hour bus service compared to a railway station within walking distance, which is perhaps a much higher share than expected beforehand. This qualitative analysis has shown some strong location factors (price of land and investment subsidies) and some weak ones (public transport services)[5]. One should be aware, however, that these results depend on the reference values chosen (see Table 11.8).

A quantitative statistical analysis may provide more insights by offering underlying structures. This analysis will be carried out in the next subsection by means of a logit model.

11.4.2 Quantitative Analysis

The statistical techniques of categorial data analysis - logit and log-linear modelling provide very powerful tools for the analysis of survey data

[5] It is important to note that in our study the importance of accessibility-related factors is smaller than those of financial factors such as cost of land and cost of capital. This means that our results are more in line with those mentioned by the report of the European Commission (E.C., 1990a) than the ones reported by Healey & Baker (1996) and NSS (1991) (see Chapter 4).

(Leitham, 1996). We use a logit model in which it is assumed that entrepreneurs value the location factors according to a certain utility function. We assume that this utility function is linear for all entrepreneurs. For entrepreneur i, the utility $U_{i,n}$ of a location n depends on the location factors of location n according to the criteria indicated in Table 11.8:

$$U_{i,n} = a_0 + a_1 z_{n1} + a_2 z_{n2} + a_3 z_{n3} + a_4 z_{n4} + a_5 z_{n5} + a_6 z_{n6} + a_7 z_{n7} + e_{i,n} \quad (1)$$

In this function, the factors a_j represent the weights (to be estimated) of the location factors whilst the factors $z_{n,j}$, represent the values of alternative n according to criterion j. The error term $e_{i,n}$ indicates that the preference of an entrepreneur also depends on many other factors which are not included in the analysis. Therefore, this term presents the uncertainty of the model. Usually, when two alternatives are compared, not all entrepreneurs will have the same preference; this appears clearly from the analysis in section 4.1.

In the survey, questions were asked on several alternative location types. In the above discussed utility function, this implies that the respondents indicate whether $U_{i,1}$-$U_{1,2}$ is larger or smaller than zero. When there is a sufficient number of observations, a statistical analysis (in this case a logit model) can be used for estimating the coefficients a_j. In this contribution, we will not give a technical description of the model and method used. For the interested reader, we refer to Maddala (1983) and Cramer (1991). Based on the data presented in Table 11.7, the logit model results in estimated coefficients as presented in Table 11.9[6].

All coefficients have the expected sign. For most explanatory variables this sign is negative, which is as expected, because the variables can be interpreted as 'costs' (distances, prices). The only positive coefficient is found for investment subsidies, which is as expected, because this variable can be interpreted as a 'benefit'. It also appears that five of the six location factors differ significantly from zero. The only exception is the distance to the

---

[6] In the equation used it is not possible to estimate the constant term a_0 of the utility function, because the estimations are based on the comparison of alternatives. The constant term disappears in these equations. Therefore, the logit model is estimated without a constant term. Moreover, it appears that given the chosen values of the scores of the location factors, there exists an independency in the matrix with the observations of the explanatory variables. The resulting multicollinearity problem is solved by giving the coefficient of the least important location factor - frequency of bus services - the value zero.

nearest railway station, which is as expected negative, but not in a significant way. Most significant are the price of land and investment subsidies, while the access to the highway is the third most significant variable.

Table 11.9 Results logit model

	coefficient	t-value
distance to a highway access	-.1637	-10.1
price of land	-.0235	-15.7
investment subsidy	.0645	11.7
distance to suppliers/customers	-.0510	-7.6
distance to a large city	-.0468	-7.0
distance to a railway station	-.0219	-1.5

- log likelihood: 1221.9
- log likelihood (all coefficients equal to zero): 1437.6

In this specification of the model, it appears that 69.7 % of the responses are correctly predicted. Compared to other analyses with this specification, this is a fairly good result, although a specification with random answers already would result in 50 % of correctly predicted answers.

The interpretation of the coefficients a_j in Table 11.9 is 'utility' per 'unit z_j'. Because we are not primarily interested in the value of the coefficients themselves, but in the trade-offs (relative weights) between the factors, the ratios of the coefficient may provide interesting insights. When we take the distance to a highway access as a starting point, the following trade-offs are found[7]. *To keep the level of utility constant, a 1 kilometre reduction of the distance to the nearest highway access should be accompanied by one of the following changes:*

* a rise of the price of land of about dfl. 7.0 per square meter;

[7] In the calculations the relations between the location factors z_j and utility are supposed to be linear. This means that an increase in distance to a highway from 1 to 2 kilometres is supposed to be equal to an increase from 9 to 10 kilometres. Of course other specifications are possible here.

* a reduction of an investment subsidy of about 2.5 %;
* an increased distance to the most important customers/suppliers of about 3.2 kilometres;
* an increased distance to the nearest large city of about 3.5 kilometres;
* an increased distance to the nearest railway station of about 7.5 kilometres.

For the results discussed until now, we used the most simple specification of the model. In this model no variables were included indicating the specific features of a company: all respondents were supposed to use the same weights. In order to broaden the scope of the analysis, we repeated the analysis and added some company specific features (amongst others: size, economic sector and current location vis a vis a highway). For each of these features an extra variable has been added, which is defined as the product of the initial value and a dummy[8]. For example, for the variable 'sector', a dummy variable has been added which has the value '1' when the company is a manufacturing company and the value '0' in all other cases. The dummy variable for 'scale' received the value '1' when the company has more than 50 employees, and the dummy 'current location' indicates whether the company is located more than 7.5 kilometres away from a highway access. In this way it can, for example, be analyzed to which extent small companies give different weights to location factors than large companies do.

From the estimation results it appears that in general these differences are not very large. The coefficients of the dummy variables are in most cases not significant. The largest differences are found in the sector-specific estimation (see Table 11.10). In Table 11.10, the upper part of the coefficients relate to non-manufacturing firms. Note that these values are rather close to those in Table 11.9. The lower part of the coefficients shows the shifts in the importance of the location factor when we compare manufacturing with non-manufacturing firms. It appears that manufacturing companies value the distance to customers/suppliers and the nearest city significantly lower (this can be inferred from positive coefficients found and the high t-values).

---

[8] In a technical sense we introduce interaction effects by using this specification. The introduction of an extra main effect is of no use in this situation, because such an effect will disappear when pairs of alternatives are compared. The reason is that the estimation concerns the differences in utility values of two alternatives as judged by a specific entrepreneur. In such a situation it is not possible to estimate a constant term.

Table 11.10 Results logit model (dummies for economic sectors included)

	coefficient	t-value
distance to a highway access	-.1892	-8.6
price of land	-.0236	-11.8
investment subsidy	.0651	8.9
distance to suppliers/customers	-.0696	-7.6
distance to a large city	-.0635	-6.9
distance to a railway station	-.0288	-1.5
dummies for the manufacturing sector:		
distance to a highway access	.0507	1.5
price of land	-.0010	-0.3
investment subsidy	.0017	0.2
distance to suppliers/customers	.0412	3.0
distance to a large city	.0375	2.8
distance to a railway station	.0150	0.4

- log likelihood: 1207.7
- log likelihood without industry dummies: 1221.9

We restrict ourselves to a qualitative presentation of the results for the other estimations. The size of a company has no significant impact on the weights given to the distinct location factors a_j. Entrepreneurs who are now located at a distance of more than 7.5 kilometres away from a highway access, however, value this criterion significantly lower. This is a nice result since it implies consistency between 'revealed preference' information, as indicated by the actual location of firms, and stated preferences as investigated here. Firms that are currently located farther away from highways apparently have managed to adjust to such a location type so that their valuation of this location aspect is relatively low. If they really would have attached a high value to highway accessibility they would never have settled at this location, or they would have already moved. Furthermore, it appears that companies which are located relatively far from a highway access give a relatively low value to the location factors 'distance to a large city' and 'distance to a railway station'. Conversely, it appears that these companies are relatively sensitive to financial variables such as price of land and investment subsidies.

11.5 Conclusions

In this Chapter a brief overview has been presented on research on location behaviour of firms in which stated preferences are used. Those surveys are used for three purposes:

1. to rank an entire list of locational factors,
2. to investigate the impact of a major change in one or more locational factors on the location and/or the development of companies,
3. to measure preferences for location profiles and locational factors.

In this Chapter a survey of the third type is used to analyze the relocation pattern - both actual and preferred - and to calculate trade-offs among a number of locational factors.

The relocations in the period 1980-1994 have led to a concentration of economic activities in the suburban areas. The share of firms in suburban areas has increased from 53.7 to 66.6 percent over this period. The share of the firms located in the old urban area (inner city and residential quarters) decreased from 34.1 to 23.3 percent. The share of the firms in the countryside also decreased slightly from 12.4 to 10.2 percent. This relocation towards the suburban areas from both the old urban area and the countryside is also found - even to a larger extent - when the preferred locations are analyzed. Since 60 % of the firms prefer another type of location, there seems to be an enormous latent potential for additional relocations. Through the analysis of potential relocations, two tendencies in the preferences of the entrepreneurs appear. First, there is a desire to upgrade the location profile. For instance, firms located on sites for heavy industry prefer a location on a site for light industry. Second, there is a strong desire to relocate to a site near to a highway ramp or - to a lesser extent - a railway station. This last tendency is more strongly present in the preferences of entrepreneurs than it is found in the actual relocation pattern over the period 1980-1994. It is possible that in the actual relocation decision other location factors are more important than distance to a highway or a railway station. Therefore, we have used a stated preference technique to measure the relative importance of a number of locational factors.

A problem with many location studies is that the results cannot be interpreted in terms of trade-offs between location factors. Therefore, we conduct a pairwise comparison approach which enables one to investigate trade-offs.

The relative importance of seven locational factors is investigated in a pairwise way using stated preference techniques.

It appears that locational factors with direct financial implications, such as price of land and investment subsidy, are the most important factors. The distance to a highway access, distance to a large city and distance to customers and clients are second order factors. The factors concerning public transport (distance to a railway station and the frequency of bus services) are least important.

The logit analysis allows the investigation of trade-offs among the seven locational factors. These trade-offs make it possible for instance to calculate the cost effectiveness of different regional policy measures, as far as they concern the interest of firms for newly developed industrial sites. By using these trade-offs, national or regional governments can compare the pros and cons of investments in public transport, or in highways, the provision of more investment subsidies, or the offering a lower price of land on industrial sites.

The research area in which the survey was carried out concerns the eastern part of the Netherlands, an area not plagued by congestion and approximately 100 km from major metropolitan areas in the Netherlands (Randstad) and Germany (Ruhr-area). This may explain the relatively low importance attached to accessibility aspects by the entrepreneurs in their responses. It is in such regions that financial instruments seem to be more effective in influencing locational behaviour of firms compared with accessibility related measures.

PART D

Infrastructure Supply Studies

Introduction to Part D

In this part the interrelationship between infrastructure supply and the space economy is examined by two case studies. In Chapter 12 the mutual impact between the historical development of the Dutch rail network and urban growth is studied. It is followed by Chapter 13, where we analyze the distribution of highways and railways among regions in Europe.

The difference between Part C and Part D of this book concerns the direction of causality. Where in Part C we assumed infrastructure to be the driving force of spatial development, in Part D the reverse direction is studied: regional and urban development are used as starting points in the analysis of the evolution of transport networks.

Chapter 12 starts with a theoretical introduction on the interrelationship between network development and urban development. This Chapter pays extra attention to network design modelling. We begin with a survey of cost-related network location models (cost minimization algorithms), then we introduce demand aspects; finally, rates of return on investments in the networks are added to the model. This resultant model is used to analyze the historical development of the Dutch railway network. After a description of the historical development of the Dutch railway network, the actual network development is compared with the network evolution as simulated by the most refined model. About 40 % of the first 20 links were predicted correctly by the model. Then the reverse relationship is investigated; we try to explain the variance in growth rates of cities by factors such as city size, population growth, and the increase in accessibility by the railway network.

In Chapter 13 we investigate the factors which explain the supply of highways and railways among European regions. Special attention is paid to network density in European border regions. The regional supply of infrastructure is measured as the length of the network per square kilometre. The regional differences in network supply appear to be relatively large. The attempt is made to explain these differences in a model at the NUTS II level by factors such as regional and interregional demand, financing possibilities, costs of construction and maintenance, regional policies, and barrier effects either by national borders or sea coasts. The supply of infrastructure mainly depends on two determinants: regional demand (population size) and financing possibilities (gross domestic product). In the model attention is given to border effects in transport infrastructure supply, because one might expect that national planning of infrastructure has led to a neglect of cross-border links in the past. It is found that negative border effects do exist in the field of infrastructure supply, but only at a relatively small spatial scale.

12 Evolution of the Dutch Railway Network and Urban Development

12.1 Introduction

Transport networks can be defined in terms of links and nodes. In this Chapter we address the issue of the mutual impacts of links on nodes and nodes on links. Most studies on this theme focus only on one of the two relationships. We try to carry out a more integrated approach to modelling transport networks. This will be done in the context of railway systems, where links can be interpreted in terms of railway lines, and nodes in terms of railway stations, or in a broader sense, the cities served by these stations (see Rietveld & Van Nierop, 1995a, 1995b).

The structure of the Chapter is as follows. In section 2 we discuss some relevant theoretical concepts on the interrelationship between networks and urban growth. Section 3 is devoted to a presentation of factors relevant for railway development in the 19th century in the Netherlands. Section 4 contains an algorithmic approach to network formation. Based on the criterion of net returns on investment, a sequential approach to the construction of railway lines is given. The network evolution simulated by the algorithm is compared with the actual development of the network in section 5. In section 6 the reverse relationship is investigated: the question addressed is to what extent the variance in growth rates of cities can be explained by factors such as the increase in accessibility. Finally, concluding remarks are given in section 7.

12.2 The Interrelationship between Network Development and Urban Development

As a point of departure for our analysis we take the scheme shown in Figure 12.1. The development of transport networks has a certain impact on the growth of cities (see Chapter 3). The supply of infrastructure has an impact on the productivity of other production factors such as private capital and labour because of possibilities of substitution. Therefore, places where infrastructure is improved experience an improvement in factor productivity which offers opportunities for the expansion of markets. Since transport infrastructure networks are localized and immobile, they lead to a differentiation in the spatial productivity landscape: some locations (cities) benefit more than other locations. Therefore, the construction of transport infrastructure networks may have polarizing effects in space.

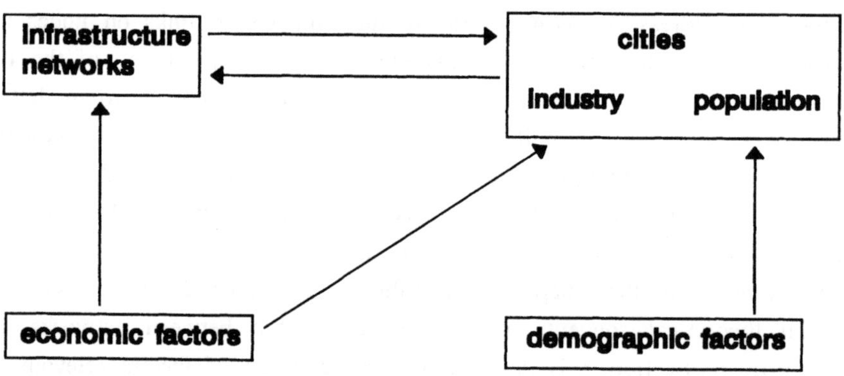

Figure 12.1 Mutual relationship between network development and urban growth

The potential degree of polarization depends on the density of nodes in a network. When the density of nodes is high (as is the case in road networks), the potential degree of polarization is low. On the other hand, when the density of nodes is low, as is usually the case with railway or aviation networks, potential degrees of polarization are higher.

Of course, transport infrastructure networks are not the only factors influencing the economic development of cities. There are several other important influential factors including knowledge infrastructure, local entrepreneurship and urbanization economies. Urbanization economies imply that economies of scale (both internal and external) exist. Such economies of scale imply that the spatial division of labour is much less subject to change than is often assumed. A consequence, as explained by Myrdal (1957) and Krugman (1991), is that historical conditions can have a strong influence on urban developments, and that imbalances in regional or urban conditions may be long lasting. Another important implication is that old infrastructure networks, even if they are no longer used, may remain to have an impact on urban development via the urbanization pattern they had influenced in the past. Voigt (1973) argues that infrastructure types important in the starting phases of urban development (railways, for example) had a strong impact on spatial patterns of urbanization. But infrastructure types that become important at a later stage (for example, the highway system) have smaller impacts, because the urbanization economies lead to more stable spatial patterns. This is one of the reasons why the evolution of the railway network during the 19th century was chosen for this Chapter. At this stage of urban development one may expect considerable impacts of infrastructure provision.

In the above discussion we focused on the impact of transport infrastructure on the development of cities. However, one should be aware that the construction of transport infrastructure does not occur in an autonomous way. If the private sector is active in providing infrastructure (as it was during the first phases of railway development), it will undoubtedly place a burden on expected economic performance of the infrastructure, given the demand for transport services. When the public sector provides infrastructure, a different situation emerges. In this case, the market for transport services may play a smaller role. Even when cost-benefit analysis is used in the preparatory phases in many countries, it is no exception the political rationality in the decision-making process leads to an other outcome than what economic rationality would dictate. Thus, an in-depth investigation of the actual allocation of funds for infrastructure towards modes and regions would be necessary (see Chapter 13 for an effort in this respect).

As indicated in Figure 12.1, there are other relevant factors along with transport infrastructure when one studies urban growth. General economic conditions and demography are among these. Urban growth will be studied by means of population totals. The reason is that data on urban employment are very incomplete for the 19th century. Whenever one may assume that

Map based on Jonckers Nieboer (1938)

Figure 12.2 The Dutch railway system in 1847

population growth in urban areas is indeed driven by employment opportunities, such an approach can certainly be defended.

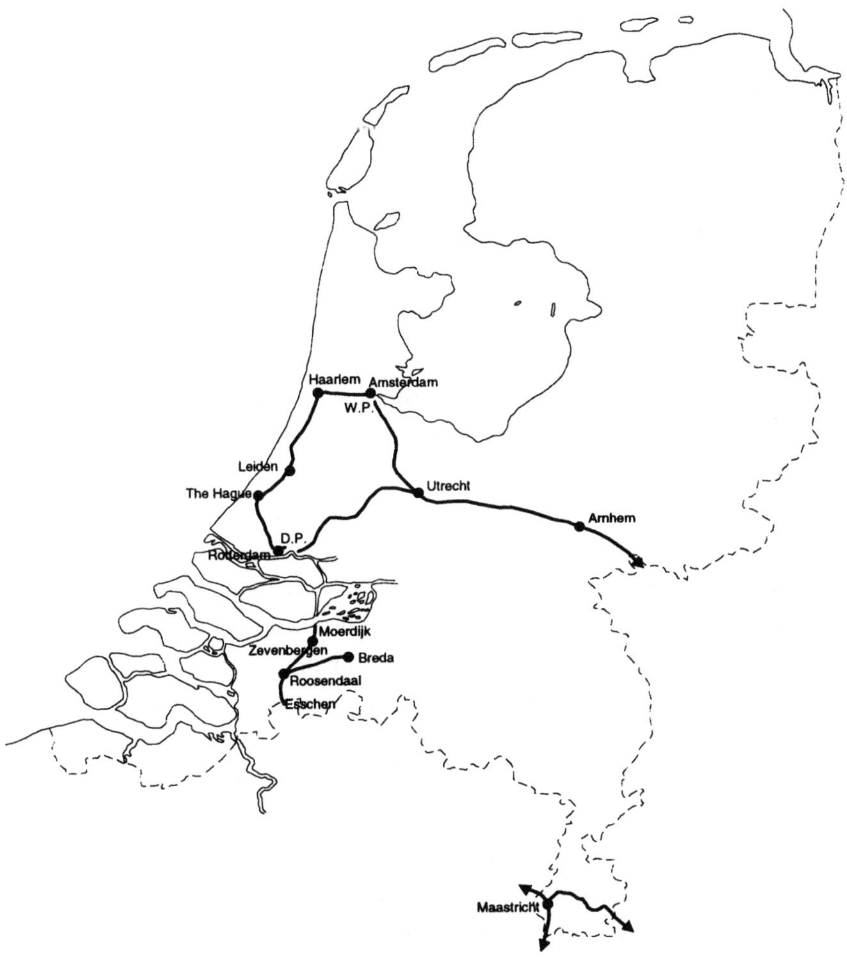

Map based on Jonckers Nieboer (1938)

Figure 12.3 The Dutch railway system in 1862

12.3 Development of the Dutch Railway System

12.3.1 Historical Description

The evolution of the Dutch railway system began in 1839 with the opening of a line between Amsterdam and Haarlem, two important cities in the western part of the Netherlands. Within eight years this line was extended to the cities of Leyden, The Hague and Rotterdam. During the same period, a line was opened connecting Amsterdam with Utrecht and Utrecht with Arnhem near the German border. Thus, a considerable number of major cities of the Netherlands already had a railway connection in 1847 (see Figure 12.2).

After 1847 the development slowed for about 15 years. Many plans were made to extend the system, but most did not materialize. There was fear that the Netherlands would lag behind its neighbours Belgium and Germany, where railway networks were much denser at that time. Indeed, most expansions during that period took place in border areas where Dutch regions were connected to the networks of the neighbour countries, not to the national system (see Figure 12.3).

An important reason why railway development was fairly slow during this period concerns the existence of a dense system of canals and rivers, which allowed adequate services by means of inland navigation, thus making the demand for another transport mode small. Another reason was that industrialization started rather late in the Netherlands, and this suggests that an important segment of the transport demand market was weakly developed.

During this period railways were constructed and operated by private companies. The low level of expansion of the railway system induced the government to enter the market with a state railway company which would build and operate lines next to those of the private railway companies. The combination of the entry of the state railway company and the increased investment efforts of the private companies led to rapid expansion of the railway system after 1863. In 1875 the Dutch railway system already consisted of a rather dense network connecting all important parts of the country.

The factors that governed the design of the Dutch railway network have been studied by Dijksterhuis (1984). A distinction can be made between factors at the demand side and the supply (cost) side. These factors will be summarized in the following two sections.

12.3.2 Factors at the Demand Side

Connecting seaports with industrial hinterlands. There were no mining activities in the Netherlands and industrial development began late. As a consequence, the major economic sector influencing railway development was transport itself. The ports of Rotterdam and especially Amsterdam had a long tradition as transshipment points of international and intercontinental trade flows. The rapid development of the Ruhr area led to an increase in the demand of transport via these transshipment points. For Amsterdam this was particularly important, since unlike Rotterdam, it did not have a direct water-connection with the Ruhr area. It is no surprise therefore that Amsterdam was most active in establishing a railway connection with Germany. Although this strategic argument played a major role in discussions about railway development in the Netherlands, it would last until 1856, 17 years after the construction of the first railway line, when Amsterdam finally achieved its direct railway connection with Germany. In the meantime, in the west of the Netherlands a railway network had already existed for over 10 years. It should be noted that the strategic importance attached to freight transport in the planning of the Dutch railway network was not reflected in the actual transport flows. The share of passenger transport in total transport was larger than expected and passenger transport experienced much higher growth rates than freight transport.

Connection of larger cities in the Netherlands. The first railway line connecting the cities of Amsterdam and Haarlem was actually meant to be a test for the feasibility of a longer freight-oriented railway line connecting Amsterdam with Germany. However, the line proved to be successful as a passenger line, and this led to the extension with Rotterdam via The Hague and Leyden. After the construction of lines between Amsterdam and Utrecht, and Utrecht and Rotterdam, a circle structure evolved which has reinforced the structure of the system of cities in Holland as a Randstad structure. It is this mechanism of connecting larger cities in the country that will be studied more in depth in this Chapter. The question addressed in section 12.3.2 is what economic criteria seem to have directed the decisions of building railway lines.

Connections with neighbour countries. It was not long before certain peripheral parts of the Netherlands were connected to neighbour countries. One might have expected that the system would first be developed as a national system after which international connections would be added (see Rietveld, 1993). But this was not true. Private initiatives in border regions soon led to railway lines in these regions that were connected with lines in Belgium and Germany. The major railway developments in the 1850s took place in the border regions. The idea of a nation state was not yet strong at that time (Knippenberg & de Pater, 1990). Although the Netherlands is a small country, there were very large cultural differences between the provinces and it would require a lengthy unification process before these differences were reduced. The railways certainly have played a large role in this after the 1860s. Another factor that may have played a role is that Belgium and the Netherlands only existed as separate countries beginning in 1839. The national boundaries were probably considered as rather artificial in the following decades. On the eastern side, the unified German state only started to exist in 1870. Before that, the Dutch eastern border did not separate two nations. Also language differences at most played a minor role at that time.

Services to smaller cities and villages. During the first phases of the evolution of the railway network the major intention was to connect the larger cities. But of course, wherever possible, stations were added near smaller cities and larger villages in order to increase effective demand. The priority given to large cities not only depended on the amount of people living there, but also on the spending capacity of urban consumers. In addition, freight transport mainly took place between cities, because it was here that industry and ports were located.

Other demand factors. The above demand factors relate to segments of the demand for transport that can be served by railway lines. Another segment that became relevant in the course of the evolution of the railway system concerns commuting. Commuter lines were constructed around some of the larger cities. Because of the relatively small size of the major cities in the Netherlands, the importance of railway lines built mainly to serve commuters was limited.

Still another market segment that induced the construction of some railway lines concerned the connections of major ports such as Rotterdam and Amsterdam with outports located at the seacoast (Hoek van Holland, Den Helder, IJmuiden). Such railway lines were useful for transport motives where speed was important, such as mail and passenger traffic.

A final demand side factor which has influenced the development of Dutch railway networks and which should be mentioned in this context is competition between railway companies. During the first 50 years of railway development, many different companies existed. Competition among these companies sometimes has led to the construction of competing connections between major cities.

The aforementioned demand related factors are not entirely independent from each other, since transport networks are usually polyvalent: railway lines can be used for several segments on the transport market simultaneously. For example, a major connection between two cities can be used for both passenger and freight transport; in addition, it can be used to connect smaller municipalities located along the line, and also to provide commuter services.

12.3.3 Cost Factors

In addition to demand related factors, cost factors also played a role in the design of railway lines. Most of these factors were relevant at the local spatial level, but some of them were also important at a higher spatial level.

River crossings. The spatial structure of the Netherlands is strongly determined by a number of large rivers mainly flowing from East to West. These rivers provided ample opportunities for transport in the East-West direction. However, land transport was problematic in the North-South direction and was apparently a substantial barrier. This also holds true for the railway system. In the initial phases of railway construction, major developments occured in the East-West direction. Thus, the railway system was mainly parallel to the existing system of waterways so that the new system reinforced the existing spatial structure. Even the very first railway line between Amsterdam and Haarlem ran completely parallel to an existing canal along this trajectory.

The large rivers were considered as unavoidable barriers during the first part of the evolution of the railway network. This is one of the reasons why the Dutch network developed in a rather fragmented way. It was only around 1870 that the first North-South connection was realized. After this, river crossings continued to play a role as a cost factor in railway construction. These crossings were much more expensive per kilometre than ordinary tracks. In addition, the river courses sometimes had a more local impact on the route of the crossing railway lines, because points of crossing were sought in such a way as to minimize construction costs. See also Haggett et al. (1977) for a discussion on the 'law of refraction in the construction of transport links'.

Physical conditions: soil, slopes. Soil conditions in the Netherlands were often difficult for railway construction, because in several regions one finds major amounts of moor. Dijksterhuis (1984) indicates that in such regions routes of railway lines were sometimes slightly adjusted to use sand layers. Steep slopes do not exist in most parts of the Netherlands. Only in the province of Limburg, has the hilly structure had some impact on the routing of railway lines.

Spatial structure of existing cities. The earlier phase of development of the railway system took place in a period when most cities still were walled and had a very compact structure. As a consequence, there was little space for a railway station within the city itself. The costs of realizing a railway station within the walled area of the city would be very high. In addition, because of the small spatial area of most cities, there was also little need for such a location of railway stations. Therefore, in many cities, railway stations were created outside the city, near the walls which had started to lose their strategically defensive importance during the first phase of railway construction.

12.3.4 Military Considerations

A final remark is in order about the military aspects of railway construction. Without exception in every country, infrastructure development is to a considerable extent driven by military rather than economic considerations. In the 19th and early 20th century, the railway system was generally regarded as extremely important for military purposes. In the Netherlands military considerations also played a role, but to a lesser degree than in some other countries. Military considerations about railway construction mainly concerned the construction of fortresses to prevent possible enemies from using the lines to penetrate the major cities. In addition, military considerations sometimes played a role in the determination of railway routes. Since railway lines in some regions run parallel to rivers, one had to choose on which side of the river the line had to be built. Military considerations were sometimes significant here, depending on the direction from where a future enemy attack was most plausible.

12.4 Network Design Methods

A mutual relationship exists between network development and urban development. In the present section we focus on the impact of cities as the nodes of networks on the construction of railway links. The aim of this section is to investigate the development of the Dutch railway system in the 19th century and see the extent to which this development can be replicated by means of network location models. We start with a survey of cost-related network location models. Then we introduce demand aspects; next, rates of return on investments in the networks are added to the model. Finally, the model is tested in a case study of the development of the Dutch railways in the 19th century.

12.4.1 Network Design: Cost Minimization Algorithms

The following definition of a network will be used in this section:
A *network* N is a pair N = (V,A), where V is a set of nodes i with weight m_i, (i = 1, 2, ..) and A is a set ordered pairs of nodes a_{ij} called arcs with length d_{ij}[1].

The practical implementation to railway networks is obvious: each city is represented by a node with a weight equal to the number of inhabitants of the city, and each link between two cities is represented by an arc with a (symmetric) length equal to the distance between these cities. There are of course many drawbacks in this representation, since it neglects many social, economic and geographic factors, but even in this simple case, the problem of building an optimal network appears to be quite untractable.

The first step in building an optimal network is to answer the question what is the objective function that has to be maximized or minimized, and what are the constraints to which this optimization is subjected. When a cost minimizing approach is followed we have several choices:

If the objective is to minimize *user costs* of a network, the resulting *maximum connected network* problem can be defined by:

---

[1] For a number of definitions of special types of networks and of properties of networks in the context of graph theory, we refer to Appendix 12.I.

(MCN) $\min \sum_i \sum_j p_{ij}$

where:

p_{ij} is the length of the shortest path between node i and j. p_{ij} can recursively be defined by:

$$p_{ij} = \min_{k,\, a_{kj} \in A} (p_{ik} + d_{kj});$$

if nodes i and j are not connected, $p_{ij} = \infty$.

The solution of this unconstrained problem is trivial: a direct link between each pair of nodes, resulting in a complete network. Note that we assume user costs to be proportional to the distance travelled, and that the demand is equal among all nodes. If construction of a link is cheap in relation to the user costs, such a complete network can be desirable. In most practical applications, the MCN is subjected to a budget constraint:

$\min \sum_i \sum_j p_{ij}$

sub

$\sum_i \sum_j \lambda_{ij}\, d_{ij} \leq B;$

$\lambda_{ij} = \lambda_{ji}, \forall i, \forall j;$

$p_{ij} \leq \infty, \forall i, \forall j;$

where

p_{ij} is the length of the shortest path between node i and j;

λ_{ij} is a binary variable, assuming values 0 and 1;

d_{ij} is the physical distance between nodes i and j and

B is the budget available.

When it is the objective to minimize the construction costs of a network connecting all the nodes, we have a *Minimum Spanning Tree* problem:

(MST) $\min \sum_i \sum_j \lambda_{ij}\, d_{ij}$

sub

$\lambda_{ij} = \lambda_{ji}, \forall i, \forall j;$

$p_{ij} < \infty, \forall i, \forall j;$

where

λ_{ij} is a binary variable assuming values 0 and 1;

p_{ij} is the length of the shortest path between node i and j.

In this formulation production costs are assumed to be proportional to the length of the trajectories built. The resulting network is a tree. A tree is a subnetwork of nodes and arcs, such that for every pair of nodes there is exactly one path connecting them. A spanning tree is a tree that contains all the nodes of a network. Spanning trees are used in the design of communication networks in which each node must be able to communicate with every other node. If the communication links are expensive, then it is desirable to have just one path between each pair of nodes so that the resulting network is a spanning tree. The Minimum Spanning Tree problem can be solved easily with a Greedy Algorithm (Kruskal, 1956). The idea of the Greedy Algorithm for solving the Minimum Spanning Tree problem is quite straightforward:

Given a network $N = (V,A)$ start building a subnetwork $N' = (V',A')$ with any arbitrary node $i \in V$ and add the nearest node $j \in V$ to N', as well as the arc a_{ij}. In the next step, consider all the arcs $a_{kl} \in A$, which connect a node $k \in V'$ with a node $l \in V \backslash V'$. Let a_{kl} be the arc with minimum length, then add the node l and the arc a_{kl} to the subnetwork N'. Repeat this step until $V' = V$.

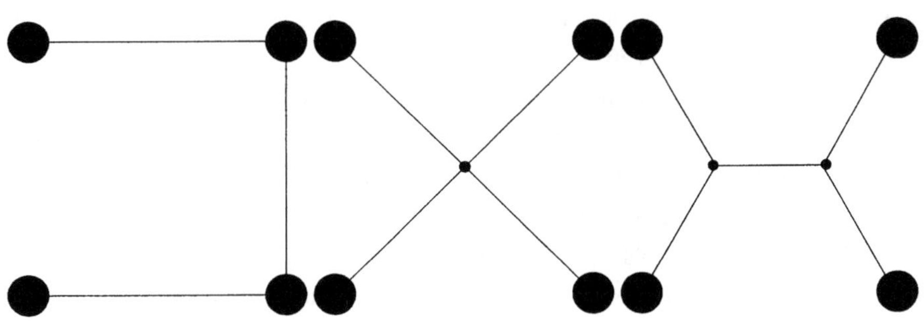

Figure 12.4 The effect of auxiliary nodes on a network

If it is permissible to add new auxiliary nodes to the network, an important improvement can be achieved with regard to the original minimum spanning tree problem. The minimum spanning tree on the 4 nodes in Figure 12.4 has length 3. When one auxiliary node is added, the total length decreases to $2\sqrt{2} \approx 2.82$. A second auxiliary node results in a total length $1 + \sqrt{3} \approx 2.73$. Unfortunately, it is very difficult to determine the optimal position and optimal number of the additional nodes. This problem is called the *Steiner problem* or the *Floating Point Location problem* and is well-solved for the case of three nodes only. In that case, the optimal location of a node added to the network is that place where the angle of the links between the new node and each pair of the old nodes is 120°. It is known that the maximum number of additional nodes needed to construct an optimal network on a network with $n = |V|$ nodes is $(n - 2)$ (Miehle, 1958). In most cases this number is smaller since additional nodes can coincide with the original nodes. The structure of 120^0 angles still holds in more complex networks, but the main problem is which nodes should be added. In an n-node network there are $\sum_{m=1}^{(n-2)} 2^{-m} \binom{n}{n+2} (n+m-2)!/m!$ arrangements possible. This can be solved by a brute-force enumerative algorithm, but the time this requires increases exponentially with the size of the problem. The Steiner problem is recognized to be *NP*-hard, that is: there is no efficient algorithm known, that solves the Steiner problem and it is very unlikely that a good solution method providing an optimal solution in a reasonable amount of time will ever be found. For an extensive treatment of *NP*-hardness and the complexity of (general) optimization problems see, for instance, Garey and Johnson (1979) or Papadimitriou and Steiglitz (1982). Some heuristic algorithms for the Steiner Problem are developed by Morgan (1967).

The former two approaches minimize either user costs or construction costs of a network. Usually however, and particularly in transportation networks, we want to take both into consideration. Minimizing a weighted sum of construction costs and discounted user costs is one possibility. But several other formulations are also possible. For example:

a. minimize construction costs subjected to a maximum detour factor S for each pair of nodes:

$\min \sum_{ij} \lambda_{ij} d_{ij}$

sub

$p_{ij} \leq Sd_{ij}, \forall i, \forall j;$

$\lambda_{ij} = \lambda_{ji}, \forall i, \forall j;$

where

d_{ij} is the physical distance between i and j;

λ_{ij} is a binary variable, assuming values 0 and 1;

p_{ij} is the length of the shortest path between i and j and

S is the maximum detour factor for each pair of nodes ($S \geq 1$).

It is not difficult to see that when the detour factor is finite, the resulting network is connected.

b. minimize user costs subjected to a budget constraint for the construction of links:

$\min \Sigma_i \Sigma_j p_{ij}$

sub

$\Sigma_{ij} \lambda_{ij} d_{ij} \leq B;$

$\lambda_{ij} = \lambda_{ji}, \forall i, \forall j;$

$p_{ij} \leq \infty, \forall i, \forall j;$

where

p_{ij} is the length of the shortest path between node i and j;

λ_{ij} is a binary variable assuming values 0 and 1;

d_{ij} is the physical distance between i and j and

B is the budget available.

In the latter problem the sum of the lengths of the shortest paths is minimized under the condition of a limited budget. However, this problem cannot be solved by a brute-force enumerative algorithm, since in a network on $n = |V|$ nodes, there are $2^{n(n-1)/2}$, that is exponentially many, configurations possible. Although some work is done in order to improve exact algorithms, we still depend on heuristic approaches. One such heuristic algorithm has been developed by Scott (1969) and MacKinnon and Hodgson (1970) for a slightly different problem formulation: the *Forward Approximation Algorithm*:

1 Determine which links of the possible network N = (V,A) are very unlikely to occur in an optimal network and remove them.

2 Let N′ = (V′,A′) be a (the) minimum spanning tree on N = (V,A).

3 Add that arc a_{ij} to the network that results in the largest improvement of the objective function, provided that the constraint is not violated.

4 Find an 1-optimal network by systematically removing each arc $a_{ij} \in V'$ and substituting it with an unused arc $a_{kl} \in V \backslash V'$ if this would result in an improvement of the objective function and the constraint is not violated.

Repeat 3 & 4 until no improvement can be made.

Step 1, although it improves the efficiency of the algorithm, is quite arbitrary, since it is hard to give a general criterium which decides whether or not an arc is implausible in an optimal network. It is obvious that the resulting network is not necessarily optimal. The results are assured to be closer to the optimal solution when a more sophisticated interchange step 4 is used in order to build a 2-optimal or a 3-optimal network. The problem is that this step is the most time consuming step of the algorithm, and interchange procedures which account for the substitution of three or more arcs at once are unfortunately far too slow.

Another approach suggested by Scott (1971) works the other way around: it starts with a complete network and successively eliminates nodes until a feasible solution is obtained.

12.4.2 Network Design: Introduction of the Demand Side

The problems as defined thus far are not very realistic, because they assume that each node has the same size. When a link is to be made to transportation networks, it is obvious that a link between two big cities is much more desirable than a link between two small cities, when the distances are more or less equal. Another point that must be considered is that the demand for transportation between two cities strongly depends on the length of the shortest path between them. A more demand orientated formulation such as maximizing the total use of the network is preferred above a formulation in terms of minimizing costs. The demand for transportation measured as the number of trips between two cities i and j can be approximated by the following gravity-type model:

$$D_{ij} = c\ m_i m_j \ / \ p_{ij}^{\alpha}$$

where

c is a constant value and

α is a distance decay parameter

m_i, m_j are masses of cities i and j

A drawback of this demand function is that every link added to the network results in a rise of the total demand for transport, while it is reasonable to assume that the total demand for transport in a city is limited. The unconstrained gravity model does not include the possibility of substitution between destinations. To allow for such a possibility, a constrained gravity model can be used. In that case the variables A_i and B_j are added to the demand function as balancing factors, resulting in the following double-constrained model:

$D_{ij} = c\ A_i B_j m_i m_j\ /\ p_{ij}^{\alpha}$

where

c is a constant;

$A_i = (\Sigma_{j \neq i}\ B_j m_j\ /\ p_{ij}^{\alpha})^{-1}$;

$B_j = (\Sigma_{i \neq j}\ A_i m_i\ /\ p_{ij}^{\alpha})^{-1}$ and

α is a distance decay parameter

The total use of the network can be measured by the number of trip kilometres:

$$\Sigma_i \Sigma_j\ c\ A_i B_j m_i m_j p_{ij}\ /\ p_{ij}^{\alpha}$$

In the case of a railway company with uniform tariffs this is proportional to the total receipts. If we also assume that operational costs are proportional to the number of trip kilometres, the above criterion is proportional to total receipts net of operational costs. The resulting network optimization problem is:

$$\max\ \Sigma_i \Sigma_j\ A_i B_j m_i m_j p_{ij}^{1-\alpha}$$

sub

$$\Sigma_i \Sigma_j\ \lambda_{ij} d_{ij} \leq B$$

where

$\lambda_{ij} = \lambda_{ji}$, $\forall i$, $\forall j$;

p_{ij} is the length of the shortest path between node i and j;

λ_{ij} is a binary variable assuming values 0 and 1;

B is the budget available.

This problem is more complex than the problems mentioned above and since there is no efficient algorithm known, we shall focus on heuristic methods. The Forward Approximation algorithm as proposed in section 2 can, in an adapted form, be used again and now consists of a repetition of steps 3 and 4.

In the present context, the algorithm can be formulated as follows:

input: a network N = (V,Ø) and a set A′, where V is a set of nodes i with weight m_i, the set A of arcs a_{ij} between pairs of nodes of V is empty and A′ is a set of all possible arcs a_{ij} between pairs of nodes in V.

output: a 1-optimal network N′, on the nodes N subjected to the budget constraint $\Sigma_i\Sigma_j\ \lambda_{ij}d_{ij} \leq B$. Thus, there is no link $a_{ij} \in A$, and a link $a_{kl} \in A'$ such that substitution of the two arcs both improves the objective function and satisfies the budget constraint

1 $\lambda_{ij} := 0, \qquad \forall a_{ij} \in A'$

2 $p_{ij} := \infty, \qquad \forall a_{ij} \in A'$

3 for all $a_{ij} \in A$, determine what would be the value of $\Sigma_i\Sigma_j A_i B_j m_i m_j p_{ij}^{(1-\alpha)}$ after adding a_{ij} to the network, and add that arc to the network for which the value of this objective is maximal, provided it does not violate the budget constraint.

4 for all $a_{ij} \in A$, examine whether there exists an arc $a_{kl} \in A'$ such that the network $N'' = (V,\{a_{kl}\} \cup A\backslash\{a_{ij}\})$ satisfies the budget restriction and improves the objective function.

5 if such a pair a_{ij} and a_{kl} exists then
$A' := \{a_{ij}\} \cup A'\backslash\{a_{kl}\}$ and
$A := \{a_{kl}\} \cup A\backslash\{a_{ij}\}$ and repeat 4
if there does not exist such a pair, go to 6.

6 if the network N′ is a maximum connected network or if no arc a_{ij} can be added without violating the budget constraint then: STOP.
The resulting network N′ is the approximation of the optimal network.
otherwise go to 3.

The computation time of this algorithm is dependent to a considerable extent on the efficiency of the algorithm used to find the shortest path between each pair of nodes. Floyd (1962) has developed an efficient algorithm that finds the shortest path between each pair of nodes in a network simultaneously. In the network optimization algorithms, in each iteration the shortest paths are recomputed several times after adding or deleting one edge to the existing network. It is not necessary to repeat the computation completely. Algorithms that use information of the already existing network are developed by Loubal (1967) and Murchland (1970). For an overview of shortest path algorithms, see Steenbrink (1974) and Tulp (1991).

Other approaches of the unconstrained version of this problem are made by Boyce et al. (1973), with a branch-and-bound algorithm and by Hodgson (1974, 1975), who uses dynamic programming.

12.4.3 The Rate of Return on Investment in Network Infrastructure

Consider the following objective function:

$$\max\ s = (\Sigma_i \Sigma_j A_i B_j m_i m_j p_{ij}^{(1-\alpha)})/(\Sigma_i \Sigma_j \lambda_{ij} d_{ij})$$

The numerator is proportional to the number of passenger kilometres travelled and is assumed to represent receipts of the railway company net of variable costs. The denominator is the length of the entire network and is assumed to be proportional to the total investment sum. Thus, the criterion is proportional to the return on investment. Decision making behaviour on the basis of the rate of return on investment can be assumed to occur in the following sequential way.

At the start of the process, that pair of nodes can be determined which leads to the highest criterion values of s. In a second step, one can determine the next pair of nodes which leads to the highest value of s, given the arc that already exists, etc. Railway construction stops when the return on adding a link in the network is lower than the rate of return outside the railway sector.

The resulting network is not required to be a connected network. So it is possible that a choice is made for linking two major cities, while there already exists a path between them, instead of connecting a small village with the remainder of the network.

Consider the question what is the first link to be built in the network. Based on the criterion of the rate of return on investment, one can rank all the potential links in order of attractiveness. Suppose that the link with the best performance is built. Given this fact, then one can again regenerate a rank order of links in terms of which one is the most attractive to build as the second link. Next, because of the interdependencies within networks, there is no guarantee that the second ranking corresponds with the first ranking. The realization of a certain link may have an impact on the rates of return of other possible links.

Network interdependencies also have other implications. For example, if a certain link a_{ij} is the most attractive link to be built as a first link, it is not assured that it will be part of the optimal network if one considers a network consisting of two links. It may appear that the optimal two-link network does not contain the first mentioned link. If decision making takes place in a consecutive way, one may easily arrive at such suboptimalities leading to path dependence. In an algorithmic sense, this problem can be overcome by using a 1-optimal (or i-optimal; see preceding sections) approach so that one considers that an interchange of links in a network may lead to a higher performance. Another consequence of network interdependencies is that the rate of return on investment on one link is influenced by whether or not another link is built. This is especially important when more than one railway company is involved in constructing and operating links. This occurs for example, in international networks; in the early stages of network development many countries have witnessed the existence of several railway companies in the same territory. Situations may occur where a link a_{ij} does not yield the requested market rate of return to firm 1, while the same holds true of link a_{kl} for firm 2 when only one of the links is built. But because of network interdependencies, it is possible that if both links were built, the rate of return of the combination a_{ij} and a_{kl} would be above the requested rate of return. If this holds true for both firms, there is no basic problem preventing the construction and operation of both links. However, if for one firm the rate of return remains below the requested level if both links are built, whereas it increases above that level for the other one, a problem emerges. In that case the first mentioned firm needs a compensation from the other one to induce it to build and operate the link concerned.

12.5 Application: the Development of the Dutch Railway System

In the application of the algorithm we have assumed that the costs of construction of the system are equal everywhere with one major exception. Crossing of long rivers such as the rivers Rhine, Waal, IJ, IJssel and Meuse are supposed to give rise to an additional cost equivalent to constructing 20 km of an ordinary connection. The value of the parameter α has been chosen as 2.

The 35 largest Dutch municipalities (see Appendix 12.II) are taken into consideration, together with Antwerp, Brussels, Aachen and Cologne. By also considering these big foreign cities, some of the developments in the peripheral areas of the country can be explained. Cologne is chosen as the centre of the Ruhr-area. Population totals have been used for 1860. The borders of the municipalities have been based on the 1970 definition. By doing so we include population of surrounding villages which merged with larger centres between 1860 and 1970 (see Van der Knaap, 1980).

Table 12.1 The simulated network

#	simulated link	f_n	#	simulated link	f_n
1	Rotterdam - Schiedam	0	11	Leyden - Gouda	.45
2	Middelburg - Vlissingen	0	12	Rotterdam - Dordrecht	.42
3	Amsterdam - Haarlem	.33	13	Deventer - Zutphen	.39
4	Delft - Schiedam	.25	14	Utrecht - Amersfoort	.43
5	The Hague - Delft	.2	15	The Hague - Schiedam	.4
6	Amsterdam - Zaandam	.66	16	Utrecht - Leyden	.44
7	The Hague - Leyden	.57	17	Amsterdam - Rotterdam	.41
8	Amsterdam - Leyden	.5	18	Veendam - Hoogezand-Sapp.	.39
9	Amsterdam - Utrecht	.55	19	Groningen - Hoogezand-Sapp.	.42
10	Heerenveen - Sneek	.5	20	Dordrecht - Breda	.4

The first 20 links generated by the railway system development model presented in section 3.3 can be found in Table 12.1. Maps of the simulated network are given in the Figures 12.5-12.8. When we consider the performance of the model, we notice that it tends to build short lines before long lines between major cities. For example, the second link generated is Middelburg - Vlissingen (6 km). These municipalities are 17th and 28th respectively in the ranking of the municipalities according to their number of inhabitants. The link between two major cities as Amsterdam (no. 1) and Leyden (no. 5) (89 km) is the 9th line generated. There are some isolated edges, but in the historical development of the network, there too were isolated clusters. The development of the network is concentrated in the western part of the country. This conforms to the historical development. The path between Amsterdam and Rotterdam is built more or less the same as the historic development. The model builds the path from Rotterdam to Amsterdam, whereas the historical path is built from Amsterdam to Rotterdam. The model generates a

Figure 12.5 The Dutch railway system simulated after 5 links

link between Amsterdam and Leyden, but it was impossible to build that link before 1852. In that year the Haarlemmermeer was impoldered. The link Haarlem - Leyden however, was finished in 1842, so Amsterdam and Rotterdam were already connected. The model can be improved by taking this into consideration. It is interesting whether the improved model will first add the link Haarlem - Leyden or Leyden - Utrecht. Both result in a connection of Amsterdam and Rotterdam.

Besides the double-constrained model, a simulation with the unconstrained

Figure 12.6 The Dutch railway system simulated after 10 links

modal is also made. As expected, the unconstrained model has a strong tendency for building links between large cities: of the first 10 links, 5 links were between Amsterdam and other big cities. An noteworthy point of the unconstrained model is that very early on it generates links from Arnhem and Maastricht to the German border.

It is not easy to judge whether the development of the railway system as simulated by the models developed in section 13.4 is similar to the actual development of the railway system. In order to measure the correspondence

Figure 12.7 The Dutch railway system simulated after 15 links

between the simulated and the actual network, one can proceed as follows. Let B_1 denote the first arc realized in the system. It is followed by the arcs $B_2;..B_N$. Similarly, C_1 is the first arc generated by the algorithm. This arc is followed by the arcs $C_2;..C_N$. Clearly, when $B_1 = C_1$, $B_2 = C_2$, etc., there is a complete correspondence between the actual and the simulated development of the system. However, this will be an exceptional case. As a measure of the correspondence between $\{B_1;..;B_N\}$ and $\{C_1;..;C_N\}$, we propose to use the share f_n of the number of nodes in $\{C_1;..;C_N\}$ which correspond with a node

Figure 12.8 The Dutch railway system simulated after 20 links

in $\{B_1;..;B_N\}$. This can be computed for any value of n = 1;..;N. Thus, if the first arc was predicted incorrectly $f_1 = 0$. Further, if $B_1 = C_2$ and $B_2 = C_1$, then $f_2 = 1$, etc.

The f_n function of the double-constrained model of course alternates strongly for low values of n, after 20 links it is .4, so about 40% of the links is rightly predicted. The values of f_n can be found in Table 12.1.

12.6 Impacts of Railway Development on Urban Development

In this section we discuss the other aspect of the mutual relationship between urban development and network development, i.e., the impact of railway construction on urban development (see Figure 12.1). There is a wide variety of factors that play a role in urban growth. In the present section we will analyze urban growth from the viewpoint of railway infrastructure. Clearly, this is an analysis with a limited scope which can only yield first approximative results. One of the difficulties with a more complete analysis is that the collection of 19th century data at the urban level is not easy. A useful source is the census which has been carried out every ten years and yields data on population totals. Data on economic structure are difficult to obtain. Thus, we used population growth as the main indicator of the development of urban growth. The assumption is that population developments reflect the economic dynamics of the cities concerned. The following statistical analysis actually does not deal with an explanation of the level of urban growth at the macro level. Instead, it addresses the variation in urban growth rates among cities for a series of decades and tries to explain the variation in these growth rates by means of a limited number of factors:

- The growth rate of the city population during the preceding decade. This factor represents the impacts of unobserved city specific factors which have a longer run impact on urban growth potentials. This variable will be denoted as g(t-2,t-1).
- The size of the city at the beginning of the decade. This factor represents the possible urbanization economies which give large cities an advantage over small cities. This variable is denoted as M(t-1).
- The number of direct railway connections the city has with other municipalities at the beginning of the decade considered. This is measured as the sum total for each city of the connectivity matrix, and will be denoted as C(t-1). Thus, after the completion of the first line between Amsterdam and Haarlem, the value of C equals 1 for these cities and zero for all other cities. The extension of the line through the connection between Haarlem and Leyden leads to a C score equal to 2 for Haarlem. Note that the accessibility concept here is of the type acc4 mentioned in Chapter 2.
- The change in the number of direct railway connections during a decade

C(t)-C(t-1), or if one assumes that adjustments take time: C(t-1)-C(t-2). If a linear relationship is used, we arrive at the following specification:

$$g(t-1,t) = a_0 + a_1{*}g(t-2,t-1) + a_2{*}M(t-1) + a_3{*}\{C(t)-C(t-1)\} + a_4{*}C(t-1)$$

The estimation results can be found in Table 12.2.

Table 12.2 Explanation of urban population growth over five decades in 44 Dutch cities

dependent variable	1840-50 (t=1850)	1850-60	1860-70	1870-80	1880-90
constant	.052	.058	.012	.056	.065
	(2.22)	(3.75)	(.59)	(3.51)	(1.91)
growth rate in (t-2,t-1)	.109	.648	.473	.686	.263
	(.55)	(3.78)	(3.61)	(5.59)	(1.21)
population in t-1/1000	-.00022	-.00015	-.00004	.00066	.00051
	(-.59)	(-.52)	(-.14)	(3.57)	(1.73)
number of lines in t-1	.0399	.00024	.028	-.00076	.0089
	(.70)	(0.02)	(2.34)	(-.11)	(.89)
change in number of lines (t-1,t)	.0069	.0462	.017	-.014	.051
	(.49)	(1.95)	(2.23)	(-1.38)	(2.21)
R^2	.022	.295	.351	.577	.276

During the first period (1840-50) none of the explanatory variables specified plays a significant role. The value of R^2 is extremely low. A possible explanation of this result is the low quality of urban population data during this period.

During the decade between 1850 and 1860 two explanatory variables have a significant impact on urban growth. The lagged growth rate is the first one: the coefficient has a value of .65, which indicates that there is indeed a strong intertemporal relationship in urban growth. This may be a reflection of omitted variables, such as sectoral composition, which continue to influence urban growth over several decades. The other significant variable is the change in the number of railway lines entering the cities. Thus, we find an indication that (change in) the railway system has indeed affected the growth of cities.

Between 1860 and 1870 three significant explanatory variables are found: lagged population growth, the number of railway lines entering the city, and the change in the number of railway lines. Up to this period urbanization economies did not play a role: no significant coefficients were found for the lagged population size. This changes after 1870.

Between 1870 and 1880 two significant variables are found: lagged urban growth and the urban population size. During this period urbanization economies start to play a role, but the railway system does not seem to have an impact. For this period the highest level for R^2 is found: .58.

For the final period considered (1880-1890), the impact of lagged growth is smaller than in the preceding decades. The significant variables relate to urbanization economies and the change in the railway system.

The conclusion is that the pattern of urban growth seems to have changed considerably from decade to decade. Railways did play a role in the evolution of the Dutch urban system, but its role should not be overemphasized: in only three of the five decades considered, the (change in the) number of connections had a significant influence on urban growth. The strongest impact has been found for the decade between 1860 and 1870. Alternative specifications with a lag in the change of the number of railway lines did not yield better results.

The significant coefficients for railway lines found in Table 12.2 are all positive. This means that in the present study no empirical support is found for the possibility that improvement of infrastructure has an adverse effect on the urban economy. The reason is most probably that in the case of railways some cities may indeed experience backwash effects due to a connection with the railway system. However, these possible backwash effects are more than compensated by the improvement of the competitive position of newly connected cities compared with smaller cities and villages that are not at all connected. Since our analysis was confined to the set of larger municipalities, it is no surprise to observe mainly positive impacts of railway connections.

12.7 Concluding Remarks

The mutual relationship between urban growth and railway development has been studied in this Chapter.

To study the impact of urban development on network development we used a model for network construction based on the evaluation of rates of return for different alternatives. This model has been used to simulate railway developments in the Netherlands. The simulated results indicate that most of the early railway developments take place in the western part of the Netherlands. This is indeed what happened in reality. About 40% of the first 20 links were predicted correctly. It is probable that a further calibration of the model will lead to better results.

In the second part of the study we tested the hypothesis that railway development has an impact on urban growth patterns. The conclusion is that indeed during three of the five decades studied significant impacts can be shown. However, the impact of railway lines on urban development should not be overemphasized. The railway system was just one among several factors that played a role in urban development during the 19th century.

The present analysis can be extended into several directions. The transport demand model can be refined by introducing other transport modes (walking, coaching, but especially also water transport). The number of nodes can be extended by taking into account more than the set of largest nodes used up until now. Another interesting extension would be to pay explicit attention to the fact that different railway companies operated in the 19th century: because of network externalities one may expect that the supply of links will be influenced by the number of suppliers and the networks they already operate.

Appendices

Appendix 12.I. Definitions of properties of graphs

In order to formalize the concept of a network, some tools of graph theory are used. Graph theory has appeared as a tool to solve a large class of mathematical problems. The language of graphs and networks and the results of graph theory will play an important part in this Chapter. Therefore, some terminology of graph theory is presented in this Appendix. An extensive treatment of graph theory with respect to optimization problems is given in, for example, Nemhauser and Wolsey (1988).

A *graph* $G = (V,E)$ consists of a finite nonempty set $V = \{1;...;n\}$ of *vertices* and a set $E = \{e_1;...;e_m\}$ of unordered pairs of different vertices, called *edges*. Thus, $e_k = \{i;j\}$, $i \neq j \in V$.

An edge e_k is incident to a vertex i if $i \in e_k$. Notation: $e_k Ii$. The graph $G = (V,E)$ can be represented by its incidence matrix

$A = (a_{ij})$

where

$$a_{ij} = \begin{cases} 1, & \text{if } e_j Ii \\ 0, & \text{else} \end{cases}$$

The graph $G = (V,E)$ can also be represented in a plane by points and lines: The points or vertices are arbitrarily placed in the plane. A line connects points i and j if $\{i;j\} \in E$. Thus, the matrix A and the picture in Figure 1 are two representations of the same graph. The number of edges incident to a vertex i is called the degree of i. Notation: d(i). Two vertices i and j are adjacent if $\{i;j\} \in E$. Notation: iAj.

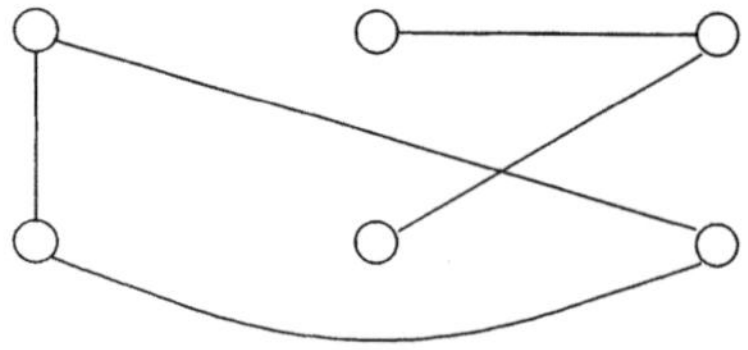

Figure 1 graphical representation of G

$$A = \begin{bmatrix} 1 & 0 & 0 & 1 & 0 & 0 \\ 0 & 1 & 1 & 0 & 0 & 0 \\ 1 & 0 & 0 & 0 & 0 & 1 \\ 0 & 0 & 0 & 1 & 0 & 1 \\ 0 & 0 & 1 & 0 & 1 & 0 \end{bmatrix}$$

Matrix A incidence matrix of G

A graph G = (V,E) is called complete if $E = \{\{i;j\} \mid \forall i \neq j\}$. Notation: $G = K_{|V|}$, where $|V|$ is the cardinality of V, i.e. the number of vertices in V.

If $V' \subseteq V$ and $E' \subseteq E(V')$, then $G' = (V',E')$ is said to be a subgraph of G. $G' = (V',E(V'))$ is the subgraph of G = (V,E) induced by $V' \subseteq V$. Notation: $G' = \langle V' \rangle$.

A sequence 0;...;k, $k \geq 1$, of vertices is called a 0-k *walk* if $\{i-1;i\} \in E$, $\forall i \in Z_k$, $Z_k = \{1;...;k\}$. A walk can also be represented by its edge sequence. The length of a walk is the number of edges in it. A *path* is a walk without vertex repetition. There is a unique partition of the vertices of the graph G = (V,E) into subsets $V_1;...;V_p$, such that vertices i and j are in the same subset V_k if and only if G contains a i-j path. Let V_k be a subset of the partition. The subgraph $G' = \langle V_k \rangle$ is called a *component* of G. Two vertices i and j are *connected* when they belong to the same component. The graph G = (V,E) is said to be connected if it has only one component.

The notion of a graph can be extended by allowing multiple edges, i.e. a pair {i;j} occurring several times in E, loops, i.e. a pair {i;i}, or by defining a weight m_i on each vertex i and a length d_{ij} on each edge e_{ij}. It can be restricted by assuming the edges to be ordered pairs of vertices. In the latter case the graph is called a directed graph or digraph. By convention, the vertices in a directed graph are called nodes and the edges are called arcs.

In graph theory a network N = (V,A) is a directed graph together with a source $s \in V$ and a terminal $t \in V$ and a capacity or length b on each arc. In this Chapter however, we use a slightly different definition of a network.

Appendix 12.II: The 35 major Dutch municipalities in 1860: population size.

1	Amsterdam	257078	19	Deventer	16346
2	Rotterdam	131663	20	Gouda	15980
3	The Hague	80521	21	Tilburg	15854
4	Utrecht	57115	22	Den Helder	15202
5	Leyden	41883	23	Kampen	13902
6	Groningen	39250	24	Zutphen	13728
7	Maastricht	31282	25	Heerenveen	13174
8	Leeuwarden	30868	26	Enschede	12848
9	Haarlem	30594	27	Amersfoort	12683
10	Arnhem	25810	28	Vlissingen	12596
11	Dordrecht	24521	29	Veendam	12576
12	's-Hertogenbosch	23404	30	Sneek	11996
13	Zwolle	23303	31	Zaandam	11676
14	Breda	22855	32	Drachten	11559
15	Delft	22101	33	Apeldoorn	11302
16	Nijmegen	21493	34	Eindhoven	11128
17	Middelburg	17198	35	Hoogezand/Sappermeer	11084
18	Schiedam	16838			

13 Determinants of Infrastructure Supply in European Regions

13.1 Introduction

In studies on the impact of transport infrastructure on the economy, infrastructure is usually considered as exogenous. The starting point is the public sector, which provides transport infrastructure at certain locations or between certain locations. Then various ex-ante and ex-post approaches can be followed to analyze how such a given public sector investment will have an impact on transport flows and the spatial economy (see Figure 13.1).

A well developed body of tools exists to analyze the impacts of transport infrastructure investments on transport flows, for example, in terms of modal choice and route choice (see e.g. Ortuzar & Willumsen, 1993). A variety of approaches can also be found on the link between transport infrastructure and spatial economic development (see Chapters 3 and 4), although it should be noted that here the degree of uncertainty of the results is usually much larger due to the long-run nature of the adjustment processes in space.

It is important not to overlook the *mutual* dependence between transport infrastructure and spatial economic development. There will be feed-backs from spatial economic developments on transport infrastructure (see Figure 13.2). These feed-back links may assume various forms. Sometimes, the public sector may decide to expand the infrastructure stock in order to *accommodate* ongoing or expected spatial economic developments in certain regions. In other cases the public sector *responds to bottlenecks* in infrastructure use. Another form would be that the public sector wants to *stimulate* certain lagging regions by providing it with better infrastructure, even though the present level of infrastructure use does not indicate the existence of bottlenecks.

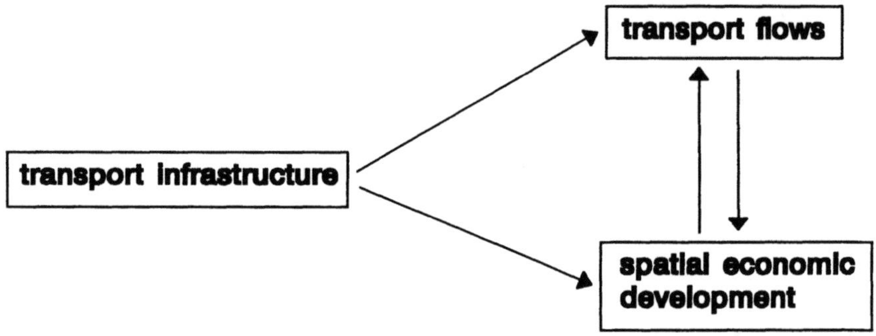

Figure 13.1 Impacts of transport infrastructure

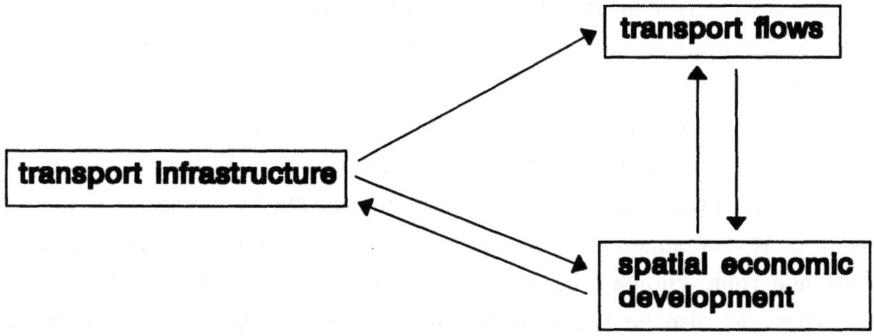

Figure 13.2 Mutual dependence between infrastructure and spatial economic development

In this Chapter we will discuss certain aspects of the feed-back relationships between spatial economic development and transport infrastructure. This feed-back relationship is important for at least two reasons. Firstly, a proper understanding of investment behaviour in the public sector is important for its own sake. The long-run dynamics of economies strongly depend on investment behaviour, and in this respect both private and public investments deserve our attention. Secondly, a deep understanding of public sector decision making may also be important for assessing the impacts of infrastructure investments on economic development. The reason is that ex-ante studies on impacts of infrastructure projects will often depend on results of ex-post experiences with earlier projects. The aspects taken into consideration at these earlier occasions will have influenced the selection of projects at that time and also the specific impacts observed.

The economic theory on the public sector supply of network infrastructure has a strong normative orientation. This was already present in the work of Dupuit (1844), who made use of the notion of consumer surplus to analyze the benefits of building canals - at that time the most dynamic type of network infrastructure. The consumer surplus concept has become a major element of social cost benefit analysis, a technique used in various European countries to assist public sector decision making in the field of infrastructure. Social cost benefit analysis is essentially a normative instrument. It tells policy makers which investment plans are most attractive and which plans should certainly not be implemented.

Public sector decision makers often do not follow the recommendations implied by social cost benefit analysis, however. There may be good reasons for this, since the focus of social cost benefit analysis is somewhat one-sided in its emphasis on the efficiency issue. Equity among groups affected by infrastructure projects is difficult to take into account with social cost-benefit analysis. In addition, it appears that environmental issues are usually only taken into account rather superficially in applications of cost benefit analysis (cf. Munda, 1993, and Van Pelt, 1994). In addition, there may be other reasons why actual infrastructure policies are not always based on the normative outcomes of cost benefit analysis. For example, governments used infrastructure investments in the past as an instrument to stimulate the economy via the demand side in the short-run without paying attention to the long-run benefits. Another example very common at the beginning of the 1980s, is that governments solve budgetary problems by decreasing the level of public investments, while other types of expenditures (salaries of civil servants, social welfare payments) are kept constant since political pressures to safeguard the latter types of public expenditures are strongest. Clearly, political rationality does not always coincide with the economic rationality implied by cost benefit analysis.

The question then arises as to what are the factors explaining public sector decisions on infrastructure. This question can be studied in various ways, for example, by looking at the total level of infrastructure investments, the distribution among infrastructure types, and the spatial distribution of infrastructure investments.

In the present Chapter we will address the last issue mentioned. We will investigate the factors which explain the supply of transport infrastructure among regions (see Rietveld & Boonstra, 1995). More particularly, we will deal with the distribution of highways and railways among regions in Europe. In section 2 we first give a short discussion on infrastructure policies in Europe. A model for infrastructure supply is formulated in section 3. In section 4 empirical results are presented. In section 5 a more detailed analysis is given of infrastructure supply in border regions. Concluding remarks are formulated in section 6.

13.2 The Spatial Distribution of Infrastructure Networks

In the 1970s spatial equity played an important role in regional development policies in European countries. It was generally hoped that concentrated infrastructure investments in lagging regions would reduce interregional disparities.

In the 1980s most countries show a different policy direction. Bottlenecks become evident in infrastructure in many urban areas. The emphasis in infrastructure policies shifts toward the international level. European integration means that many other traditional instruments of national economic policy lose importance. Infrastructure is increasingly considered as a major instrument for national governments that want to ensure a national economy strong enough to face international competition.

Regional equity considerations have become less important in the national policies of the 1980s. Some countervailing force has been exerted by the EU which has launched the European Fund for Regional Development (EFRD) for lagging regions which has a strong infrastructure component (cf. Armstrong & Taylor, 1993).

Another theme which has gained importance concerns the international level. The planning of infrastructure networks in the past was dominated by national perspectives. In the 1980s the international perspective began exert an influence. This can be seen in the examples of the Eurotunnel-project connecting France and the UK, plans to improve Denmark's connections with Sweden and Germany, and the plans to extend the French TGV network to Belgium, Germany and the Netherlands. Dominance of a national perspective in infrastructure planning may easily lead to neglect of the situation in border regions. This raises the issue of the position of border regions in transport networks. To what extent can we observe that border regions were under-supplied with transport infrastructure in the past? This will be one of the questions explored in the next sections.

In most European countries rail related investments have been more dynamic than investments in other transport infrastructure types during the last decade (see Chapter 2) and it is expected that this trend will continue over the next decades. This will reinforce the position of large cities in Europe for two reasons. The first reason is that transport systems in larger urban agglomerations will be improved by the (further) introduction of light rail. The second reason concerns the introduction of high-speed rail connections which will favour large cities where demand is highest. These rail investments have a polarizing impact on spatial development (see Chapter 5). One should not exaggerate this impact, however, since investments in road continue to be much bigger than investments in rail throughout Europe (see Chapter 2).

13.3 A Spatial Model of Infrastructure Supply

Regions differ considerably in the amount of network infrastructure available to them. In this section we will formulate a model on the supply of the highway and railway network. For both networks the main actor functions at the national level. It is the national government which directly or indirectly decides (via the national railway company) on the shape and spatial distribution of these networks. For railways it has not always been like that. In the 19th century, railway lines began to be supplied by private companies with a regional orientation (see Chapter 12). But in most European countries it did not take long before national governments started to control or nationalize railway companies.

The regional supply of network infrastructure will be measured in terms of the length of the network (in km) per km^2 per area of the region. One would of course prefer to take into account the capacity per link, (e.g., the number of lanes of a highway), but data on capacities are not available at a European level. The supply of infrastructure concerns a stock which has usually been developed over a long period. For example, most of the current canal system in Europe dates back to the first half of the 19th century. For rail the historical component is also considerable.

When modelling the regional supply of network infrastructure, the following factors are assumed to influence network density:

- regional demand
- interregional demand
- possibilities of finance
- costs
- barriers and borders
- regional policies

Network infrastructure serves both *regional and interregional demand.* The relative importance of the two depends of course on the spatial scale considered and the type of infrastructure. For roads the *regional* demand is usually predominant: the large majority of road network users only make short distance trips. Demand will be measured by means of population density. It is expected that high population densities induce high network densities. There may be a reason to investigate a non-linear relationship between the two, since high densities may give rise to relatively low distances travelled per inhabitant, and hence to a relatively smaller infrastructure need.

From a theoretical viewpoint one might prefer to use car density as a representation of demand for infrastructure. However, regional data on car ownership are not complete at a cross national level.

Interregional demand can be taken into account by incorporating population density in neighbouring regions. The higher the population density in the neighbour regions, the larger the demand for infrastructure for interregional use in the region itself.

Possibilities for *financing* will play a role in the supply of network infrastructure. As indicated in Button and Rietveld (1993), the ways of funding infrastructure are rather different among European countries, but a common trend towards a stronger involvement of the private sector is evident. In this study we will use the level of gross domestic product per capita as an indicator of finance possibilities. Institutional factors designate whether it is the *national* or the *regional* level of GDP per capita which is most relevant as a determinant of regional infrastructure density. In most European countries the planning of the highway and railway system involve the mobilization of funds at the national level (cf. Stoffelsma, 1992), so that one would expect that the national GDP per capita is the most appropriate candidate.

Costs of construction and maintenance may play a role in the supply of infrastructure. These costs have a national component (e.g., related to the level of labour costs in a country) as well as a regional one. Regional physical conditions (e.g. presence of mountains) may make construction so expensive that priority is given to other regions where conditions are more favourable. In our study we measure the costs by means of a dummy variable: regions where the highest altitude is at least 2000 m receive the value 1, other regions receive the value 0. One might prefer more refined data (e.g., on slopes) in this context, but this appears to be difficult to obtain at the regional level.

Barrier effects on infrastructure densities may occur for both physical and non-physical reasons. Water is an example of a physical barrier. A region bordering on the sea will have less need for transit traffic by road or rail. Thus, one would expect that its infrastructure density is lower compared with other regions. A countervailing force may be that regions with major sea ports require a high infrastructure density for the road and rail sections of the logistic chain.

National borders are an example of non-physical barriers to infrastructure networks. One may expect that the lack of international coordination in the planning of infrastructure networks as well as the relatively low demand for cross-border mobility of the past has led to a neglect of infrastructure development in border regions.

Finally, *regional policies* may be expected to have a positive impact on infrastructure supply, since these policies are sometimes used to provide regions with higher levels of infrastructure than would be warranted if only regional demand were to be taken into account. A regional policy dummy is introduced such that regions receiving more than a certain critical level from the European Regional Development Fund - more than 50 ECU per capita between 1975 and 1983 (see Armstrong & Taylor, 1985) - receive a value 1, and the others a value 0.

On the basis of the above considerations, we have formulated the following equation for the supply of infrastructure per region:

$$INF/AREA = a_o + a_1 (POP/AREA) + a_2 (POP/AREA)^2 + a_3 (GDP/POP) + a_4 COST + a_5 SEABOR + a_6 (NTLBOR) + a_7 (REGPOL)$$

where:

INF/AREA: regional infrastructure density of railroads or highways (km per square km)

POP/AREA: regional population density (persons per square km)

GDP/POP: regional or national gross domestic product per capita

COST: dummy variable which attains the value 1 when the highest altitude in a region is at least 2000 m, and which is zero in all other cases

SEABOR: share of total border of a region which borders the sea

NTLBOR: share of total border of a region which is also a national border

REGPOL: dummy variable which attains the value 1 if a region receives heavy regional development support, and which is zero in all other cases

The coefficients a_1, a_3 and a_7 are expected to be positive. The others (except a_0) are expected to be negative.

An alternative formulation where infrastructure demand in neighbouring regions is taken into account in a more complete way would be:

$$(INF/AREA)_r = a_0 + a_1 (POP/AREA)_r + a_2 (POP/AREA)^2_r + a_3 (GDP/POP)_r + a_4 COST_r + a_5 \sum_{r' \varepsilon S_r} sh_{r'} (POP/AREA)_{r'} b_{r'} + a_6 REGPOL$$

Where S_r is the set of all regions contiguous to region r. The factor $sh_{r'}$ is the share of region r' in the total border length of region r. When a region borders on the sea, the corresponding population density in r' is set equal to zero. The factor $b_{r'}$ is added to take into account the influence of a border effect. If r and r' are part of the same country, $b_{r'}$ is equal to 1. If r and r' are separated by a border, it is equal to a factor b, which is expected to be smaller than 1.

The last term of the above equation can be rewritten as:

$$a_5 \sum_{r' \varepsilon S_r} sh_{r'} (POP/AREA)_{r'} + (b-1)a_5 \sum_{r' \varepsilon T_r} sh_{r'} (POP/AREA)_{r'}$$

where T_r is the set of regions contiguous to region r which are located in another country (T_r may be empty). The first part of the above formula describes the impact of all contiguous regions (including the sea and foreign regions), whereas the second part describes the impact of national borders. Since b is expected to be smaller than 1, the contribution of the last term to infrastructure supply is expected to be negative.

13.4 Estimation of the Infrastructure Supply Model

In our analysis we used Eurostat data. Table 13.1 contains results for a selected number of European regions defined at the NUTS-II level. The Table shows that there is a clear relationship between population density and infrastructure density. In all countries considered, regions with a low population density tend to have a low infrastructure density. Differences among countries are clear: highway densities in the Netherlands, Belgium and Germany are high compared with France, Italy and Spain. The differences among regions are clearly smaller for rail densities than they are for highway densities. This may be related to the difference in life cycle between rail and road (cf. Grübler & Nakicenovic, 1991). Major parts of the railway system were already developed in the 19th century, whereas the development of the highway system only started after World War II. On the other hand, the strong historical component in the railway system has, to some extent, been polished away by the process of railway closures. These closures can be interpreted as an effort to adjust the railway system to the current economic and spatial conditions.

The infrastructure supply functions formulated in the preceding section have been estimated for 92 regions in the following countries: France, Germany, Italy, Spain, Portugal, the Netherlands, Luxembourg and Belgium. A detailed

Table 13.1 Infrastructure densities in a selected number of regions (1984)

	Highway (m/km^2)	Railway (m/km^2)	Population ($persons/km^2$)
Ile de France	35.3	139.9	852
Basse Normandie	3.1	65.3	78
Nordrhein Westfalen	57.8	157.3	490
Bayern	26.8	100.1	155
former DDR	15.6	127.2	152
Lombardia	22.8	64.4	372
Basilicata	2.7	35.7	62
Cataluna	15.6	42.1	189
Andalucia	1.3	23.5	76
Brabant	77.1	160.2	661
West-Vlaanderen	53.6	96.3	348
Zuid-Holland	80.8	107.4	940
Drenthe	28.3	39.9	161

Table 13.2 Regression analysis of railway and highway infrastructure supply in European regions (1984)

	railway density		highway density	
constant	-10.3		-12.7	
population density	110	(5.45)*	120	(10.41)
population density squared	4.6	(0.41)	-33	(-5.37)
GDP per capita	.66	(7.80)	.15	(3.30)
cost dummy	-17.5	(-3.21)	-2.20	(-.74)
regional = national border (in %)	16.2	(1.43)	12.9	(2.10)
regional = sea border (in %)	-27.6	(-2.24)	(-5.87)	(-.87)
R^2	.84		.79	

* t-values in parenthesis. With 85 degrees of freedom, the critical value for t is 1.66 given a one tailed test at the 5% level of significance

account of the data is given in Boonstra (1992). Estimation results are given in Table 13.2.

Most of the factors mentioned above appear to play a role in infrastructure supply. For both railways and highways the intraregional demand as expressed by population density exerts a clearly stimulating impact on infrastructure supply. A non-linear relationship between population density and infrastructure density is found for highways: in densely populated regions, highway supply is smaller than one might expect on the basis of a linear relationship. This may be a reflection of the fact that car use in urban areas is lower than in rural areas, because car ownership is lower and trip distances are shorter because cities are more compact. Another possible explanation is related to the measurement of infrastructure capacity. We did not take into account the number of lanes of highways, only their length. Therefore, highway density does not fully reveal differences in capacity, since the number of lanes of highways is higher in urban areas than elsewhere. This measurement problem may be captured by the squared population density term.

Gross domestic product per capita measured at the national level is also a significant explanatory factor of infrastructure density. If one uses the regional GDP per capita instead of the national one, a less significant result is found (results can be found in Boonstra, 1992): infrastructure investments in highways and railways are clearly financed by a pooling of investment funds at the national level.

Two indicators of interregional aspects of infrastructure are taken into account: the share of the regional border which is also a national border, and the share which is also a sea border. The result for the sea border has the expected negative sign, but it is only significant for railways. Thus, regions bordering on the sea have lower railway densities because there are no border crossing railway lines in certain spatial directions.

It is remarkable that the estimation for regions bordering on other countries does not yield the expected negative result. For highways even a significant positive border effect is observed. There is no evidence therefore that European border regions are handicapped by a lagging supply of infrastructure. Even the reverse seems to be true. One important factor which helps to understand this result is that the regions taken here are large. For example, the region of Nordrhein Westfalia is one of these regions with a national border share of 25%. Obviously, one would not tend to consider this large region as a typical border area. One would therefore expect that if a system with smaller regions were to be used, one would find clearer signs of border effects (this is confirmed in Rietveld, 1993). This consideration of regional size can only explain why border effects are not significant, but not why they have the wrong sign. We are left with an intriguing result. Another possible explanation of this result is that border regions may be underdeveloped regions receiving special support from national government and the EC for road and rail investments. Indeed, it is true that border regions may be underdeveloped, but as can be seen from the pattern of expenditures of the European Regional Development Fund, there is not a clear correlation between border regions and development regions (cf. Armstrong & Taylor, 1985). For example, most Italian, French and Spanish underdeveloped regions are not border regions.

It must also be mentioned that the coefficient for the regional policy variable was not significant for both infrastructure types. Since the introduction of the REGPOL variable did not substantially change the coefficients of the other variables, it has been deleted in Table 13.2. The insignificant result for the REGPOL variable means that the regions receiving money from the ERFD do not perform considerably better in terms of infrastructure availability compared with other regions. This is not particularly surprising, since the investment sums involved in the EFRD are small compared with total investments in infrastructure. On the other hand, it is interesting to note that those regions receiving EFRD support do not have an underdeveloped infrastructure compared with other regions, if the other factors in the regression equation hold constant. The latter qualification is important: regions receiving EFRD support do have low infrastructure densities, but when one corrects for the other explanatory factors included in the regression equation, they are not worse-off in a specific sense.

Cost considerations appear to play a role in infrastructure supply: for railways high altitude regions have a lower infrastructure density than other regions. Also for highways a negative sign is found for the cost dummy, but it is not significant.

Estimation results for the alternative formulation mentioned in the preceding section are presented in Table 13.3.

Most of the results in Table 13.3 are consistent with those of Table 13.2:

- The coefficient of population density is positive and significant for both modes.

- The coefficient of the squared population density is negative and significant for highways.
- Railway supply is relatively low in high altitude regions where construction costs are high.
- The border effect is positive for both modes. A difference is that this effect is no longer significant for the alternative specification.
- The most notable difference in results is found for the population density in neighbouring regions variable. For highways a positive and significant coefficient is found. This variable is, to a certain extent, related to the sea border variable used in Table 13.2, since a sea border implies a zero density for the relevant neighbouring region. In Table 13.2 it was the railway system for which this variable appeared to have a significant impact.

Table 13.3 Regression analysis of railway and highway infrastructure supply in European regions, alternative formulation (1984)

	railway density		highway density	
constant	-19.7		-13.3	
population density	100	(4.42)*	102	(9.08)
population density squared	11	(0.88)	-23	(-3.99)
GDP per capita	.67	(7.23)	.10	(2.39)
cost dummy	-13.8	(-2.40)	.12	(0.04)
population density neighbouring regions	31	(1.09)	50	(3.93)
population density neighbouring foreign regions	53	(1.02)	41	(0.60)
R^2	.82		.83	

* t-values in parenthesis

The estimation results indicate that regional infrastructure supply for both railways and highways depends in a significant way on intraregional demand (as expressed by population density), and on possibilities of finance as measured by GDP per capita at the national level. Cost considerations play an important role for railway supply. Interregional demand (as expressed by population density in neighbouring regions) has a significant impact for rail or road, depending on the specification chosen. No significant effect could be found of regional policies leading to preferential treatment of underdeveloped regions. An intriguing result is that border effects are positive, though not always considerable. This means that at the spatial level chosen here, border regions are on average not suffering from a lack of infrastructure; rather one would say that the opposite is true.

In section 13.1 we have already mentioned the issue of the two causal relationships between infrastructure (network density) and regional development (population density). Actually, what has been done in this Chapter is that we estimated the supply of infrastructure without taking into account the demand function. Ignoring the issue of simultaneous equations leads to a certain correlation between explanatory variables and the disturbance terms. This leads to a bias in the estimation of the coefficients concerned. The estimation results should be used with some caution, accordingly. A possible solution would be the formulation of a complete interdependent system and the use of an econometric technique for simultaneous relationships. Another possible way out would be the formulation of a dynamic system of equations.

13.5 Infrastructure Supply within Border Regions

The empirical results shown in the above section indicate that at the NUTS II level, regions located at national borders do not suffer from low infrastructure densities, rather the opposite seems to be true. A possible explanation is that the regions considered are too large. If the analysis was carried out at a smaller spatial scale so that border regions only included areas near national borders, one would expect a negative effect on infrastructure densities.

In order to test this, we used data on Dutch border regions. The seven regions (provinces) concerned have been divided into two parts: one part consists of all municipalities located at the national border; the second part consists of the rest of the provinces. Table 13.4 contains results for relevant indicators.

Table 13.4 Indicators for infrastructure supply in municipalities located at the national border relative to the rest of the provinces in the Netherlands

	index for municipalities located at the national border relative to the rest of the provinces			
	population density	income per capita	highway density	railway density
Groningen	39	98.8	46	0
Drenthe	218	97.0	137	0
Overijssel	121	96.2	35	29
Gelderland	109	96.5	24	47
Limburg	171	100.9	172	104
Noord-Brabant	55	100.4	81	50
Zeeland	103	104.1	0	0

An interesting result is that population density in municipalities at the border is on average not lower than in the rest of the border provinces. The median value is 109, compared with 100 for the rest of the provinces. Especially the provinces of Drenthe and Limburg have high density municipalities at the border (all relative to the rest of the respective province). The median value for income per capita is slightly below 100 (98.8). Table 13.4 shows that intermunicipality income differences are small in the Netherlands. Intermunicipality infrastructure differences are much bigger, however. The median values for highways and railways in border municipalities relative to the rest of the provinces are 46 and 29, respectively. Thus, although the border municipalities have a slightly above average population density, their infrastructure densities are clearly below average.

This result for the Netherlands strongly supports our expectation that border regions defined as municipalities located at borders indeed have low infrastructure densities. This disadvantage entirely disappears when border regions are defined in terms of the much larger spatial units at the NUTS II level. Negative border effects therefore do exist in the field of infrastructure supply, but only on a fairly small spatial scale. This negative impact of low infrastructure supply in border municipalities can be felt in three ways. First, in intra-regional transport in border regions. Second, in border crossing interregional transport. Here, large detour factors may be expected, especially for trips which would be short could they have been made as the crow flies. Third, border municipalities will also have relatively bad connections with other regions within their own country. Here too, detour factors will tend to be larger than average.

13.6 Concluding Remarks

In this Chapter we have investigated the factors having an impact on the distribution of transport infrastructure among regions.

When analyzed at the NUTS II level, the supply of infrastructure depends mainly on two determinants: regional demand as represented by population size, and financing possibilities measured by the gross domestic product in the pertaining nation. The role of interregional differences in construction costs is less prominent, but this may be caused by the somewhat crude measurements. No significant impact was found of regional policies on the infrastructure stock. The spatial autocorrelation which may be expected to be relevant in an analysis of network densities has a rather limited role. At this level of spatial aggregation, the impact of surrounding regions is not always significant (for a more detailed analysis of spatial autocorrelation, we refer to Rietveld & Wintershoven, 1998). It is also remarkable that border regions, when defined at this spatial level, do not suffer from a lack of infrastructure supply compared with other regions. From Table 13.2 we infer that the negative impact on infrastructure supply for a region located at a sea coast is much clearer than the impact of being located near a national border. An analysis at a much more detailed spatial level reveals that national borders do have a negative impact on infrastructure densities, but that this is limited to a relatively small area near the border.

Several directions for further research are noteworthy. First, the network properties of railways and highways can be taken into account in a more thorough way. This would imply the representation of networks in terms of nodes and arcs which would enable one to study spatial interdependencies in a more refined way.

Another interesting direction for future research is the development of a dynamic model of network supply. The present level of the stock of network capital is the result of many decisions in the past. Especially for railway infrastructure the historical component is very important (see Chapter 12). Such a dynamic approach would also permit an integrated analysis of the mutual relationships between transport infrastructure and regional development. This would imply that in addition to an explanation of the expansion (or contraction) of networks by means of features of regions, an analysis would be given of the impacts of network development on the development of regions in the realms of employment and population.

14 Conclusions

14.1 The Infrastructure Concept

In Chapter 2 we recognize that infrastructure is an ambiguous concept. In the literature a fair number of properties of infrastructure have been formulated (such as non-excludiability and non-rivalness). However, it appears that not all types of capital usually considered as infrastructure actually share these properties. This mismatch between a formal definition of a term and the common use of the term fortunately does not appear to be a hindrance for research in this field, as long as explicit attention is paid to what one means with the terms used.

One illustration of the definitional problems focuses on some researchers who state that an essential element of infrastructure is the strong public sector involvement in its provision. However, certain types of transport infrastructure (as it is usually considered) do not share this property (cf gas or oil pipelines, some airports, sea ports). Apparantly, countries vary largely in the extent to which the public sector supplies certain infrastructure types.

With respect to this issue of public sector involvement, it seems that macro-economic data usually do not give accurate information on infrastructure as it is commonly conceived. Public sector capital and infrastructure investments only partially overlap. This implies that macro-economic data often do not give a proper view on infrastructure developments. The *overlap* between the two seems to be *decreasing*, since the most dynamic infrastructure sectors (telecommunications and aviation) take place outside the public sector. This provides a warning signal against the notion that public sector investments should be a fixed share of total investments.

A traditional way to measure infrastructure quality of regions is to compare the infrastructure stock of the region (in capacity, length or in monetary terms) to some scale indicator such as population size or surface. As noted in Chapter 2, such an approach has unpleasant limitations. It ignores the network character of transport infrastructure and spatial spill-overs are neglected. Therefore, we have placed special attention in this study on the notion of *infrastructure services. It is not the transport infrastructure stock measured in length or investment amount per se that counts, but the services that it provides to the users.*

14.2 Accessibility

A promising way to introduce infrastructure services is to use the concept of *accessibility*, which has been chosen as a key concept for this book (especially in Chapters 2, 5, 6, 7 and 10). Accessibility can be defined as the potential of opportunities for interaction. It is based on both network features and the spatial distribution of activities. Since for each activity the relevant types of destinations are different, one may have to identify different accessibility indicators. In addition, a review of the literature reveals that there are a good number of ways to operationalize the concept of accessibility. Our review in Chapter 2 contains a list of no less than 11 alternative operationalizations.

In Chapter 5 we show the fruitfulness of the accessibility concept by applying it at the European level to various infrastructure networks (road, rail, aviation). In this Chapter we use an accessibility concept that is based on a spatial interaction approach. The accessibility concept can be used to analyze the effects of changes in infrastructure networks on the positions of the respective nodes in networks. We find that accessibility not only depends on physical networks, but also on non-physical barriers to communication, such as border effects that have a strong impact on long distance spatial interaction, and hence on long distance accessibility. Neglect of these non-physical factors will lead to a distorted view of the potential services of new investments in infrastructure networks.

To broaden the scope of our analysis in Chapter 6, we compared a number of alternative approaches to the accessibility of cities in the European infrastructure network. We find high levels of correlation between the accessibility indicators of the cities in most cases. The results are rather robust with respect to a number of choices that inevitably must be made when such a concept has to be applied (selection of cities, demarcation of cities, operationalization of the concept). This leads to the comfortable conclusion that the outcomes do not depend too much on the measurement tool. An interesting result is that a comparison of *perceived* accessibility of European cities from a business travel perspective and the accessibility of cities where use is made of airline travel times and frequencies leads to a high level of correspondence (correlation). It appears that the airline is the dominant transport mode for this category, while rail and road accessibility has only a minor role in these perceptions.

An uncomfortable finding of our comparative analysis of accessibility studies concerns the equity issue. The accessibility measures can be used to study the contribution of transport infrastructure projects to the solution of regional cohesion problems at the European level. For example, an interesting question is to what extent the large-scale introduction of high-speed rail will lead to larger or smaller variations in accessibilities. We find that the result depends on the type of accessibility indicator used. Of the four studies surveyed in Chapter 6, three indicate that the European high-speed rail projects will lead to higher inequality in accessibilities of cities, whereas the fourth study indicates a decrease in inequality. This equity issue deserves more attention in future research on infrastructure services.

The accessibility concept is also used in various other Chapters. Especially in Chapter 10 it is obviously the main determinant of the locational attractiveness of Dutch cites as perceived by Dutch entrepreneurs. An important result is that including the accessibility of foreign cities gives a better explanation of the perceived locational attractiveness of cities. This means that entrepreneurs in a country do include foreign markets in their perception of locational attractiveness of cities. This is an indication that in accessibility studies the boundaries of the study region should not be too narrowly chosen.

Another result concerns the relationship between accessibility and urbanization economies. Urbanization economies reflect higher productivity owing to the proximity of markets for various inputs and outputs. Accessibility essentially does a similar thing, but at a wider spatial range by including the markets in other cities located farther away. According to this interpretation, *urbanization economies are part of accessibility.* This has two important consequences.

First, in studies on urbanization economies the location of cities farther away should also be taken into account to fully incorporate the notion of *proximity.* Why would one only consider the benefits of having a large variety of activities nearby, and ignore the benefits of having such activities at somewhat longer distances? The decisive point is of course to what extent these benefits are subject to a strong distance decay. If distance decay is very strong, one may indeed confine the analysis to a local level. But this is an empirical matter which should not be based on an a priori delimitation of the spatial range of urbanization economy effects.

Second, in studies of accessibility the position of *'internal' accessibility* deserves special attention. It is common practice that accessibility indicators at the urban level include the internal accessibility (the potential of opportunities for interaction within the city itself). There are good reasons for this: neglecting internal accessibility would lead to the unexpected result that big cities achieve low accessibility scores and small cities located near large cities have high scores. It appears that the contribution of a large city to its own accessibility may be considerable. Neglecting this part of accessibility would lead to distorted results. However, to accurately measure internal accessibility is not easy when data on the spatial distribution of activities, internal transport networks and speeds are unavailable. A promising way to arrive at better estimates of internal accessibility would be the use of GIS based approaches.

Ignoring the partial overlap between urbanization economies and accessibility means that when in an explanatory study of productivity or urban growth a significant contribution is found for accessibility. It may both reflect the existence of urbanization economies and of economies derived from cities located farther away. This would call for an in-depth analysis where internal and external accessibility are both explicitly represented.

14.3 Impacts of Infrastructure Supply

The theoretical perspectives presented in Chapter 3 indicate that *the contribution of transport infrastructure to productivity of private production factors is beyond doubt. However, the impact it has on the other production factors (labour, private capital) is subject to rather complex mechanisms with positive and negative feed-back loops, the net outcome of which is rather uncertain.*

Consider for example the production function approach. The higher productivity of private production factors owing to an increase of the infrastructure stock implies that fewer inputs are needed to produce the same level of output (assuming that the other inputs are substitutes of infrastructure). However, the decrease in production costs will lead to a decrease in prices and hence an increase in output. This would in turn lead to a higher demand for the inputs. The net balance of the two effects is uncertain. In addition, from a macroeconomic perspective, crowding out effects have to be taken into account leading to feed-backs via higher taxes or higher interest rates. Thus, the production function approach does not obtain a clear conclusion with respect to the effect of infrastructure investments on employment or private capital investments.

A similar result is found for the interregional trade approach. A decrease in transport costs due to an infrastructure project reinforces the tendency towards regional specialization. The comparative advantages of regions can be exploited to a larger extent than before the reduction in transport costs. This leads to a situation where strong sectors in the regions will expand and weak sectors in the various regions will contract. What does this imply for the workers in both regions? Most regions will have a mixture of sectors, some of which will decline and others that will expand. Hence, the balance of labour demand may be both positive and negative. Labour market flexibility is especially important here. If, in a region workers who lose their jobs in contracting sector A are not qualified for work in expanding sector B, unemployment will result, especially when labour migration is limited. Such inflexibility will also have distorting effects on wage formation in the regions. We therefore conclude that, also from the interregional trade perspective, theory predicts mixed results: employment may both increase and decrease as a result of better infrastructure.

Given this general result that the employment effects of transport infrastructure investments for regions may be both positive and negative, one may wonder to what extent more definite results can be reported in particular situations. In Chapter 3 we discuss a few special cases. Consider for example the case with three regions where an improved infrastructure link is provided between two of them. The above reasoning indicates that it is in general not possible to indicate which of these two will be the main beneficiary in terms of size of economic activity. However, compared with the third region (of which the trade flows are assumed not to benefit from the improved link), the two regions directly affected improve their joint market share in the three region economy. This means that although it is not possible to say in general which of two regions will benefit most in case of an improved link between them, the region certainly losing will be the third one. This result explains why regional authorities are often so keen on improving the quality of their external infrastructure links.

Another issue concerns the improvement of a link between two structurally different regions, for example between a 'core' and a 'periphery' region. Which of the two will benefit most: does it lead to a reinforcement of the core, or to an equalization of positions in economic space? As discussed in Chapter 3, Krugman (1991) presents a simple trade model between core and periphery under the regime of scale economies. The model leads to the conclusion that with very high transport costs trade is not profitable, so that production will be located in both core and periphery. A decrease of transport costs will lead to concentration of production in the core, because there the biggest market is found and so the costs of transport to both markets are lower from there. A further decrease of transport cost would lead to a shift of production towards the periphery when inputs are cheaper in the periphery. This simple model would lead to the conclusion that at a high level of transport costs, the improvement of transport infrastructure tends to lead to concentration of production in the core (implying an increase in interregional inequality). When the infrastructure network is already mature, centrifugal forces become stronger according to this model, thus implying an urban-rural manufacturing shift and a decrease of interregional inequality. We arrive at the conclusion that *with this particular model the core benefits in the early phases of infrastructure development and the periphery benefits in the later phases.* In more complex models however, such an unambiguous conclusion will no longer be possible.

Concerning the core-periphery distinction, we note in passing that it is a property of the accessibility concept that improvement of a link between a 'large' and a 'small' region will lead to a relatively large increase in accessibility of the small region (see also Chapter 2). *Thus, the small region will benefit most* ***in terms of accessibility*** *when infrastructure between a large and* *a small region is improved.* The background of this result is the large share of the internal accessibility of the large region. An implication is that *when a core region wants to increase its relative accessibility position in a system of regions, the best way seems to be the improvement of the* ***internal*** *infrastructure network.*

Note that the conclusions in the above paragraph have been formulated in accessibility terms only. The example provided by Krugman indicates that other factors also have to be considered to discover the direction in which economic activities shift in response to infrastructure improvements.

In the empirical work reported in this book we find a clear confirmation of the rather indeterminate results of infrastructure improvement on the location of economic activity, as long as these are measured at a regional level. This already holds true for the studies on two mega projects discussed in Chapter 4: the Channel tunnel between Great Britain and France, and a couple of major bridges/tunnels in and around Denmark. Both studies indicate modest spatial shifts in economic activity as a consequence of these large projects.

Similar results are found in the studies reported in other Chapters: highway development at the interregional level (Chapters 7 and 8), and 19th century railway construction impacts on urban development (Chapter 12).

The above conclusions are based on an analysis at a high level of spatial aggregation (regions, countries). In the next section we investigate the extent to which other spatial scales give rise to other conclusions.

14.4 Spatial Scale of Infrastructure Impacts

An important aspect of infrastructure impact studies is the spatial scale of the impacts. Improvements of infrastructure usually have effects at two spatial levels: shifts in the relative accessibility of one region compared with the other, and shifts in the relative accessibility of zones within a region. The two levels must be clearly distinguished. The discussion on shifts in the location of economic activity in this concluding Chapter has thus far focused on interregional shifts. Our empirical results (and also the theoretical backgrounds surveyed above) indicate that such interregional shifts are usually limited. A rather undisturbed spatial development at the regional level does not imply, however, that at a lower spatial level the effects are equally small. As documented in Chapters 8, 9, and 11, the local effects are clearly larger.

A major force in the changes in spatial distribution of activity within a region appears to be that dynamic successful firms need space for expansion. When this space is not available at the given location these firms usually consider locations nearby. The recent construction of infrastructure offers good opportunities for a relocation towards a zone with an attractive accessibility near a highway exit (or a railway station). Bad infrastructure usually does not function as a push factor, but good infrastructure appears to function as a pull factor (see Chapter 8).

A careful analysis of case studies (Chapter 9) reveals that the largest dynamics in terms of growth of firms does not necessarily occur in the zone where the infrastructure itself is built. Other zones may also be the main beneficiaries of the network improvement.

The differentiation of spatial scales (regional versus local) is important for the distinction between generative and distributive effects (see Chapter 3). A major part of the firms that locate near a highway exit would also have found a location in the same region if this particular road and exit would not have existed. Such firms are not part of the generative effect of the infrastructure. *Rapid growth of economic activity in zones near highway exits is indeed a clearly visible indication of the importance of infrastructure in industrial location, but the impact mainly concerns the intraregional distribution. Explicit attention upon the spatial scale of infrastructure impact is therefore needed to avoid wrong conclusions about infrastructure impacts.*

14.5 Infrastructure Impacts: Models Versus Entrepreneurial Perceptions

In our empirical Chapters we have used analyses that consist of combinations of model applications and surveys on entrepreneurial perceptions. One might argue that well-estimated models or statistical methods are to be preferred above surveys addressing perceptions of entrepreneurs that are necessarily subjective. There are several circumstances, however, where model based approaches face severe limitations. An example of such a situation is a model that has to be applied to a case that lies outside its domain. A major infrastructure improvement like the Channel tunnel may have effects that cannot be captured by any existing model. Another example occurs when multicollinearity prevents a proper analysis of the impact of infrastructure. This may happen both with time series and cross-section data. Strongly correlated explanatory variables in time or space hamper the clarification of the contribution of infrastructure to spatial development in a statistical analysis.

In our opinion the gap between the two approaches is not as large as it sometimes seems. First, we note that models can be used to analyze subjective entrepreneurial priority statements (see, for example section 11.4). Second, the actual behaviour of entrepreneurs is in many ways driven by their subjective expectations, views and priorities. Therefore, it is important to have knowledge of these aspects. Third, some models can only be applied when they are fed with inputs that may not yet be available. For example, the impact of a new bridge depends on the pricing strategies of competing ferries. Subjective information about these pricing strategies may have to be used if one wants to apply route choice models for this purpose.

Well known objections against the use of subjective information of entrepreneurs are that they are subject to biases due to strategic considerations and that perceptions are unreliable. Some of these arguments have been discussed in Chapters 8 and 11. We conclude that by taking a number of measures such biases can be reduced (see section 8.1). We also found in several empirical cases that the subjective statements and perceptions really do make sense when they are compared in a systematic way with observed data (see Chapters 6 and 11).

One of our conclusions in Chapter 11 on locational preferences reads that useful information on priorities can be obtained when questions are formulated so that respondents indicate their preferred alternative in a choice set. This information can be presented in terms of trade-offs between criteria and is therefore probably more fertile than the usual approach, where respondents are asked to give rankings of location criteria in order of importance.

When we consider the views of entrepreneurs on infrastructure improvement on the performance of the firm, we get plausible results: a large majority considers market development and internal factors of much greater importance than infrastructure. Nevertheless, depending on the regional circumstances, a certain part (10-25 percent) of the firms indicate that without the infrastructure improvements (construction of a major highway) they would have had a smaller number of workers. For investment levels even higher values are found (20-35%). These figures mean that a good share of firms do indeed consider a major highway improvement in their region as a stimulus of their activities. However, these effects are most probably mainly distributive in nature. The reason is that the benefiting firms may be rather easy to identify, because they are concentrated in a band of some 7.5 km on both sides of the highway, whereas the mechanisms of crowding out most probably function in a much more diffuse way. Crowding out will occur in a much broader region so that the negative effects will be very thinly distributed; as a consequence they easily remain unobserved by the responding firms.

14.6 Locational Behaviour

In the regional economic dynamics, transport infrastructure improvements can have different impacts on firms. First, existing firms might grow or decline; second, new firms may emerge; third, infrastructure improvements may influence the relocation decision of existing firms.

In this section we focus on the third category. As can be learned from Chapters 8 and 9, transport infrastructure becomes important after the initial relocation decision is taken. The forcing push factors are the inability to expand at the old location and to a lesser extent environmental restrictions within residential areas. Only after this relocation decision is taken does transport infrastructure become an important pull factor in the decision of where to relocate. However, the size of the playing field is bounded by the imperfect information of the entrepreneur about the opportunities available, especially when it concerns locations at an increasing distance of the present location. *Firms appear to relocate over short distances (in our case study 42 % of the relocated firms stays within the same four digit postal code area, 75 % stays within the same municipality).*

Transport infrastructure appears to play a clear role in the perception of the entrepreneurs. A good access by transport infrastructure is seen as a status determinant by the company. For example, some firms want to be located near an international airport or a high quality public transport terminal, despite wether the employees, customers or clients travel regularly via these transport modes.

In Chapter 11 we investigated the long-term locational desires of Dutch firms with respect to types of sites at the intraregional level. These sites have been defined in terms of infrastructure and locational features. The desiderata are largely consistent with the observed relocation pattern over the last fifteen years. There is an even stronger relocation potential than the actual relocations have shown from the inner city and the residential areas towards suburban sites. Those sites with an access to a highway are especially favoured. The potential for relocations is high: over 60 % of the firms prefer to be located on another type of site compared with the present site. One should certainly take into account that new developments may lead to other locational preferences. Yet these results indicate that in the long-run, we have to rely on a further spatial redistribution of economic activity within regions. This redistribution is especially oriented toward locations with high highway accessibility.

We should not forget, however, that locational inertia remains high: many firms may stay at the present location type because the advantage of moving is too small compared with the costs involved. In addition, the known preference of many firms for a location with highway access does not provide a guarantee of success for industrial or office areas built near highway access points. As we found in the case of the Amsterdam orbital motorway (Chapter 9), the largest effects did not occur in the zones where the last segments of the motorway were constructed, but instead in highway connected zones that were already in use for over twenty years.

14.7 Analysis of the Behaviour of Suppliers of Infrastructure

In the last part of our book we have paid special attention to the supply side of infrastructure. This has been an underdeveloped topic in infrastructure research. The background is that infrastructure projects are often considered as exogenous impulses. However, a proper analysis of infrastructure impacts implies that more explicit attention has to be given to the actors providing it.

These actors may be private (cf. Chapter 12, where we investigate the evolution of the Dutch railway network at the time it was primarily supplied by private firms), or they may be public (cf. Chapter 13, where infrastructure supply is analyzed in European regions).

It is important to consider the supply side of infrastructure for at least two reasons. The first is that the supply side is interesting in and of itself. Theories developed in the fields of economics and political science on the behaviour of the public sector have to be tested to arrive at a more concrete understanding of the basic forces driving public investments. The second reason is that neglect of the supply side may lead to distorted results on the impacts of infrastructure supply on the users. For example, if infrastructure investments of the public sector increase in periods of rapid growth because the funds for investments are larger, one might wrongly interpret a correlation between the two as an indication that infrastructure investment causes growth. A similar example holds for the spatial aspect: if investment projects are particularly realized by the public sector in prosperous regions to prevent congestion problems, this may lead to the spurious result that infrastructure projects generate growth. Similarly, if infrastructure were to be provided mainly to backward regions, a one-sided analysis might suggest that investments have adverse results.

In our two studies of infrastructure supply we found clear evidence that supply can be analyzed by means of economic models. For the Dutch railway case we formulated, estimated and simulated a dynamic joint model to represent supply of railway links and response of the demand side in terms of differentiated urban growth rates. We indeed find that the supplier of infrastructure responds to present market opportunities, whereas urban development is influenced by past and present investments in specific railway lines.

For European infrastructure supply at the regional level, we find clear evidence that supply is influenced by regional and national economic conditions. Because of a lack of historical data, we did not investigate the question of the extent to which the current economic conditions have been influenced by past projects.

14.8 Is Infrastructure Effective?

Throughout the book we have observed that infrastructure improvements have a wide variety of effects on firms and consumers. However, these effects are often difficult to trace because of the lengthy time period it may take before an economy adjusts to the new situation. A general result is that *the effects are often large at a local level, but at higher levels of spatial aggregation they are much smaller. The overall conclusion with respect to the infrastructure projects studied in this book is that at the regional level, the effects on the location of economic activity are usually modest.*

The effects of a project will also depend on the particular features of the *existing network* and of the regions concerned. When a network is already well-developed, an extension will not contribute much in terms of accessibility unless it concerns the construction of capacity in a congested bottleneck or a missing link.

Another important feature concerns the *structure of the regional economies* concerned. Some sectors depend strongly on particular types of infrastructure (see also Chapter 11). Thus, the position of sectors in the regional economies affected have to be taken into account.

A third consideration concerns *macro-economic conditions*. In a period of low growth or decline, investment and relocation levels will be low so that the private sector response to infrastructure provision will also be small. This underlines the importance of a long-term perspective. Macro-economic conditions are also important for the magnitude of crowding-out effects.

We conclude that there are risks in the transfer of the results of the study of a certain infrastructure project to another project. Such a transfer has to be preceded by a careful check of the comparability of the conditions (existing network, regional economy, macro-economic conditions) and of the spatial scales involved. When comparability is low, project specific impact studies are recommended.

In the present study we have only focused on a limited number of aspects of infrastructure. Much attention has been paid to the spatial and accessibility aspects of transport infrastructure and infrastructure use by firms. Much less has been said about the productivity aspects and the use by consumers. These issues will of course have to play an important role in evaluations of specific infrastructure projects. Our main conclusion that the impact of infrastructure supply on the interregional allocation of economic activity is usually rather limited should not be understood as a message that infrastructure is unimportant: from the other perspectives (productivity, use by consumers), it undoubtedly may have an important role.

14.9 Prospects

Transport infrastructure will remain a challenging subject of research. The long lifetime of the capital stock concerned, the large number of actors affected, the network features, and the spatial dimensions give rise to a complex interplay of factors which will not easily be boring for analysts.

The focus of our analysis has been on the effects of the *construction* of infrastructure. But one should not forget that after the construction phase infrastructure is also an interesting subject of research. Several other aspects of infrastructure deserve attention including organizational and institutional aspects (conditions of use, regulation) and financial aspects (how to charge for the use). The importance attached to the environmental consequences of transport infrastructure will probably lead to a higher emphasis on the efficient use of the existing infrastructure rather than on the construction of new infrastructure. This will be a promising subject of research in the future.

The network character of transport infrastructure is usually not taken into account in standard macro or regional economic models. Therefore, these models are not so suitable for an analysis of transport infrastructure impacts. Their strong point is that they allow the investigation of various feedbacks of infrastructure projects on the macro economy, including so-called crowding out effects. Such crowding out effects are usually neglected in transport models. Therefore, a fruitful direction of further research will be the combination of both types of models.

We have shown that accessibility is quite a useful tool for representing the transport network features in regional or macro economic models. An interesting spin-off of our analysis is we find that accessibility and urbanization economies are rather close concepts. Further research into this relationship is promising to help clarify the understanding of urbanization economies. This implies that accessibility also becomes a candidate as a factor that enters regional or national production functions to represent the contribution of transport infrastructure services to production processes. The use of accessibility measures is the most natural way to represent spatial spill-over effects in production.

Such a further analysis will probably help improve the computation of the so-called internal accessibility. In this respect the use of detailed spatial data and network data in geographical information systems is promising.

Causality issues in transport infrastructure supply are important. Too often infrastructure is considered as an exogenous factor. In order to arrive at an adequate analysis of transport infrastructure impacts, more research efforts should be directed toward the behaviour of suppliers of transport infrastructure.

References

ABC World Airways Guide, (1991), Dunstable, Reed Telepublishing Ltd.

Amano, K. & M. Fujita (1970), A long run economic effect analysis of alternative transportation facility plans - regional and national, *Journal of Regional Science*, Vol. 10, pp. 297-323.

Anas, A. (1982), *Residential location markets and urban transportation: economic theory, econometrics, and policy analysis with discrete choice models*, New York, Academic Press.

Anderson, D.L. (1983), Your company's logistic management: an asset or a liability?, *Transportation Review*, pp. 111-125.

Andersson, A.E., C. Anderstig & B. Harsman (1989), *Knowledge and communications infrastructure and regional economic change*, Umea, University of Umea.

Anderstig, C. & L.-G. Mattson (1991), An integrated model of residential and employment location in a metropolitan region, *Papers in Regional Science*, Vol. 70, pp. 167-184.

Armstrong, H. & J. Taylor (1985, 1993), *Regional economics & policy*, Oxford, Allen.

Arnott, R. & J.R. Stiglitz (1981), Aggregate land rents and aggregate travel costs, *Economic Journal*, Vol. 91, pp. 331-347.

Aschauer, D.A. (1989), Is public expenditure productive?, *Journal of Monetary Economics*, Vol. 23, pp. 177-200.

Bannock, G., R.E. Baxter, & E. Davis (1992), *Dictionary of economics*, London, Penguin.

Bannister, D. & K.J. Button (eds.) (1993), *Transport, the environment and sustainable development*, London, E & FN Spon Chapman & Hall.

Barra, T. de la (1989), *Integrated land use transport modelling: decision chains and hierarchies*, Cambridge, Cambridge University Press.

Berechman, J. & P. Gordon (1986), Linked models of land use transport interactions: a review, in: B. Hutchinson & M. Batty (eds.), *Advances in urban systems modelling*, Amsterdam, North-Holland, pp. 109-132.

Biehl (1993), *The role of infrastructure in regional development*, Mannheim, University of Mannheim.

Biehl, D., E. Hussman, K. Rautenbach, S. Schnijder & V. Südmeyer (1975), *Bestimmungsgr ünde des regionale entwicklungspotentials*, Tübingen, Mohr.

Biehl et al. (1986), *The contribution of infrastructure to regional development*, Brussels, European Commission.

Blum, U. (1982), Effects of transportation investments on regional growth, *Papers of the Regional Science Association*, Vol. 58, pp. 151-168.

Blum, U. (1997), Benefits and external benefits of transport: a spatial view, in: D.L. Greene, D.W. Jones & M.A. Delucchi (eds.), *The full costs and benefits of transportation*, Berlin, Springer Verlag, pp. 219-244.

Boonstra, J. (1992), *Barrieres in verkeer en infrastructuur; een internationaal perspectief*, Thesis, Economische Faculteit, Amsterdam, Vrije Universiteit.

Botham, R. (1983), The road programme and regional development: the problem of the counterfactual, in: K.J. Button & D. Gillingwater (eds.), *Transport, location and spatial policy*, Aldershot, Gower, pp. 23-56.

Boyce, D.E., A. Fahri & R. Weischedel (1973), Optimal network problem: a branch-and-bound algorithm, *Environment and Planning*, Vol. 5, pp. 519-533.

Bradley, J. (1997), *Policy design and evaluation: EU structural funds and cohesion in the European periphery*, Dublin, Economic and Social Research Institute.

Bröcker, J. (1984), How do international trade barriers affect interregional trade? in: A.E. Anderson, W. Isard & T. Puu (eds.), *Regional and industrial development theories, models and empirical evidence*, Amsterdam, North-Holland, pp. 219-239.

Bröcker, J. (1989), How to eliminate certain defects of the potential formula, *Environment and Planning A*, Vol. 21, pp. 817-830.

Bruinsma, F.R. (1990), *Infrastructuur: investeringen en werkgelegenheid*, OSA-Voorstudie V-36, Den Haag, Organisatie voor Strategisch Arbeidsmarktonderzoek.

Bruinsma, F.R. (1994), *De invloed van transportinfrastructuur op ruimtelijke patronen van economische activiteiten*, Ph.D. Thesis, Amsterdam, Vrije Universiteit.

Bruinsma, F.R. & P. Rietveld, (1991), *Infrastructuur en stedelijke ontwikkeling in Europa*, Werkstuk 27, Amsterdam/Utrecht, Vrije Universiteit/Stedelijke Netwerken.

Bruinsma, F.R., P. Nijkamp & P. Rietveld (1992), Regional economic transformation and social overhead investment, *Tijdschrift voor Economische en Sociale Geografie*, Vol. 83, pp. 3-12.

Bruinsma, F.R. & P. Rietveld (1993), Urban agglomerations in European infrastructure networks, *Urban Studies*, Vol. 30, pp. 919-934.

Bruinsma, F.R., G. Pepping & P. Rietveld (1993), *Economische uitstraling opening ringweg Amsterdam*, Rotterdam, Adviesdienst Verkeer en Vervoer, Rijkswaterstaat.

Bruinsma, F.R., S.A. Rienstra, P. Rietveld & M.T.A. Hilferink (1995a), *De structurerende werking van infrastructuur op interregionaal niveau langs verbindingsassen: een analyse op COROP-niveau*, Den Haag, Projectbureau IVVS.

Bruinsma, F.R., S.A. Rienstra & P. Rietveld (1995b), *De structurerende werking van infrastructuur op interregionaal niveau langs verbindingsassen: een ondernemerssurvey*, Den Haag, Projectbureau IVVS.

Bruinsma, F.R., G. Pepping & P. Rietveld (1996), Infrastructure and urban development: the case of the Amsterdam orbital motorway, in: D.F. Batten & C. Karlsson (eds.), *Infrastructure and the complexity of economic development*, Berlin, Springer Verlag, pp. 231-249.

Bruinsma, F.R. & P. Rietveld (1997), A stated preference approach to measure the relative importance of location factors, *International Journal of Development Planning Literature*, Vol 12, pp. 125-140.

Bruinsma, F.R., S.A. Rienstra & P. Rietveld (1997), Economic impacts of the construction of a transport corridor: a multi-level and multi-approach case study for the construction of the A1 highway in the Netherlands, *Regional Studies*, Vol. 31, pp. 391-402.

Bruinsma, F.R. & P. Rietveld (1998a), The accessibility of European cities: theoretical framework and comparison of approaches, *Environment and Planning A*, Vol. 30, pp. 499-521.

Bruinsma, F.R. & P. Rietveld (1998b), The impact of accessibility on the valuation of cities as location for firms, in: J. Egeln & H. Seitz (eds.), *Städte vor neuen Herausforderungen*, Baden Baden, Nomos, forthcoming.

Button, K.J. (1993), *Transport economics*, Aldershot, Edward Elgar.

Button, K.J. (1996), Ownership, investment and pricing of transport and communication infrastructures in: D.F. Batten & C. Karlsson (eds.), *Infrastructure and the complexity of economic development*, Berlin, Springer Verlag, pp. 147-165.

Button, K.J. & P. Rietveld (1993), Financing urban transport projects in Europe, *Transportation*, Vol 20, pp. 251-265.

Cattan, N. (1992), Air and rail accessibilities and attractivities of European metropolis, in: A. Bonnafous (ed.), *Selected proceedings of the 6th World Conference on Transport Research*, Lyon, LET Universite Lumiere Lyon, Vol. 1, pp. 43-54.

C.B.S. (1966-1992), *Statistiek van de wegen*, Voorburg, Centraal Bureau voor de Statistiek.

C.B.S. (1970-1992), *Regionaal economische jaarcijfers*, Voorburg, Centraal Bureau voor de Statistiek.

C.B.S. (1989), *Criminaliteit en strafrechtpleging*, Voorburg, Centraal Bureau voor de Statistiek.

C.B.S. (1995), *Nationale rekeningen*, Voorburg, Centraal Bureau voor de Statistiek.

Chapman, K. & D.F. Walker (1992), *Industrial location*, Oxford, Basil Blackwell.

Cheshire, P. (1990), Explaining the recent performance of the European Community major urban regions, *Urban Studies*, Vol. 27, pp. 311-333.

Christaller, W. (1933), *Die zentralen orte in Süddeutschland*, Jena, Fischer.

Ciccone, A. & R. E. Hall (1996), Productivity and the density of economic activity, *American Economic Review*, Vol. 86, pp. 54-70.

Cook, T. (1991), *European timetable*, London, Thomas Cook Publishing.

Costa, J.D.S., R.W. Ellson & R.C. Martin (1987), Public capital, regional output and development, *Journal of Regional Science*, Vol. 27, pp. 419-437.

Cramer, J.S. (1991), *The LOGIT-model: an introduction for economists*, London, Edward Arnold.

Daluwatte, S. & A. Ando (1992), Regional agglomeration and transportation improvements in postwar Japan, in: A. Bonnafous (ed.), *Selected proceedings of the 6th World Conference on Transport Research*, Lyon, LET Universite Lumiere Lyon, Vol. 1, pp. 195-206.

Dasgupta, M. & F.V. Webster (1992), Land/use transport interactions: policy relevance of the ISGLUTI-study, in: A. Bonnafous (ed.), *Selected proceedings of the 6th World Conference on Transport Research*, Lyon, LET Universite Lumiere Lyon, Vol. 1, pp. 267-278.

DATAR (1989), *Les villes Europeennes*, Montpellier, Maison de la Geographie.

Deno, K.T. (1988), The effect of public capital on US manufacturing activity, *The Southern Economics Journal*, Vol. 55, pp. 400-411.

Diamond, D. & N. Spence (1989), *Infrastructure and industrial costs in British industry*, London, The Department of Trade and Industry.

Dietvorst, A.G.J., J.A. van Ginkel, A.O. Kouwenhoven, B.C. de Pater & W.J. van Bremen (1984), *Algemene sociale geografie; ontwikkelingslijnen en standpunten*, Weesp, Romen.

Diewert, W.E. (1986), The measurement of the economic benefits of infrastructure services, in: *Lecture Notes in Economics and Mathematical Systems*, Vol. 278, Berlin, Springer Verlag.

Dodgson, J.S. (1974), Motorway investment and sub-regional growth, *Regional Studies*, Vol. 8, pp. 75-91.

Dupuit, J. (1844), De la mesure de l'utilité des travaux publics, Annales des Ponts et Chaussées; an English translation appeared in: On the measurement of the utility of public works, *International Economic Papers*, no. 2, 1952, pp. 83-110.

Dutch Ministry of Transport (1989), *Tweede Structuurschema Verkeer en Vervoer*, Den Haag, SDU.

Dijksterhuis, R. (1984), *Spoorwegtracering en stedebouw in Nederland: historische analyse van een wisselwerking: de eerste eeuw 1840-1940*, Ph.D. Thesis, Delft, Technische Universiteit Delft.

E.C. (1990a), *The regions in the 1990s*, Directorate General for Regional Policy, Brussels, European Commission.

E.C. (1990b), *The European highspeed train network*, Brussels, European Commission.

E.C. (1993), *European high speed rail network; socio-economic impact study*, Brussels, European Commission.

Echenique, M. (1986), The practice of modelling in developing countries, in: B. Hutchinson & M. Battey (eds.), *Advances in urban system modelling*, Amsterdam, North Holland, pp. 275-298.

E.C.M.T. (1988), *Investment in transport infrastructure in ECMT countries*, Paris, European Conference of Ministers of Transport.

E.C.M.T. (1997), *Statistical trends in transport 1965-1992*, Paris, European Conference of Ministers of Transport.

Elbers, C. (1992), *Spatial disaggregation in general equilibrium models*, Ph.D. Thesis, Amsterdam, VU University Press.

Engle, R.F & C.W.J. Granger (1987), Co-integration and error correction: representation, estimation, and testing, *Econometrica*, Vol. 55, pp. 251-276.

Erlandsson, U. & C. Lindell (1993), *Svenska regioners kontakt- och resemöjligheter i Europa 1992*, Rapporter och Notiser 119, Lund, Institutionen för kulturgeografi och ekonomisk geografi vid Lunds Universitet.

Eurostat (1990), *Regio's statistisch jaarboek 1989*, Luxembourg, Eurostat.

Eurostat (1997), *Regio's statistisch jaarboek 1996*, Luxembourg, Eurostat.

Evers, G.P.M., H.P. van der Meer, J. Oosterhaven & J.B. Polak (1987), Regional impacts of transport infrastructure; a multisectoral potentials approach, *Transportation*, Vol. 14, pp. 113-126.

Fayman, S., P. Metge, K. Spiekermann, M. Wegener, T. Flowerdew & I. Williams (1995), The regional impact of the Channel tunnel: qualitative and quantitative analysis, *European Planning Studies*, Vol. 3, pp. 333-356.

Floyd, R.W (1962), Algorithm 97, shortest path, *Communications of the association of computing machinery*, Vol. 5, p. 345.

Forslund, U.M. & B. Johansson (1995), Assessing road investments: accessibility changes, cost benefit and production effects, *Annals of Regional Science*, Vol. 29, pp. 155-174.

Fotheringham, A.S. & M.E. O'Kelly (1989), *Spatial interaction models*, Dordrecht, Kluwer.

Frey, B.S. (1983), *Democratic economic policy*, Oxford, Martin Robertson.

Frost, M.E. & N.A. Spence (1995), The rediscovery of accessibility and economic potential: the critical issue of self-potential, *Environment and Planning A*, Vol. 27, pp. 1833-1848.

Fujita, M. (1989), *Urban economic theory: land use and city size*, Cambridge, Cambridge University Press.

Fukuchi, T. (1978), Analyse economie-politique d'un développement régional harmonisé, *Collections INSEE*, Vol. 61, pp. 227-253.

Funck, R.H. (1996), Competition among locations: objectives, instruments, strategies, perspectives, in: H. Giersch (ed.), *Urban agglomeration and economic growth*, Berlin, Springer Verlag, pp. 227-255.

Garey, M.R. & D.S. Johnson (1979), *Computers and intractability: a guide to the theory of NP-completeness*, New York, Freeman.

Gramlich, E.M. (1994), Infrastructure investment: a review essay, *Journal of Economic Literature*, Vol. 32, pp. 1176-1196.

Granger, C.W.J. & P. Newbold (1974), Spurious regressions in econometrics, *Journal of Econometrics*, Vol. 2, pp. 1045-1066.

Grübler, A. & N. Nakicenovic (1991), *Long waves, technology diffusion and substitution*, research report 91-17, Laxenburg, International Institute for Applied Systems Analysis.

Guiliano, G. (1986), Land use impacts of transportation investments: highway and transit, in: S. Hanson (ed.), *The geography of urban transportation*, New York, The Guildford Press, pp. 247-279.

Gutiérrez, J., R. Conzález & G. Gómez (1996), The European high-speed train network: predicted effects on accessibility patterns, *Journal of Transport Geography*, Vol. 4, pp. 227-238.

Haggett, P., A.D. Cliff & A. Frey (1977), *Location analysis in human geography*, London, Arnold.

Hall, P. (1990), Keynote address on orbital motorways, in: D. Bayliss (ed.), *Orbital motorways*, London, Thomas Telford, pp. 1-31.

Hamerslag, R. (1980), Spatial development, developments in traffic and transportation and changes in the transportation system, in: J.P.B. Tissot van Patot, J.B. Polak & J.B. van der Kemp (eds.), *Changes in the field of transport studies*, Den Haag, Nijhoff, pp. 61-79.

Handy, S.L. & D.A. Niemeier (1997), Measuring accessibility: an exploration of issues and alternatives, *Environment and Planning A*, Vol. 29, pp. 1175-1194.

Hansen, W.G. (1959), How accessibility shapes land-use, *Journal of the American Institute of Planners*, Vol. 25, pp. 73-76.

Hart, R.A. & G.T. Moore (1973), The development of spatial cognition; a review, in: R.M. Downs & D. Stea (eds.), *Image and environment; cognitive mapping and spatial behaviour*, Chicago, Aldine, pp. 246-288.

Hartog, H. den, K.A. Heineken, B. Minne, R.J.J. Roemers & H.J. Roodenburg (1986), *Investeren in Nederland*, onderzoeksmemorandum 17, Den Haag, Centraal Planbureau.

Healey & Baker (1990-1996), *European Real Estate Monitor*, London, Healey & Baker.

Hirschman, A.O. (1958), *The strategy of economic development*, New Haven, Yale University Press.

Hodgson, M.J. (1974), Stating interurban highway construction by dynamic programming, *Annals of Regional Science*, Vol. 8, pp. 123-136, reply on comments: (1975) *Annals of Regional Science*, Vol. 9, pp 105-108.

Hoevenagel, R. (1994), *The contingent valuation method: scope and validity*, Ph.D. Thesis, Amsterdam, Vrije Universiteit.

Isard (1956), *Location and the space-economy*, Cambridge, MIT Press.

Isserman, A.M. (1990), *Research designs for quasi-experimental control group analysis in regional science*, Morgantown, West Virginia University.

Isserman, A.M., T. Rephan & D.J. Sorensen (1989), *Highways and rural economic development; results from a quasi-experimental approach*, Leningrad, Paper presented at the Seminar on Transportation Networks and Regional Development.

Jara-Diaz, S. (1986), On the relations between user benefits and the economics of transportation activities, *Journal of Regional Science*, Vol. 26, pp. 379-391.

Jensen-Butler, C. & B. Madsen (1998), An eclectic methodology for assessment of the regional economic effects of the Femern Belt link between Scandinavia and Germany, *Regional Studies*, forthcoming.

Johansson, B. (1992), *Infrastructure, accessibility and economic growth*, Stockholm, Royal Institute of Technology.

Johnson, E.A.J. (1970), *The organization of space in developing countries*, Cambridge, Harvard University Press.

Jonckers Nieboer, J.H. (1938), *Geschiedenis der Nederlandsche Spoorwegen*, Rotterdam, Nijgh en van Ditmar.

Kau, J.B. (1976), The interaction of transportation and land use, in: P.F. Wendt (ed.), *Forecasting transportation impacts upon land use*, Leiden, Martinus Nijhoff, pp. 112-134.

Kawashima, T. (1978), Regional impact simulation model BALAMO, in: H. Straszak & B. Wagle (eds.) *Models for regional planning and policy making*, Vienna, Proceedings of the Joint IBM/IIASA Conference, pp. 183-201.

Keeble, D., P.L. Owens & C. Thompson (1982), Regional accessibility and economic potential in the European Community, *Regional Studies*, Vol. 16, pp. 419-432.

Keeble, D., P.L. Owens & C. Thompson (1983), The urban-rural manufacturing shift in the European Community, *Urban Studies*, Vol. 20, pp. 405-418.

Kelejian, H.H. & D.P. Robinson (1997), Infrastructure productivity estimation and its underlying econometric specifications: a sensitivity analysis, *Papers in Regional Science*, Vol. 76, pp. 115-131.

Knaap, G.A. van der (1980), *Population growth and urban systems development; a case study*, Leiden, Martinus Nijhof.

Knippenberg, H. & B. de Pater (1990), *De eenwording van Nederland*, Nijmegen, SUN.

Korteweg, P.J. (1992), Kantoorruimte en segmentering, *VOGON-Journaal*, pp. 17-21.

Kramer, J.H.T. (1990), *Luchthavens en hun uitstraling*, Ph.D. Thesis, Nijmegen, Katholieke Universiteit Nijmegen.

Kroes, E.P. & R.J. Sheldon (1988), Stated preference methods: an introduction, *Journal of Transport Economics and Policy*, Vol. 22, pp. 11-25.

Kroes, E.P., A.J. Daly, H.F. Gunn & A van der Hoorn (1996), The opening of the Amsterdam Ring Road, a case study on short term effects of removing a bottleneck, *Transportation*, Vol. 23, pp. 71-82.

Krugman, P. (1991), *Geography and trade*, Cambridge, MIT Press.

Kruskal, J.B. (1956), On the shortest subtree of a graph and the travelling salesman problem, *Proceedings of the American Mathematical Society*, Vol. 7, pp. 48-50.

Lakshmanan, T.R. (1989), Infrastructure and economic transformation, in: A.E. Andersson, D. Batten & P. Nijkamp (eds.) *Advances in spatial theory and dynamics*, Amsterdam, North Holland, pp. 241-261.

Lathrop, G.T. & K.E. Cook (1990), The effects of beltways on urban development: a discussion of U.S. experience, in: D. Bayliss (ed.), *Orbital motorways*, London, Thomas Telford, pp. 143-156.

Launhardt, W. (1882), Die bestimmung des zweckm äszigten standortes einer gewerbligen anlage, *Zeitschrift des Vereins Deutscher Ingenieure*, Vol. 26, pp. 106-115.

Leitham, S. (1996), *Transport infrastructure and industrial operation and location*, Ph.D. Thesis, Edinburgh, Napier University.

Liew, C.K. & C.L. Liew (1985), Measuring the development impact of a transportation system: a simplified approach, *Journal of Regional Science*, Vol. 25, pp. 241-257.

Linneker, B.J. & N.A. Spence (1992), An accessibility analysis of the impact of the M25 London orbital motorway on Britain, *Regional Studies*, Vol. 26, pp. 31-47.

Lösch, A. (1940; 1954), *The economics of location*, New Haven, Yale University Press.

Loubal, P.S. (1967), *A network evaluation procedure*, Highway research record 205, Washington D.C., Highway Research Board.

Lowry, I.S. (1964), *A model of metropolis*, Santa Monica, Rand Corporation.

Lukkes, P. (1988), *De beste plek*, Onderzoek en advies 43, Groningen, Rijksuniversiteit Groningen.

Mackett, R.L. (1981), *The impact of transport planning policy on the city; a model-based approach to Leeds*, Ph.D. Thesis, Leeds, University of Leeds.

Mackett, R.L. (1993), Structure of linkages between transport and land use, *Transportation Research B*, Vol. 27B, pp. 189-206.

Mackett, R.L. (1996), Modelling the implications of new public transport technology: an approach using artificial intelligence, in: Y. Hayashi & J. Roy (eds.), *Transport, land-use and the environment*, Dordrecht, Kluwer Academic Publishers, pp. 297-315.

MacKinnon, R.D. & M.J. Hodgson (1970), Optimal transportation networks: a case study of highway systems, *Environment and Planning*, Vol. 2, pp. 267-284.

Maddala, G.S. (1983), *Limited-dependent and qualitative variables in econometrics*, Cambridge, Cambridge University Press.

Madsen, B. & C. Jensen-Butler (1991), The regional economic effects of the Danish Great Belt link and related traffic system improvements, in: H.G. Smit (ed.), *European Transport Planning Colloquium 1991*, Delft, ETPC, pp. 259-288.

Martellato, D., P. Nijkamp & A. Reggiani (1998), Measurement and measures of network accessibility: economic perspectives, in: M. Beuthe & P. Nijkamp (eds.), *Advances in European transport policy analysis*, Aldershot, Avebury, forthcoming.

McKinnon, A. (1996), The contribution of road construction to economic development, *Roads and economy*, Brussels, European Federation for Transport and Environment, pp. 19-26.

Meester, W.J. & P.H. Pellenbarg (1986), *Subjectieve waardering van bedrijfsvestigingsmilieus in Nederland*, Groningen, Rijksuniversiteit Groningen.

Mera, K. (1973), Regional production functions and social overhead capital, *Regional and Urban Economics*, Vol. 3, pp. 157-186.

Michelin (1988), *Wegenatlas van Europa*, London, The Hamlyn Publishing Group Limited.

Miehle, W. (1958), Link-length minimization in networks, *Operations Research*, Vol. 6, pp. 232-243.

Miller, H.J. (1998), *Measuring space-time accessibility benefits within transportation networks; basic theory and computational procedures*, Salt Lake City, University of Utah.

Mills, E. & G. Carlino (1989), Dynamics of county growth, in: A.E. Andersson, D. Batten, B. Johansson & P. Nijkamp (eds.), *Advances in spatial theory and dynamics*, Amsterdam, North Holland, pp. 195-205.

Mishan, E.J. (1976), *Cost-benefit analysis*, New York, Praeger.

Mohring, H. (1961), Land values and the measurement of highway benefits, *Journal of Political Economy*, Vol. 69, pp. 236-249.

Mohring, H. (1993), Land rents and transport improvements: some urban parables, *Transportation*, Vol. 20, 267-283.

Morgan, M.A. (1967), Hardware models in geography, in: R.J. Chorley & P. Haggett (eds.) *Models in geography*, London, Methuen.

Morisugi, H., E. Ohno & T. Miyagi (1993), Benefit incidence of urban ring road: theory and the case study of Gifu ring road, *Transportation*, Vol. 20, pp. 285-303.

Morita, K. (1992), *Impact of transportation network upon factory locations in Japan*, Tokyo, Industrial Location Policy Division.

Mueller, D.C. (1989), *Public Choise II*, Cambridge, Cambridge University Press.

Munda, G. (1993), *Fuzzy information in multicriteria environmental evaluation models*, Ph.D. Thesis, Amsterdam, Vrije Universiteit.

Munnell, A.H. (1993), An assessment of trends in and economic impacts of infrastructure investments, in: OECD, *Infrastructure policies for the 1990's*, Paris, OECD, pp. 21-54.

Murchland, J.D. (1970), *A fixed method for all shortest distances in a directed graph and for the inverse problem*, Ph.D. Thesis, Karlsruhe, University of Karlsruhe.

Musgrave, P. & R.A. Musgrave (1984), *Public finance; theory and practice*, New York, Mc Graw-Hill.

Myrdall, G. (1957), *Economic theory and underdeveloped regions*, London, Duckworth.

N.E.A. (1990), *Congestie en het functioneren van bedrijven, een case-study rondom de S20 in Noord-Brabant*, Rijswijk, NEA.

N.E.I. (1991), *Majeure ruimtelijke en infrastructurele operaties in grootstedelijke agglomeraties in Noordwest-Europa*, Rotterdam, Nederlands Economisch Instituut.

Nemhauser, G.L. & L.A. Wolsey (1988), *Integer and combinatorial optimization*, New York, Wiley.

Neusser, H.G. (1985), Die bedeutung von hemmnisfaktoren für die entwicklung des verkehrsaufkommens, *DFVLR-Nachrichten*, Vol. 45, pp. 32-34.

Newman, P. & J. Kenworthy (1989), *Cities and automobile dependence: an international sourcebook*, Aldershot, Gower Technical.

Nijkamp, P. (1986), Infrastructure and regional development: a multidimensional policy analysis, *Empirical Economics*, Vol. 11, pp. 1-21.

Nijkamp, P., P. Rietveld & I. Salomon (1990), Barriers in spatial interactions and communications, *The Annals of Regional Science*, Vol. 24, pp. 237-252.

NSS (1991), *Eisen aan de bedrijfsomgeving in de Randstad*, Den Haag, NSS beleidsonderzoek & beleidsadvies.

Nurske, R. (1952), *Problems of capital formation in developing countries*, Oxford, Basil Blackwell.

O.E.C.D. (1986), *Infrastructure investments and their contribution to employment and economic growth*, Paris, OECD.

Ortuzar, J. & L.G. Willumsen (1993), *Modelling transport*, London, Wiley.

Papadimitriou, C.H. & K. Steiglitz (1982), *Combinatorial optimization; algorithms and complexity*, London, Prentice Hall.

Pellenbarg, P.H. (1982), Subjectieve waardering van vestigingsmilieus, *Geografisch Tijdschrift*, Vol. 16, pp. 219-226.

Pellenbarg, P. (1985), *Bedrijfslokatie en ruimtelijke kognitie*, Ph.D. Thesis, Groningen, Rijksuniversiteit Groningen.

Pellenbarg, P.H., A. Pruntel & W.G. van der Weij (1993), *Ruimtelijke cognitie en ruimtelijk gedrag van ondernemers in Nederland*, Groningen, Rijksuniversiteit Groningen.

Pelt, M.J.F. van (1994), *Ecological sustainability and project appraisal*, Aldershot, Avebury.

Perroux, F. (1958), *La coexcistence pacifique*, Paris, Presses Universitaires de France.

Pierce, D.A. & L.D. Haugh (1977), Causality in temporal systems, characterizations and a survey, *Journal of Economics*, Vol. 5, pp. 265-293.

Pirie, G.H. (1979), Measuring accessibility: a review and proposal, *Environment and Planning A*, Vol. 11, pp. 299-312.

Plassard, F. (1992), *The relationship between regional policy and transport and communication networks*, Paris, European Conference of Ministers of Transport.

Pluym, W.K. & S.Y. Roosma (1984), *Economische betekenis van transport-infrastructuur*, Groningen, Federatie van Noordelijke Economische Instituten.

Poppelaars, J.A.A. (1993), *Sociaal-culturele voorzieningen als vestigingsplaatsfactor*, Den Haag, Instituut van Onderzoek voor Overheidsuitgaven.

Predöhl, A. (1925), Das standortproblem in der wirtschaftstheorie, *Weltwirtschaftliches Archiv*, Vol. 21, pp. 294-331.

Prud'homme, R. (1996), Assessing the role of infrastructure in France by means of regionally estimated production functions, in: D.F. Batten & C. Karlsson (eds.), *Infrastructure and the complexity of economic development*, Berlin, Springer Verlag, pp. 37-47.

Rich, D.C. (1980), *Potential models in human geography*, Geoabstracts Catmog 26, Norwich, University of East Anglia.

Rienstra, S.A. & P. Rietveld (1997), *Spatial economic impacts of developing international top office locations; a case study for Amsterdam South*, Rome, Paper presented at the 37th European Regional Science Association Congress.

Rietveld, P. (1989), Infrastructure and regional development, *Annals of Regional Science*, Vol. 23, pp. 255-274.

Rietveld, P. (1993), International transportation and communication networks in Europe; the role of barrier effects, *Transportation Planning and Technology*, Vol. 17, pp 311-317.

Rietveld, P. & L. Janssen (1991), Telephone calls and communication barriers, *Annals of Regional Science*, Vol. 24, pp. 307-318.

Rietveld, P & J. van Nierop (1995a), The evolution of transport networks: the case of the Dutch railway system, in: D. Banister (ed.), *European transport and communication networks*, London, Wiley, pp. 121-140.

Rietveld, P. & J. van Nierop (1995b), Urban growth and the development of transport networks; the case of the Dutch railways in the nineteenth century, *FLUX*, no. 19, pp. 31-43.

Rietveld, P. & J. Boonstra (1995), On the supply of network infrastructure, highways and railways in European regions, *Annals of Regional Science*, Vol. 29, pp. 207-220.

Rietveld, P. & P. Wintershoven (1998), Border effects and spatial autocorrelation in the supply of network infrastructure, Papers in Regional Science, forthcoming.

Rothengatter, W. (1994), Do external benefits compensate for external costs of transport? *Transportation Research A*, Vol. 28A, pp. 321-328.

Rouwendal, J. & P. Rietveld (1994), Commuting distance distributions of Dutch households, *Urban Studies*, Vol. 31, pp. 1445-1457.

R.P.D. (1991), *Perspectieven in Europa*, Den Haag, Rijksplanologische Dienst.

Rijkswaterstaat (1992), *Effects of the opening of the Amsterdam orbital motorway: final report phase 1*, Rotterdam, Transportation and Traffic Research Division, Rijkswaterstaat.

SACTRA (1994), *Trunk roads and the generation of traffic*, London, HMSO.

Sasaki, K., S. Kunihisa & M. Sugiyama (1995), Evaluation of road capacity and its spatial allocation, *Annals of Regional Science*, Vol. 29, pp. 143-154.

Sasaki, K., T. Ohashi & A. Ando (1997), High-speed rail transit impact on regional systems: does the Shinkansen contributie to dispersion?, *Annals of Regional Science*, Vol. 31, pp. 77-98.

Schmenner, R.W. (1982), *Making business location decisions*, Englewood Cliffs, Prentice Hall.

Scott, A.J. (1969), The optimal network problem, *Transportation Research*, Vol. 3, pp. 201-210.

Scott, A.J. (1971), *Combinatorial programming, spatial analysis and planning*, London, Methuen.

Seitz, H. (1993), *The impact of the provision of urban infrastructures on the manufacturing industry in cities*, Moscow, Paper presented at the 33rd European Regional Science Association Congress.

Seitz, H. (1995a), The productivity and supply of urban infrastructures, *Annals of Regional Science*, Vol. 29, pp. 121-141.

Seitz, H. (1995b), Public infrastructure, employment and private capital formation, in: OECD, *Investment, productivity and employment*, Paris, OECD, pp. 123-150.

Seo, J.G., J.E. Moore II & P. Gordon (1996), A sequential land use/-transportation model with externalities: linking the dynamics of regional economic growth and urban spatial structure, in: Y. Hayashi & J. Roy (eds.), *Transport, land-use and the environment*, Dordrecht, Kluwer Academic Publishers.

Sheffi, Y. (1985), *Urban Transportation Networks*, London, Prentice-Hall.

Snickars, F. (1989), Effects of infrastructure provision on urban development, *Tijdschrift voor Economische en Sociale Geografie*, Vol. 80, pp. 106-123.

Snickars, F. & A. Granholm (1981), *A multiregional planning and forecasting model with special regard to the public sector*, Laxenburg, International Institute for Applied Systems Analysis.

Song, S. (1996), Some tests of alternative accessibility measures: a population density approach, *Land Economics*, Vol 72, pp. 474-482.

Spiekermann, K. & M. Wegener (1996), Trans-European networks and unequal accessibility in Europe, *European Journal of Regional Development*, Vol. 4, pp. 35-42.

Steen, P.J.M. van & B.M.R. van der Velde (1993), *Expansieruimte voor bedrijven*, Onderzoek en advies 69, Groningen, Rijksuniversiteit Groningen.

Steenbrink, P.A. (1974), *Optimization of transport networks*, Ph.D. Thesis, Utrecht, NV Nederlandse Spoorwegen.

Steward (1994), *Econometrics*, New York, Philip Allan.

Stoffelsma, R.J. (1992), *Centrale overheid en regionale welvaart*, Ph.D. Thesis, Groningen, Rijksuniversiteit Groningen.

Sturm, J.E. & J. de Haan (1995), Is public expenditure really productive? New evidence for the USA and the Netherlands, *Economic Modelling*, Vol. 12, pp. 60-72.

Thünen, J.H. von (1821), *Der isolierte staat in beziehung auf landwirtschaft und nationalökonomie*, Jena, Fischer.

Treyz, G.I. (1980), Design of a multi-regional policy analysis model, *Journal of Regional Science*, Vol. 20, pp. 191-206.

Tulp, E. (1991), *Searching time table networks*, Amsterdam, Vrije Universiteit.

Verhoef, E.T. (1996), *The economics of regulating road transport*, Cheltenham, Edward Elgar.

V.G.M. (1990), *VastGoedMarkt*, no. 1.

Vickerman, R.W. (1995), The regional impacts of Trans-European networks, *Annals of Regional Science*, Vol. 29, pp. 237-254.

Vickerman, R.W., K. Spiekermann & M. Wegener (1998), Accessibility and economic development in Europe, *Regional Studies*, forthcoming.

Voigt, F. (1973), *Verkehr*, Berlin, Springer Verlag.

Voogd, H., P. Nijkamp & P. Rietveld (1990), *Multicriteria evaluation in physical planning*, Amsterdam, North Holland.

Vuren, T. van (1997), Gegenereerde vraag: schijn of werkelijkheid?, *Verkeerskunde*, Vol. 48, pp 26-29.

Weber, A. (1909; 1929), *A theory of the location of industries*, Chicago, Chicago University Press.

Webster, F.V., P.H. Bly & N.J. Paulley (1988), *Urban land-use and transport interaction*, ISGLUTI, Aldershot, Avebury.

Weibull, J.W. (1980), On the numerical measurement of accessibility, *Environment and Planning A*, Vol. 12, pp. 53-67.

Wegener, M. (1986), Transport network equilibrium and regional deconcentration, *Environment and Planning A*, Vol. 18, pp. 437-456.

Wegener, M. (1994), Operational urban models: state of the art, *Journal of the American Planning Association*, Vol. 60, pp. 17-29.

Wegener, M. (1996), Reduction of CO_2 emissions of transport by reorganisation of urban activities, in: Y. Hayashi & J. Roy (eds.), *Transport, land-use and the environment*, Dordrecht, Kluwer Academic Publishers, pp. 103-124.

Williams, H.C.W.L. (1977), On the formation of travel demand models and economic evaluation measures of user benefit, *Environment and Planning*, Vol. 9A, pp. 285-344.

Williams, H.C.W.L. & M.L. Senior (1978), Accessibility, spatial interaction and the spatial benefit analysis of land use-transportation plans, in: A. Karlqvist, L. Lundqvist, F. Snickars & J.W. Weibull (eds.), *Spatial interaction theory and planning models*, Amsterdam, North Holland, pp. 253-287.

Willeke, R. (1992), *Benefits of different transport modes*, Paris, European Conference of Ministers of Transport.

Wilson, A.G. (1982), *Transport location and spatial systems*, Leeds, School of Geography.

Wilson, F., A. Stevens & T. Holyoke (1982), *Impact of transportation on regional development*, Transportation research record, Washington D.C., National Academy of Sciences.

World Bank (1994), *World development report 1994: infrastructure for development*, Oxford, Oxford University Press.

Youngson, A.J. (1967), *Overhead capital*, Edinburgh, Edinburgh University Press.

FSC
www.fsc.org
MIX
Papier aus verantwortungsvollen Quellen
Paper from responsible sources
FSC® C105338

If you have any concerns about our products,
you can contact us on
ProductSafety@springernature.com

In case Publisher is established outside the EU,
the EU authorized representative is:
Springer Nature Customer Service Center GmbH
Europaplatz 3, 69115 Heidelberg, Germany

Printed by Libri Plureos GmbH
in Hamburg, Germany